D1032859

PRENTICE-HALL FOUNDATIONS OF MODERN BIOLOGY SERIES

William D. McElroy and Carl P. Swanson
EDITORS

THE CELL

Fourth Edition

Carl P. Swanson

Ray Ethan Torrey Professor of Botany
University of Massachusetts

Peter L. Webster

Associate Professor of Botany
University of Massachusetts

Prentice-Hall, Inc., Englewood Cliffs, New Jersey 07632

Library of Congress Cataloging in Publication Data

Swanson, Carl P
The cell.

(Foundations of modern biology series)
Bibliography: p. 292
Includes index.
1. Cytology. I. Webster, Peter L., (date)–
joint author. II. Title.
QH581.2.S89 1977 574.8'7 76-25952
ISBN 0-13-121707-0
ISBN 0-13-121699-6 pbk.

Printed in the United States of America

10 9 8 7 6 5 4 3 2 1

Prentice-Hall International, Inc., London
Prentice-Hall of Australia Pty. Limited, Sydney
Prentice-Hall of Canada, Ltd., Toronto
Prentice-Hall of India Private Limited, New Delhi
Prentice-Hall of Japan, Inc., Tokyo
Prentice-Hall of Southeast Asia Pte. Ltd., Singapore
Whitehall Books Limited, Wellington, New Zealand

To D.N.S. and M.A.W.

Contents

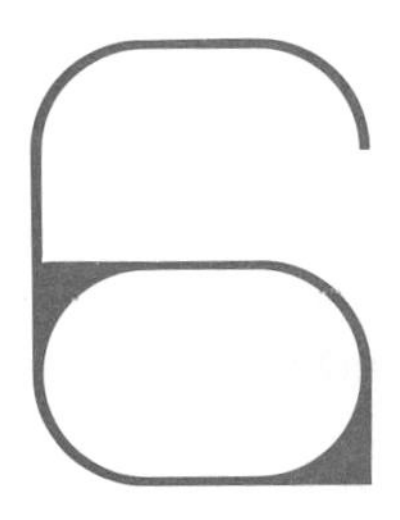

Comparative cytology 134

The cell in reproduction 161

Meiosis— the transmission of information 181

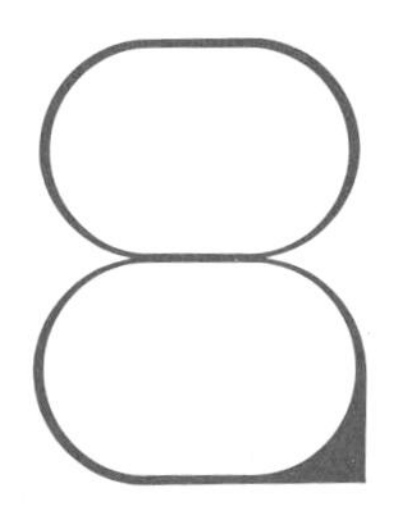

The cell in development 221

The cell in death 243

Cells through time 259

Appendix A

Appendix B

Appendix C

Preface

This fourth edition of *The Cell* introduces a number of changes that should distinguish it from its predecessors. The changes do not, we believe, greatly alter the thrust previously expressed, but rather they deepen and broaden a consideration of the place of the cell in modern biology. The changes, therefore, although extensive, are more in the nature of reinforcements and not of shifts to new or different directions. The volume is obviously lengthier than earlier editions. This has resulted from a greater emphasis given to bioenergetics and the mechanisms whereby the cell manipulates matter and energy; from an increased volume and improved quality of illustrative material; from much revised and updated information on organelles and their roles in cellular affairs; from a new chapter on the evolution of cells; and from a doubling of the authorship, not by replication, which would mean more of the same, but by addition, thereby bringing, hopefully, an element of hybrid vigor to all facets of the volume.

We wish to acknowledge our indebtedness to those who have, in one way or another, assisted in bringing this volume to its present state; to those who have generously allowed use of their illustrative materials; and particularly to Professor H. T. Yost of Amherst College, whose critical reading of the entire manuscript has been of much aid to us. Any errors, however, remain our responsibility.

Carl P. Swanson
Peter L. Webster

1 The cellular basis of life

Every science has one or more periods of time when it seems to burst its bounds and exhibit an extraordinary pattern of growth. The stimulus for such change is usually a new discovery or a recently advanced theory which poses new problems or allows old problems to be viewed in a totally different light. Since the early 1950s, this kind of change has characterized biology, and it would not be unreasonable to say that a veritable explosion of biological knowledge has occurred, with every facet of the science being affected. Further, it is not simply that the volume of biological knowledge has increased tremendously, although this is the case; the character of the information generated also has been different, and it in turn has altered the structure of the biological sciences. New instrumentation and techniques, more appropriate experimental organisms, and, more importantly, new ways of thinking and of asking critical and testable questions have all contributed to this burst of knowledge, and have influenced the direction, character, and promise of the biological sciences. Much of this recent activity falls within the area of cell biology.

The significance of the information generated during the past quarter century is as great philosophically as it is scientifically. As human beings we have an interest in knowing who we are, where we exist in space and time, and how we came to be what and where we are. We recognize individual uniqueness, not only in ourselves and our fellow human beings, but in the plants and animals around us; the biological sciences can now supply a rational explanation for this uniqueness, not only in more readily understandable morphological and behavioral terms, but also in some instances in unambiguous, concrete, molecular terms. The purpose of this volume is to explore the cell, which we now know to be the organizational basis of this uniqueness.

It is through examination of the intimate details of nature that the unity, continuity, and diversity of the living world are made manifest. It is, however, one thing to recognize details and quite another to understand them. Understanding requires that we ask questions capable of being answered, devise critical experiments to prove or disprove some point of ambiguity or uncertainty, and construct testable hypotheses and theories that bind our isolated facts and observations into meaningful wholes. As a result of these observations, experiments, and speculations, we have come to recognize that the world around us, living and nonliving, is an ordered system governed by certain rules that we call natural laws. Science, in fact, makes sense only if it is assumed that nature is orderly. The function of science, therefore, is to uncover that order and attempt to explain it.

Our contact with the world around us is through our five senses of touch, taste, hearing, smell, and sight. They enable us to apprehend, and to visualize to our own satisfaction, the world of reality external to us, and we come to realize that this world is made up of matter and energy, which we detect, define, describe, and sometimes measure in specific ways. With our unaided senses and in unsubtle ways, we usually have no difficulty in distinguishing the sky and the land from the water, a gas and a solid from a liquid, the living from the nonliving. On a more refined level, we can distinguish degrees of roughness, intensity, and shade of color (if we are not color-blind), and an acid taste from one that is salty, sweet, or bitter. But human powers of sensory discrimination are limited. We all know, for example, that water, steam, and ice are made up of molecular H_2O, and that they have different characteristics, but our ordinary senses cannot tell us *why* this is so except to indicate that the differences are correlated with temperature. We hear only within a certain range of sound waves, and see only that portion of the light spectrum called the visible region (Figure 1.1). When we

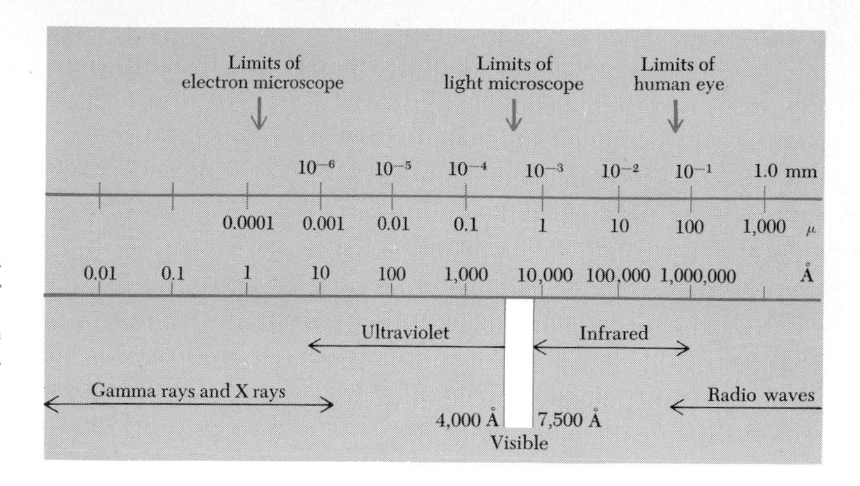

Figure 1.1
The electromagnetic spectrum on a logarithmic scale, measured in millimeters, microns, and angstrom units: 1 μm (micrometer) = 0.001 mm (millimeter) = 10,000 Å (Angstroms). The approximate lower limits of resolution of the human eye, the light microscope, and the electron microscope are given.

try to go beyond these limits, we can no longer directly comprehend the physical nature of things and must resort to instruments, experiments, and speculation to penetrate areas outside of the sphere naturally circumscribed by our senses.

Try to imagine how much of your knowledge of yourself and of the universe around you has been gained *only* through use of your five senses, as compared to that derived through instruments and/or experiments. With knowledge only from your senses, you would find your world shrinking appreciably in size and diversity. The heavens, for example, would be a blue vault studded with stars at night, but at distances that could only be guessed at; light-years, galaxies, black holes, and neutron stars would be terms without meaning. The world of cells and microorganisms would disappear from your frame of reference, and the structure of all other organisms, including yourself, would be understood only in terms of larger, readily visible portions. We do, however, possess the means whereby we can extend the use of our senses beyond their normal capabilities. On a clear night, for example, the naked eye can discern stars that number only in the several thousands. But powerful telescopes show that the Milky Way, the galaxy of which our solar system is a part, contains billions of stars similar to our sun, while the universe as a whole contains billions of comparable galaxies. Light and electron microscopes, on the other hand, are concerned not with great distances but rather with the world of the minute, the microcosm of viruses, bacteria, cells, and large molecules, all of them ordinarily invisible because of their size. Similarly, photographic plates, more sensitive to certain portions of the electromagnetic spectrum (Figure 1.1) than are our eyes, extend the limits of our "visibility" so that we can detect the long infrared rays and radio waves on one side of the spectrum and the short ultraviolet rays, X rays, and gamma rays on the other.

Whatever means we use, we attempt to define the "things" we observe in terms of the units of which they are composed, and the more refined our knowledge, the more powerful and discriminating our instruments and techniques, the more precise become our definitions of these units—their limits, basic nature, and modes of aggregation into larger units. It would, indeed, be impossible for you to read these pages without understanding letters, the basic symbolic units of our language, the numbers that make up our decimal or metric systems, or the rules by which they are put together to express a thought or a quantitative value. One of the first goals of a science, therefore, whether it be physics, chemistry, or biology, is to determine the uniqueness of the units with which it is concerned, for unless such units are understood and accepted by everyone in a particular field, meaningful communication is difficult and scientific knowledge in that field cannot progress.

The cell doctrine

The basic unit of life, the *cell,* is a physical entity; life, with its ability to reproduce, mutate, and respond to stimuli, does not exist in smaller units of matter. We can break cells apart and, by centrifugation, extract selected portions for study much as the physicist breaks up atoms. We find that these cellular fragments can carry on many of their activities for a time; they may consume oxygen, ferment sugars, and even form new molecules. But these activities individually do not constitute life, any more than the behavior of a subatomic particle is equivalent to the behavior of an intact atom. The disrupted cell is no longer capable of continuing life indefinitely; we, therefore, conclude that the cell is the most elementary unit that can sustain life, even though, as we shall see, the cell is highly complex. Viruses, on the other hand, are smaller and less complex than cells, but they cannot maintain life independently of the cells they parasitize (page 19).

Compared to the atom and the molecule, the cell is a unit of far greater size and complexity. It is a microcosm having a definite boundary, within which constant chemical activity and a flow of energy proceed. At ordinary temperatures, a chemically quiescent cell is dead. The cytologist (one who studies cells), therefore, seeks to identify the kinds of cells that exist, to understand their organization and structure in terms of their activities and functions, and to visualize the cell not only as a total entity (as, for example, the unicellular bacterium), but also as an integral part of the elaborate organs and organ systems of multicellular plants and animals.

The now familiar idea that the cell is the basic unit of life is known as the *cell doctrine*. It is essentially a statement of fact, not a theory of a debatable or controversial nature. It developed gradually through microscopical observations of the structure of many plants and animals, and eventually the presence of cells as a common structural feature of all biological organization was recognized. Englishman Robert Hooke first saw the remains of dead cells in 1665 in a piece of cork as he was using his newly invented microscope (Figure 1.2), and he coined the word "cell" to describe the tiny structures, thinking that they resembled the

Figure 1.2
Robert Hooke's microscope with which he observed the microscopic structure of cork and his drawing of it (in circle). Here, in his own words, is a description of his experiment: "I took a good clear piece of Cork and with a Pen-knife sharpen'd as keen as a razor, I cut a piece of it off, and thereby left the surface of it exceeding smooth, then examining it very diligently with a Microscope, me thought I could perceive it to appear a little porous; but I could not so plainly distinguish them as to be sure that they were pores. . . . I with the same sharp pen-knife cut off from the former smooth surface an exceeding thin piece of it, and placing it on a black object Plate. . . . and casting the light on it with a deep plano-convex Glass, I could exceedingly plainly perceive it to be all perforated and porous, much like a Honey-comb, but that the pores of it were not regular . . . these pores, or cells, were not very deep, but consisted of a great many little Boxes, separated out of one continued long pore by certain Diaphragms . . . Nor is this kind of texture peculiar to Cork only; for upon examination with my Microscope, I have found that the pith of an Elder, or almost any other Tree, the inner pulp or pith of the Cany hollow stalks of several other Vegetables: as of Fennel, Carrets, Daucus, Bur-docks, Teasels, Fearn . . . & c. have much such a kind of Schematisme, as I have lately shewn that of Cork."

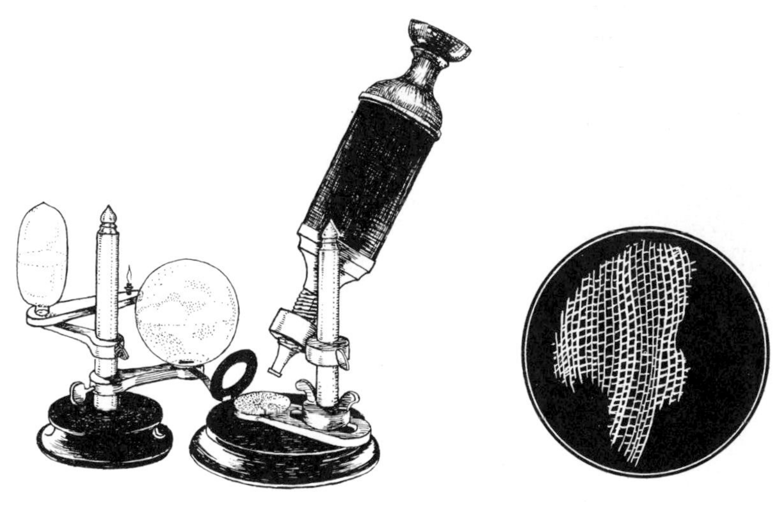

Figure 1.3
A microscope made about 1770 by George Adams for King George III of England, an active patron of the arts and sciences. Compare this with the microscopes of Robert Hooke and of today. An equally ancient microscope, made of bronze and once the gift of the French King Louis XV to his mistress Madame de Pompadour was sold at auction in 1976 for $74,000. (Photo courtesy of the Museum of the History of Science, Oxford University.)

unadorned cells occupied by monks. By 1800 good miscroscopes (Figure 1.3) were becoming available, as were techniques for the fixation and staining of cells, and there was general acceptance of the idea that organisms were cellular, but there was also a good deal of confusion over the definition of cells, the significance of their contents and cell walls, their mode of origin, and their role in organization. In 1838 and 1839, two German scientists, botanist M. J. Schleiden and zoologist Theodor Schwann, took the loose threads of ideas and observations available to them and wove them into a doctrine, which stated that cells with nuclei were the structural and functional basis of all living things. Many of their ideas concerning cellular structure, function, and origin have been proved erroneous, but by emphasizing the importance of the cell, they gave coherence to the biological thought of their time and focused attention on the one structure that had to be understood if biology was to advance beyond its purely descriptive stage. We now recognize that a study of the cells of such very different organisms as bacteria, orchids, and humans aids in the understanding of the structure and function of all organisms, an impossibility prior to the acceptance of the universality of the cell doctrine. Who would have thought, before this period, that a human being and an orchid had anything in common? It is now considered that the cell doctrine, however vague its beginnings and however long it took for formulation, ranks with Darwin's *theory of evolution through natural selection* and the *theory of the gene* as one of the foundation stones of modern biology.

Some 20 years after the announcements of Schleiden and Schwann, Rudolf Virchow, a great German physician, made another important generalization: *cells come only from preexisting cells.* When biologists further recognized that sperm and ova are also cells that unite with each other in the process of fertilization, it gradually became clear that life from one generation to another is an uninterrupted succession of cells. Growth, development, metabolism in all of its aspects, inheritance, evolution, disease, aging, and death are, therefore, but varied aspects of cellular behavior, even though each of these phenomena also can be viewed at higher and lower levels of biological organization.

Most generalizations have exceptions that cast doubt on their universal validity. This is true for the cell doctrine, but we shall not consider these exceptions until after we have examined cellular structure in some detail. Let us now see what the cell doctrine, as presently interpreted, embodies in the way of solid ideas. There are essentially three:

First, as we have already mentioned, the cell doctrine states that life exists only in cells; organisms are, therefore, made up of cells; the activity of an organism is dependent on the activities of cells, individually and collectively; and the cell is the basic unit through which matter and energy are acquired, converted, stored, and utilized, and in which biological information is stored, manipulated and expressed. Second, as a direct corollary of the first generalization, the cell doctrine has embodied within it the idea that the continuity of life has a cellular basis, which is another way of stating Virchow's generalization. Now, however, we can be more explicit, adding that genetic continuity in a very exact sense includes not only the cell as a whole but also some of its smaller components, such as genes and chromosomes, as well as the hereditary mechanism for transmitting its genetic substance to the next generation. The nature of viruses, as we shall see later, also reinforces the concept that the whole cell is the basic unit of heredity. A virus possesses genes and a chromosome, but it cannot reproduce itself without the aid of the cell it infects. Third is the idea that there is a relationship between structure and function.

This has been called the principle of complementarity; it means, briefly, that orderly behavior and orderly structures are intimately related to each other, and that within the domain of cells the biochemical activities of cells occur within, and indeed are determined by, structures organized in a definite way. We shall encounter this idea again in our discussion of cellular components.

Andre Lwoff, the French microbiologist, has expressed the cell doctrine in yet another way:*

> When the living world is considered at the cellular level, one discovers unity. Unity of plan: each cell possesses a nucleus imbedded in protoplasm. Unity of function: the metabolism is essentially the same in each cell. Unity of composition: the main macromolecules of all living beings are composed of the same small molecules. For, in order to build the immense diversity of living systems, nature has made use of a strictly limited number of building blocks. The problem of diversity of structures and functions, the problem of heredity, and the problem of diversification of species have been solved by the elegant use of a small number of building blocks organized into specific macromolecules . . . Each macromolecule is endowed with a specific function. The machine is built for doing precisely what it does. We may admire it, but we should not lose our heads. If the living system did not perform its task, it would not exist. We have simply to learn how it performs its task.

Cells and energy

In the preceding statement Lwoff describes the cell as a "machine" that performs a task. More than likely, the word "machine" will evoke an image that is more a reflection of our technological and mechanical age than of the living world as we know it. We touch a switch to flood a room with light, and we know, if we stop to think about it at all, that the switch is somehow connected to an outside source of electrical energy, with that energy being converted, by resistance in a filament, to incandescent heat, which we then see as radiant energy. A comparable thing happens when we start an automobile: a turn of a switch sets into motion a chain of events that converts the chemical energy of gasoline into heat energy, which, through expansion of gases, is translated into the mechanical energy for driving the gears and wheels of the machine. Switches, filaments, motors, and

* Andre Lwoff, *Biological Order* (Cambridge, Mass: M.I.T. Press, 1962), pp. 11, 13.

wheels are devices through which energy flows, or by which energy is controlled for a given purpose.

There is no switch that can turn life on or off; life is not that simple or that obviously mechanical. We may not think ourselves, or any other organism, to be as predictable or as automatic in response or performance as an electrical circuit or a gasoline-driven motor, but living systems, like machines, manipulate energy in a controlled way. Also like machines, living organisms are characterized by a high degree of order of structure and behavior, and it is this orderliness that either permits, or is necessary for, the controlled manipulation of energy. The maintenance of this necessary order in turn requires a continual input of energy; life, therefore, is basically an energy-converting and energy-consuming process, and it can continue only as long as a supply of appropriate kinds of energy is available (Figure 1.4).

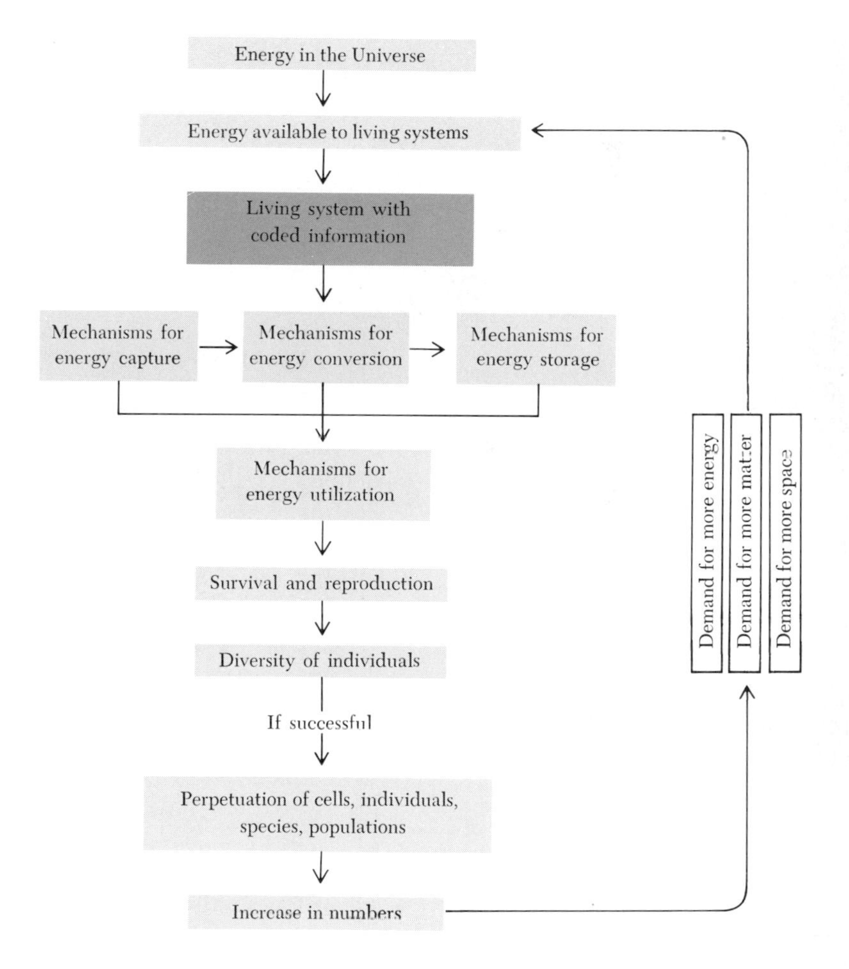

Figure 1.4
Dynamics of living systems viewed as an energy problem, with the coded information within the cell governing the movement and manipulation of energy through ordered channels. It should be recognized that since structure and behavior are intimately related, the mechanisms governing the flow of energy are as much a reflection of the coded information of the cell as is the flow of energy itself. It should also be realized that the diagram reflects hierarchies of explanatory convenience more so than physical realities, although an attempt has been made to depict the flow of energy as sequentially accurate as possible. (After a diagram provided by Dr. H. H. Hagerman.)

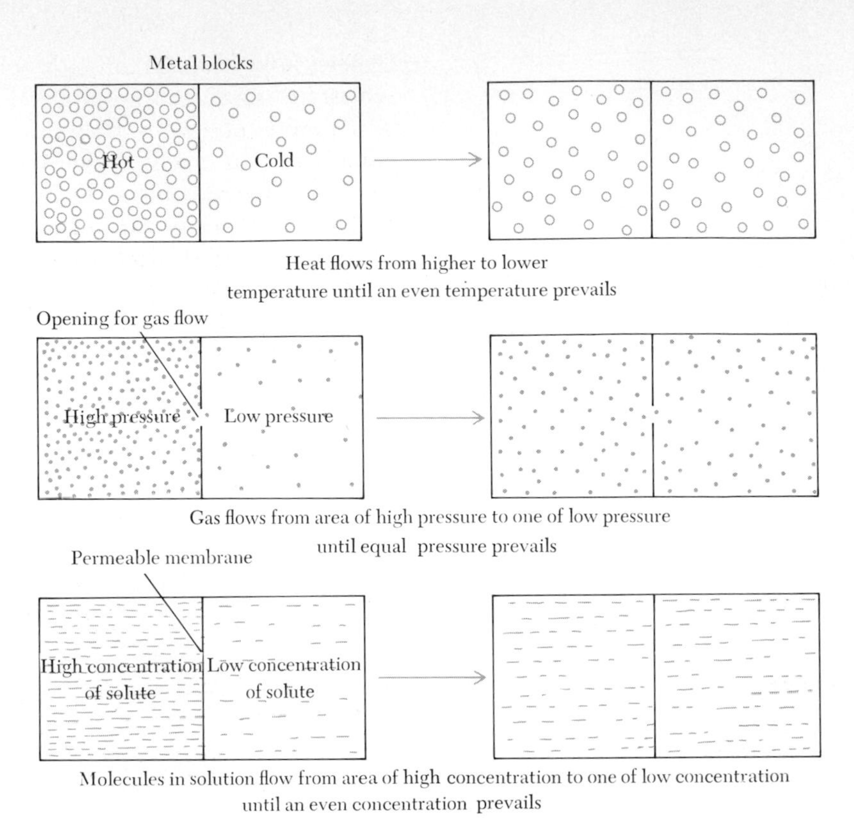

Figure 1.5
Examples of processes tending toward equilibrium, and hence toward increasing entropy.

The First Law of Thermodynamics states that energy (often defined as the capacity for doing work) cannot be created or destroyed. It can, however, be changed spontaneously into forms less able to do work. The Second Law of Thermodynamics tells us that all physical and chemical processes tend toward equilibrium, that is, they proceed from an ordered to a disordered state. Once a system reaches its equilibrium state, the energy within it is maximally randomized; such randomized energy is known as *entropy,* and is no longer available to do work. Thus we can consider our universe to be "running down" in the sense that the energy that can be used to perform work is becoming less and less available. The component of the total energy that is available to perform work—that is, the useful energy—is known as *free energy.* From what we have said, we can see that processes tend toward a minimum free energy state and a maximum entropy state (Figure 1.5).

The following forms of energy are listed from highest to lowest in terms of free energy (or from lowest to highest entropy): gravitation, nuclear reactions, sunlight (radiant energy), chemical reactions, waste heat, and microwave radiation. The primary energy of the universe, from which all other kinds of energy are derived, is that of gravitational attraction, leading to the contraction of massive objects and, through contraction, to the degradation of this energy into chemical reactions, motion, light, and heat. The entropy of the universe as a whole is increased by this degradative process, but life is an incidental beneficiary since it is the energy of sunlight that enabled living systems to come into being, and to continue to build order out of disorder through the process of *photosynthesis* and the manipulation of chemical energy. The green plant, for example, maintains a high free energy content by trapping the external radiant energy of sunlight to build chemical bonds, which in turn lead to the formation of complex but orderly molecules and structures; animals maintain a similarly high free energy status, but they do so by utilizing a lower form of energy, that bound up as chemical energy in the organic molecules of their diet. They consequently are not only less efficient as energy utilizers when compared to green plants, but they also contribute more significantly to an increase in the entropy of the environment (Figure 1.6). Man is a prime example of this fact, particularly in the present human technological state with its high rate and inefficient use of energy. Life in general, however, delays an increase in entropy for a small fraction of the substance of this planet, and so long as sufficient sunlight is available it should be able to do so indefinitely. This may be a minor feature of no

Figure 1.6
A diagram to illustrate the place of the green plants, animals, and humans (with their machines of civilization) in the energy picture of this planet. Gravitational energy, which produced the sun and governs its activities, has the lowest level of entropy; entropy increases as one proceeds toward the right of the diagram, until the point is reached where the character of the energy precludes the further performance of work.

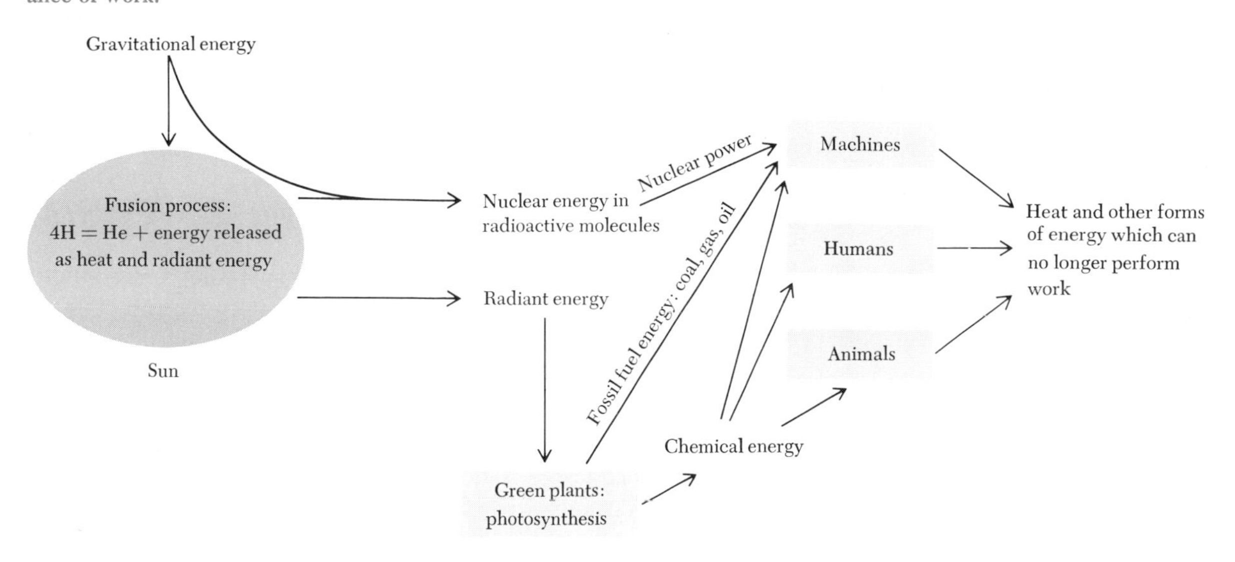

consequence in the total energy picture of the universe, but it is obviously of supreme importance to us as part of the living world.

Physical and chemical processes that proceed toward equilibrium are thought of as closed systems, although a completely closed system of change in which no energy is gained or lost is an ideal, not an actual one (with the possible exception of the universe itself). By comparison, life is an open system. It does not tend toward equilibrium for it is constantly energy-consuming. Life achieves a steady state—a dynamic situation far from equilibrium in that there is a continual interaction with the environment, with energy and matter being taken in to be converted into living substance, and energy and matter being given off, generally in a different form.

The remarkable feature of life, therefore, is that it seems to function in a qualitatively different manner from the remainder of the universe, in that living systems are order-creating both as to their structure and, intimately related to this, their management of energy. Life consequently increases the free energy state of its own substance and lowers its entropy, a seeming contradiction of the laws of thermodynamics. Functioning as they do, however, living systems do not violate any physical law, since the order represented by the maintenance and reproduction of life is "paid for" by energy commanded from the environment, and at the cost of a similar or greater increase in the entropy of the environment.

What is true for life itself is true for the cell, since it is the basic unit of living organization. In a very general way, Figure 1.4 diagrams the channels of energy flow within and between cells, and between cells and their environment. Each step is an orderly process associated with particular and highly ordered intracellular structures, and we shall find that their organization and behavior are intimately related to the kind and manner of energy being managed. The diversity of organisms in the living world, with most possessing many kinds of cells, suggests that there are likely to be variations among cells as to the patterns of energy flow and the structures associated with this flow. Variations have been found, mainly in structures, but the basic energy patterns are surprisingly similar in all forms of life.

A last feature indicated in Figure 1.4 is that the energy management of living systems is not a haphazard one; rather it is directed by information coded within a particular portion of the cell. Life, with all of its varied attributes, was made possible in the long-distant past when matter somehow acquired a particular pattern of organization within which energy from the environment could be acquired and manipulated. Once life arose, therefore, it acquired a measure of control over its environment, and in a sense took command of certain external energy sources and directed their use to its own advantage. Evolution, which through the course of time has produced the diversity of cells and organisms now alive, can be thought of as a continuing experiment in the creation of new structures for, and new ways of, energy management, with each new experimental approach being determined by changes that take place in the informational code. We will be dealing with all of these aspects in subsequent chapters, but we might well, with good reason, subscribe to the words of the poet William Blake—"Energy is Eternal Delight."

Cells in general

Our purpose then, in the chapters that follow, is to examine cells by whatever techniques or instrumentation are needed to uncover and to understand their structure and function. There is no single ideal cell that serves our purpose, so whatever cell most appropriately illustrates the point under discussion will be used. This can be done without fear of bias or distortion because of the unity of plan, function, and composition of which Lwoff speaks: what is learned from one kind of cell can be applied, sometimes directly, sometimes with modification, to other kinds of cells. We do, on the other hand, recognize two general classes of cells: the *prokaryotic cell* typical of bacteria and blue-green algae, and the *eukaryotic cell* found in all other organisms, plant and animal.

The prokaryotic cell is the simplest kind known and, from what the fossil record can tell us, probably the first to come into existence perhaps 3 to 3½ billion years ago. As a rule, these cells are small in dimensions: from 0.1 to 0.25 μm (micrometers) among the mycoplasmas, the smallest cells known (Figure 1.7); a few micrometers in length and somewhat less in width among the bacteria (Figure 1.8); and a bit larger in the blue-green algae (Figure 1.9). The living portion of the cells of bacteria and blue-greens is limited externally by a *plasma membrane,* outside of which a more or less rigid cell wall and a jellylike, mucilaginous capsule or sheath are present. The composition of the wall, which determines the infectivity of some pathogenic groups, varies with the particular prokaryotic species; the bacterial wall, for example, contains lipids, carbohydrates, and complexes of mucopeptides derived from amino acids and amino sugars, while the wall of the blue-greens tends more toward that of the eukaryotic cell in that it incorporates some cellulose.

The cellular contents consist of a less electron dense nuclear

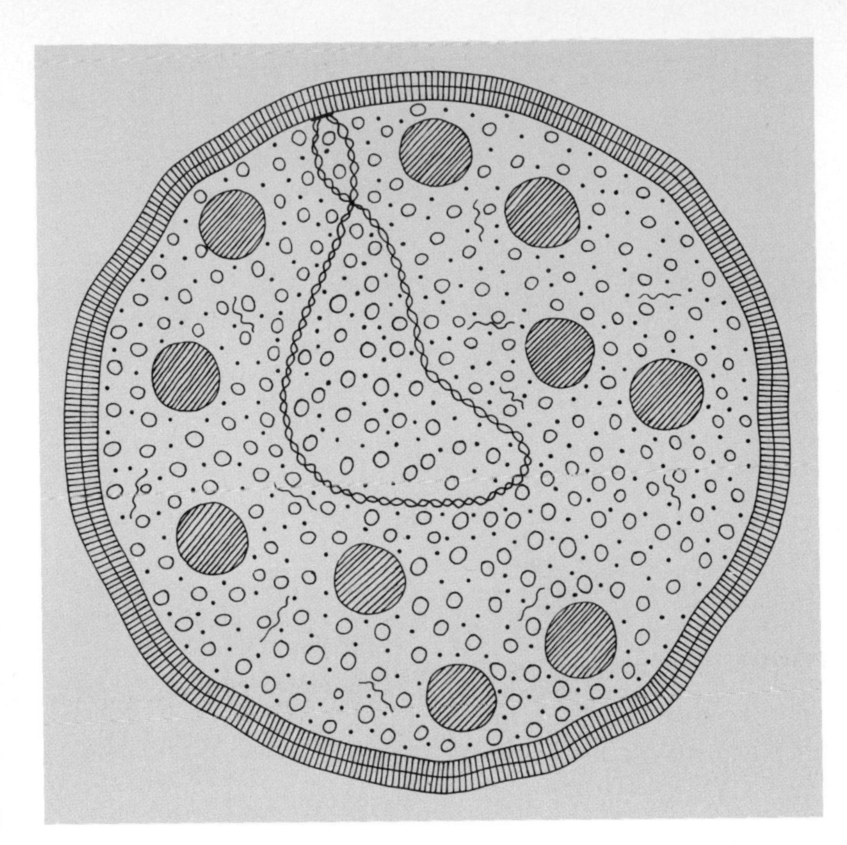

Figure 1.7
Schematic drawing of the morphology of a cell of mycoplasma, one of the smallest cells known. It consists of an outer plasma membrane, a circular double helix of genetic material attached to a structure in the plasma membrane called the mesosome, ribosomes (larger spheres), and other dissolved substances (smaller spheres and threads). The diameter of the cell is about 0.1 μm.

Figure 1.8
Cells of bacteria. Lower left: cells of *Bacillus cereus,* stained to reveal the nuclear areas under the light microscope. The cells will divide by fission, a process of pinching the cell in two between the nuclear areas (see also Figure 11.6). (Courtesy of Dr. C. F. Robinow.) Lower right: electron micrograph of a dividing cell of *Bacillus megatherium* showing the rather heavy cell wall, the plasma membrane that lies just beneath the wall, the light nuclear areas containing thin, twisted strands of DNA, the relatively undifferentiated cytoplasm, which is, however, very rich in ribosomes, and the mesosomes (the large membrane-bound bodies), which are formed by in-foldings of the plasma membrane. (Courtesy of Dr. Stanley C. Holt.)

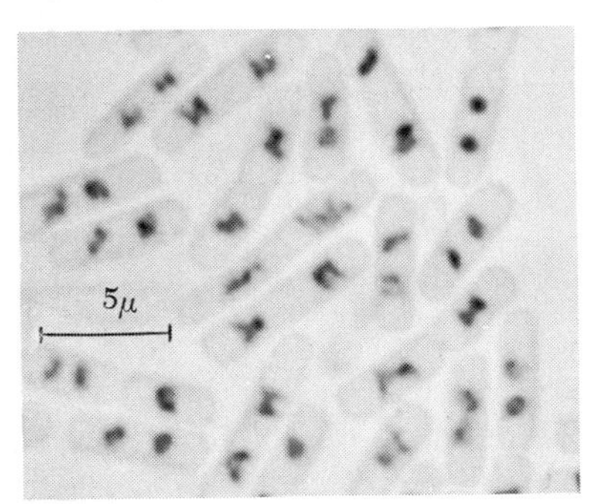

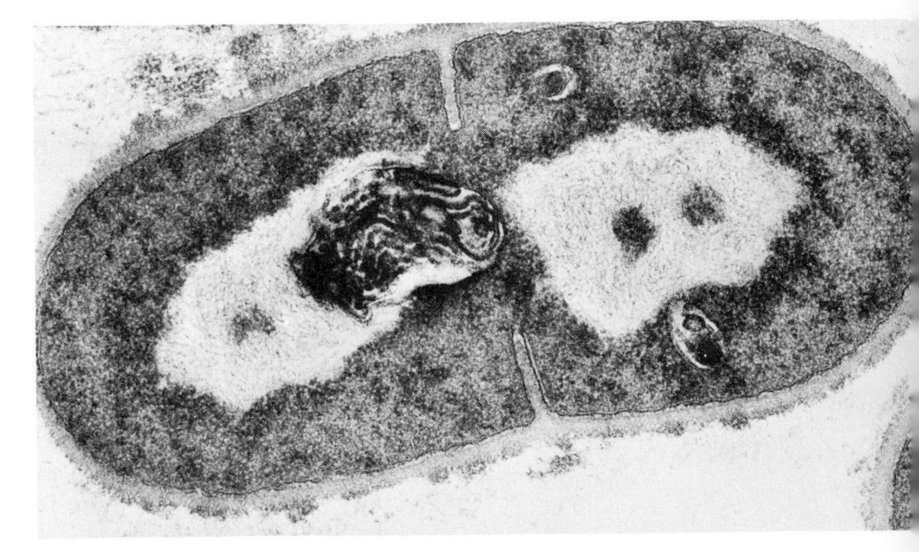

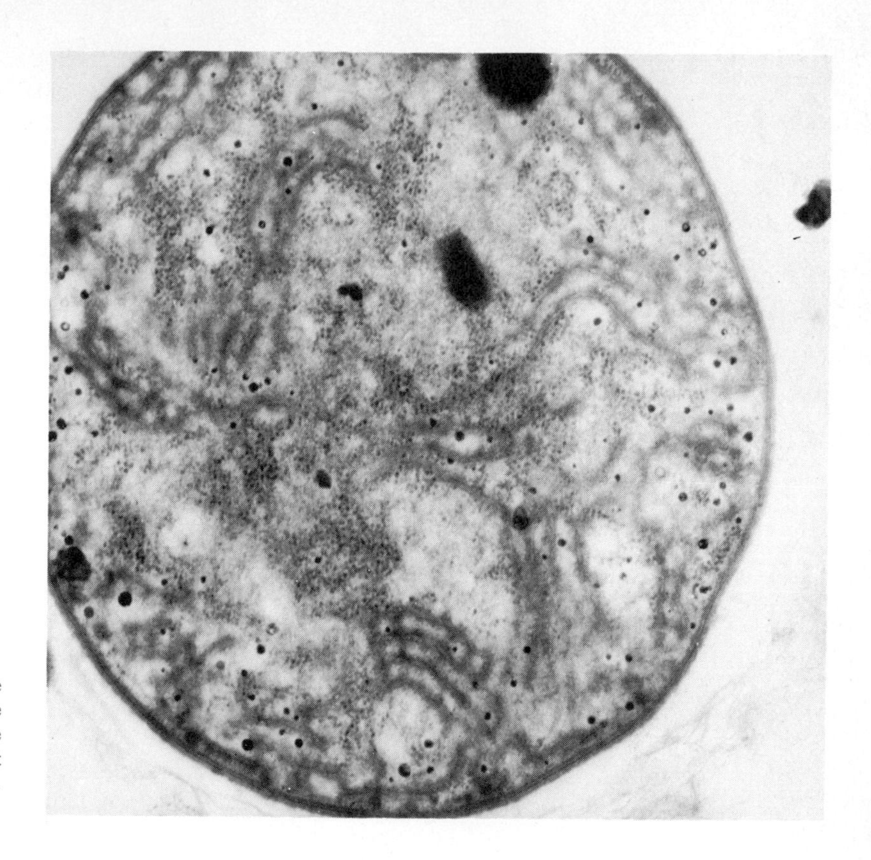

Figure 1.9
Electron micrograph of a cell of the blue-green alga, *Nostoc*, showing the photosynthetic membranes and free ribosomes. The nuclear area is not clearly defined in this illustration. (Courtesy of Dr. J. D. Duckett.)

area and a very dense cytoplasm. Thin, tangled fibers, about 30 to 50 Å (Å = angstrom, 1/10,000th of a micron) in diameter, traverse the clear nuclear area. They are readily extractable, at least from bacteria, and it is known that they are strands of deoxyribonucleic acid (DNA), the hereditary material of the cell. The blue-green algae and some bacteria possess layered membranes that are involved in photosynthesis and that appear to be derived from infoldings of the plasma membrane. The photosynthetic pigment in bacteria is bacteriochlorophyll; the blue-green algae possess chlorophyll-a and phycocyanin, the latter a blue pigment. *Ribosomes*, small, rounded structures about 150 to 200 Å in diameter, are also found in the cytoplasm, but no other internal structure is characteristic of the prokaryotic cell. Some species possess flagella, but their internal structure is quite different from that of a eukaryotic flagellum, lacking, as they do, the 9 + 2 fibrillar structure (see Figures 5.13 and 5.15, Chapter 5). The lack of visible structure in the cytoplasm may be somewhat misleading, however, for the prokaryotes as a group are versatile organisms, deceptively complex as to their biochemical activities and fully capable of carrying on an independent existence in the right environment.

The bacteria divide by simple fission (Figure 1.8). The DNA of the nuclear area appears to be attached to a mesosome, a structural feature of the plasma membrane. When the cell is about to divide, this is initiated by division of the mesosome; the two halves then separate and DNA, the hereditary material, is pulled apart into two masses, which will form the nuclear areas of the two daughter cells. Whether this is also the manner of division of all of the prokaryotes is not known.

Bacteria and blue-greens show an additional similarity in their ability to form resting spores under adverse conditions. A spore coat, at least in bacteria, forms around the nuclear area and a small amount of cytoplasm, and the cell can remain in a state of almost suspended animation until favorable conditions cause the spore to germinate. As a spore, the bacterium is most resistant to adverse conditions, even surviving temperatures of up to 120°C for a brief period. The blue-greens, on the other hand, may actually thrive in hot springs with temperatures ranging up to 70°C; they are responsible for the colored formations seen at such springs, the colors coming from varied accessory pigments.

The eukaryotic cell is a far more elaborately structured and partitioned unit than is the prokaryotic cell from which it is presumably derived (Figure 1.10). An internal division of labor has taken place, accomplished by the use of membranes. The exterior of the cell is bounded by a plasma membrane, to which, in the case of plant cells, an outer wall of cellulose and other materials has been added; the hereditary material is enclosed in a *membrane-bound nucleus,* and is segmented into complex nucleoprotein bodies, or *chromosomes,* the number of which is characteristic for each species; *mitochondria* convert the chemical energy of carbohydrates and fats into energy forms which the rest of the cell can use when needed; the complex cytoplasmic membrane systems of the *endoplasmic reticulum* and *Golgi apparatus* are concerned with the synthesis and packaging of the macromolecules needed for cellular structure and function; sunlight is trapped by green plants in membrane-bound and internally layered *chloroplasts,* and converted into, and sometimes stored as, the chemical energy of carbohydrates; and *vacuoles, lysosomes,* and *peroxisomes,* all membrane-enclosed, play additional roles in the life of eukaryotic cells. Only the ribosomes, chromosomes, *microtubules,* and *microfibrils* are not membraneous in nature. Each of these structures mentioned will be examined in detail in subsequent chapters, but while it is obvious that membranes and membrane-

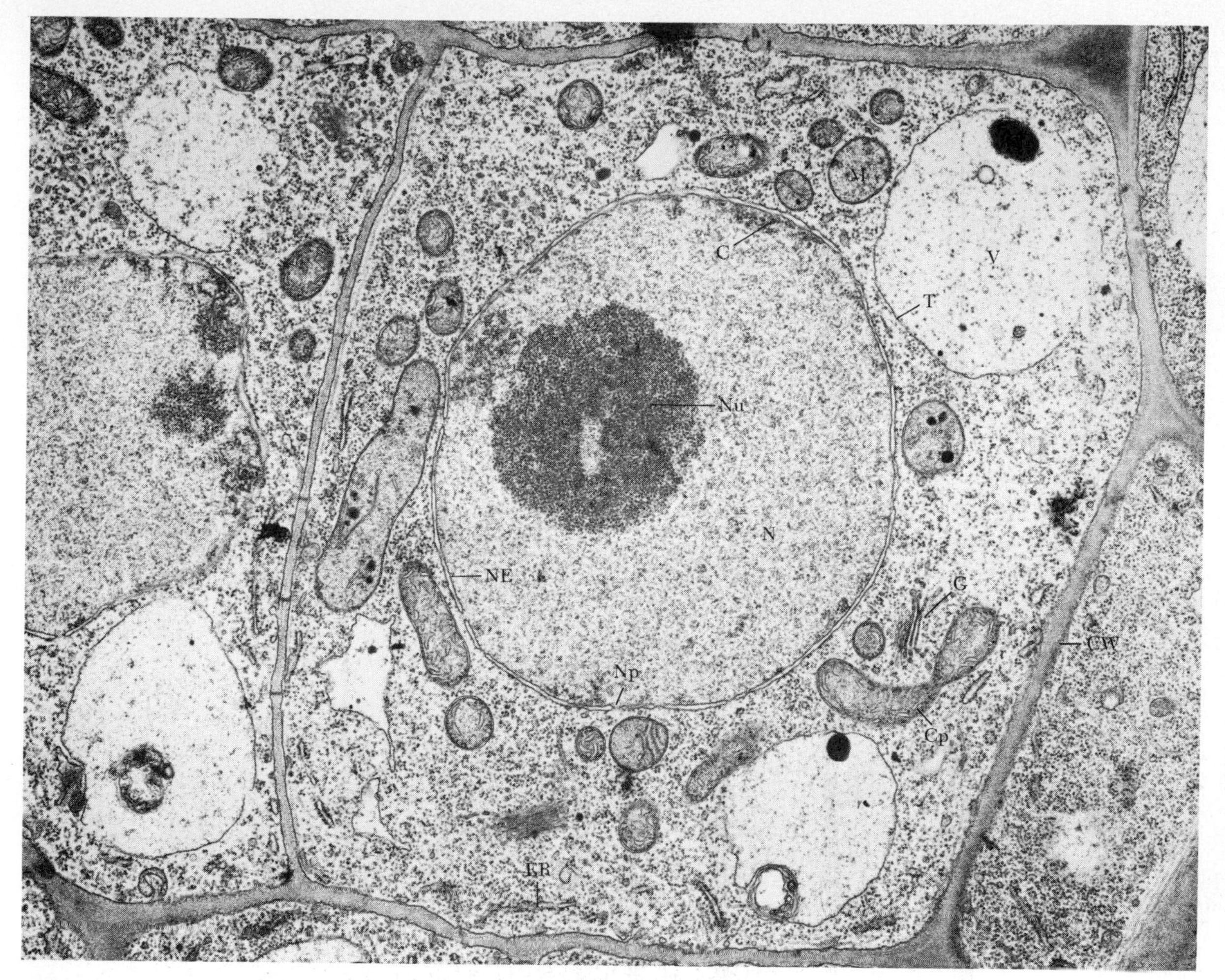

Figure 1.10

Electron micrograph of a eukaryotic cell (plant) showing most of the membranous structures of the cell. N: nucleus; Nu: nucleolus; NE: nuclear envelope; Np: pore in the nuclear envelope; C: chromatin concentrated at the nuclear envelope (the granular aspect of the nucleus is also due to the presence of chromatin); M: mitochondrion; Cp: chloroplast; G: Golgi bodies; ER: endoplasmic reticulum; CW: cell wall; V: vacuole; T: tonoplast, the membrane surrounding the vacuole. The granular aspect of the cytoplasm is due to the presence of many ribosomes. (Courtesy of M. C. Ledbetter.)

bound structures are concerned with energy manipulations in a direct way, so too are all other structures with the exception of the chromosomes. They are bearers of cellular information, but even they are concerned with energy, although as somewhat remote directors rather than manipulators. The expression of chromosomal information takes many forms, so we should anticipate that cellular structures must be flexible as to use and diverse in character to participate in the many reactions that take place in the eukaryotic cell. We need to remember, however, that the simplicity of the prokaryotic cell's structure is not a reflection of simplicity of biochemical activity.

Table 1.1 summarizes the existing differences between prokaryotic and eukaryotic cells. There is, so far as we know, no graded

Table 1.1
Features of Prokaryotic and Eukaryotic Cells.

Feature	*Prokaryote*	*Eukaryote*
Plasma membrane	Present	Present
Nuclear membrane	Absent	Present
Mitochondria	Absent	Present
Endoplasmic reticulum	Absent	Present
Golgi apparatus	Absent	Present
Ribosomes	Present	Present
Cell wall	Present, composed of amino sugars and muramic acid	Lacking in animals; present in plants, with cellulose a major component
Capsule	When present, of mucopolysaccharides	Absent
Vacuoles	Absent	Present (particularly in plants)
Lysosomes	Absent	Present
Chromosomes	Single naked structure composed only of DNA	More than one present, and composed of DNA and proteins
Photosynthetic apparatus	Membranes with chlorophyll-a, and phycocyanin in blue-greens, bacterio-chlorophyll in bacteria	Chloroplasts with chlorophyll a and b in stacked grana
Flagella	Present in some species, but lacking 9 + 2 fibrillar structure	Present in some species, but possessing 9 + 2 fibrillar structure
Division	Simple fission	Spindle of microtubules

series of cells that extends from the prokaryotic mycoplasmas to the complex eukaryotic cells of higher plants and animals. It is probable that the latter were derived evolutionarily from the former, or, more likely, that those forms existing today derived from a common ancestor, but if so the intermediate stages are missing or unknown. We will see, however, in subsequent chapters that both chloroplasts and mitochondria have their own DNA and ribosomes, and hence a degree of hereditary independence from the remainder of the eukaryotic cell. It has been suggested that both chloroplasts and mitochondria, on structural and functional grounds, are invaders that took up residence in other cells, and that during the course of evolution have become adapted to a symbiotic existence. We will examine this question more fully in the last chapter.

Viruses

Viruses are not cells in a structural sense, lacking cytoplasm as well as the organelles of other cells. However, so much information about how organisms function in an hereditary way has been derived from viruses that they warrant consideration in any discussion of cells.

Viruses exist in a wide variety of shapes, sizes, and complexity, and they infect or inhabit, in highly specific patterns, a wide range of host cells. Some of them, the bacteriophages (literally, bacteria-eaters), infect bacteria (Figure 1.11). Some restrict themselves to plant species, such as TMV, the tobacco mosaic virus, or the tobacco necrosis virus. The mosaic virus exists as long rods; the necrosis sometimes forms a crystalline complex of many individual viruses when precipitated from solution (Figure 1.12). Still others infect only animal cells. Smallpox, chicken pox, measles, mumps, and poliomyelitis are among the diseases of humans caused by animal viruses, while still others appear to be responsible for the transformation of normal into malignant cells.

As a group, the viruses exhibit certain characteristics. Some

Figure 1.11
Electron micrographs of bacterial viruses. Both of these DNA-containing bacteriophages, P2 on the left and T6 on the right, attack the colon bacterium, *Escherishia coli.* Their detailed structure is indicated in Figure 1.13.

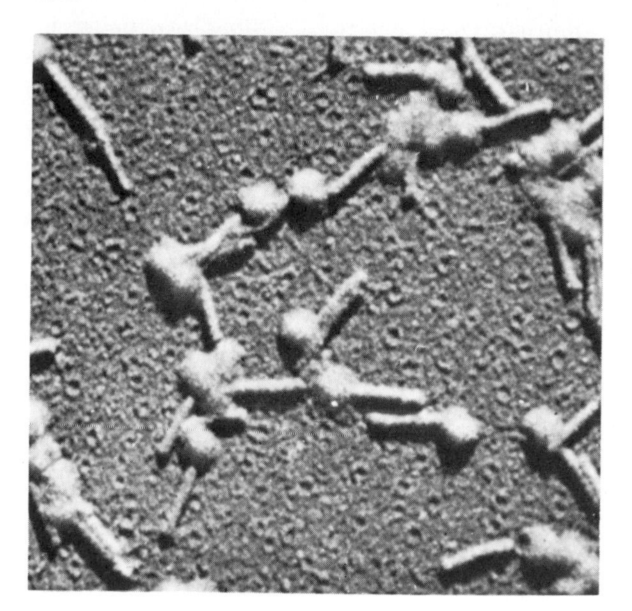

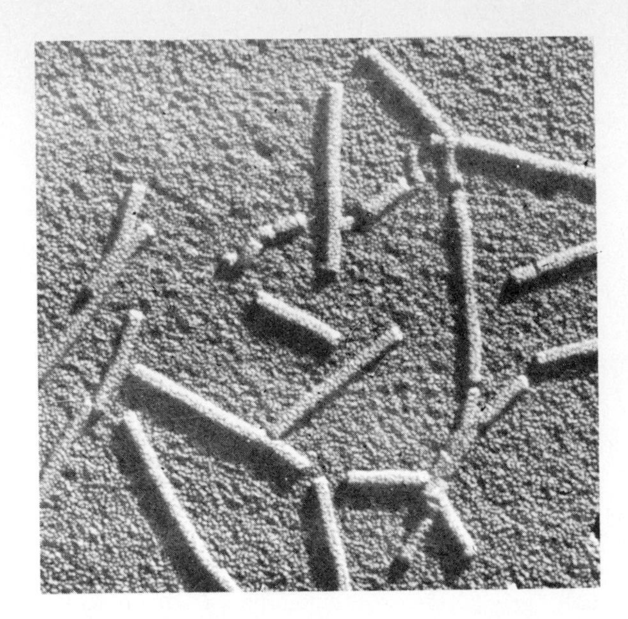

Figure 1.12
Electron micrographs of two RNA-containing viruses, which infect plants. Left: tobacco necrosis virus. The individual viral particle is spherical and about 250 Å in diameter, but when precipitated with ammonium sulfate they characteristically form a crystalline structure. Right: tobacco mosaic virus. Each virus is rod-shaped, with the protein on the outside and RNA on the inside (see Figure 1.14 for detailed structure). (Courtesy of Dr. L. W. Labaw.)

characteristics set them apart from cells, others enable an investigator to treat them experimentally as if they were cells or organisms. First, they are extremely small, able to pass through a filter that would strain out even the tiniest prokaryotic cells. Their presence in a filtrate that was infective led to their discovery long before they were pictured in an electron microscope. Thousands may be formed within a single bacterium, although in being so prolific they may lyse (kill) the cell at the time of their release. Others, such as RNA tumor viruses, are released without disrupting the cell. There are, in fact, many molecules of biological origin that are larger than the very small viruses.

Second, as their size would suggest, the viruses are very simply constructed compared to even a mycoplasma cell. Basically, they consist of a central core of nucleic acid (either DNA or RNA, a related ribonucleic acid) and an outer covering of protein. The nucleic acid represents their hereditary potential, while the protein serves two purposes: a protective covering for the nucleic acid when outside of the host cell, and a means of specificity for recognizing the kind of cell it can infect. The different ways of packing the two kinds of macromolecules provide the viruses with their distinctive shapes and sizes (Figures 1.13 and 1.14).

Third, viruses differ from cells, whether prokaryotic or eukaryotic, in that they are unable to lead an independent existence. They must infect a cell before reproduction can occur; otherwise, they are simply inert chemicals packaged in a special way. Their de-

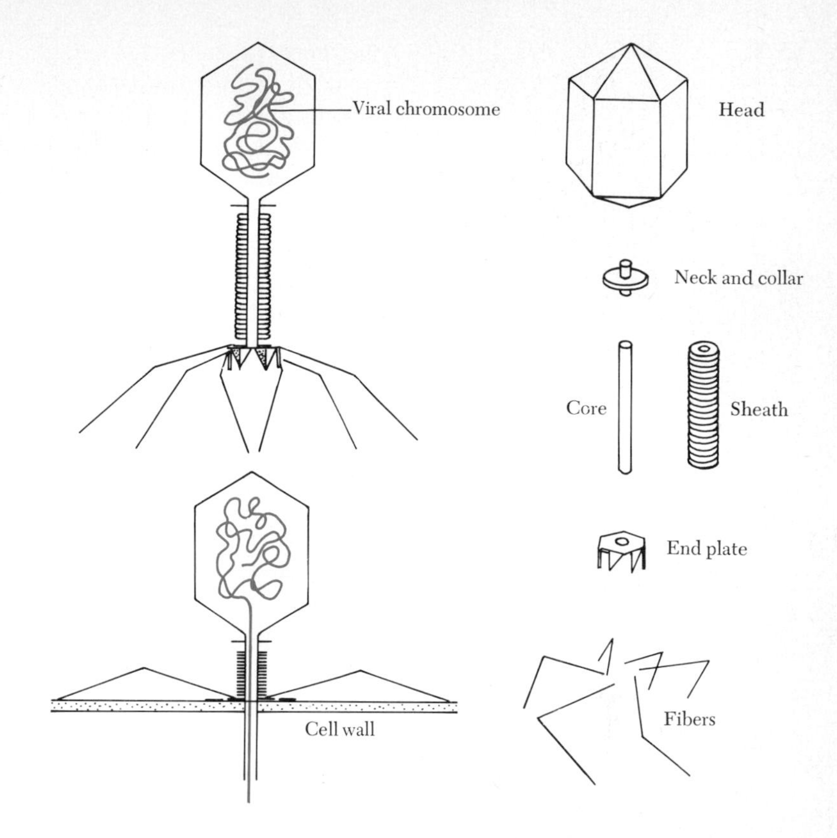

Figure 1.13
Diagrammatic representation of the structure of a T-even (T2, T4, T6) bacteriophage, with the intact virus, its component parts, and as it appears attached to the bacterial wall. The DNA is contained in the protein head. After attachment to the bacterial wall by the action of the end plate and fibers, the protein sheath contracts, driving the core through the wall; the DNA then passes from the head, through the hollow core, and into the interior of the cell where it can either multiply or become attached to the bacterial chromosome. If it multiplies, the bacterial contents will be used up in the process, and the wall will burst to release the virus particles into the medium where they can infect other cells. If the viral DNA attaches itself to the chromosome, it becomes an integral part, replicating along with the remainder of the chromosome, but without interfering with the behavior or integrity of the bacterial cell.

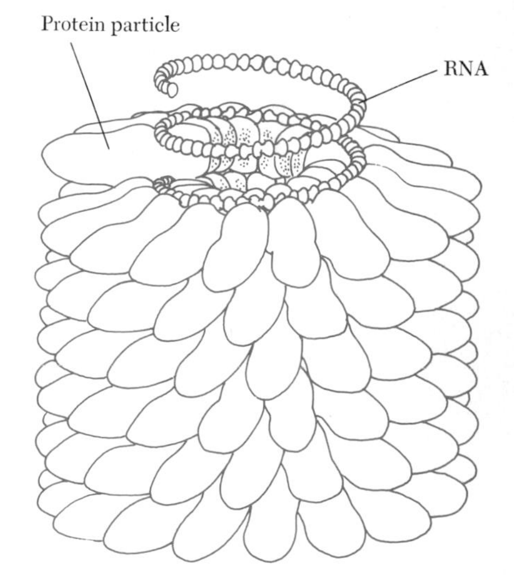

Figure 1.14
Diagrammatic representation of the structure of the tobacco mosaic virus. The protein particles, each having a molecular weight of 17,000, are assembled helically around the RNA molecule, eventually forming a hollow rod about 3,000 Å long, and a diameter of about 170 Å. It is possible to dissociate the protein and RNA. The RNA of the virus is infective by itself, the protein is not; the RNA, therefore, is the crucial hereditary molecule, the protein is more or less protective. If the RNA and the protein are added separately in solution, they will aggregate spontaneously to form typical TMV rods.

pendent existence as obligate cellular parasites suggests that to reproduce they must somehow commandeer the machinery of the invaded cell and make use of it for their own purposes. The host cell and the invading virus are, therefore, intimately related to each other, although how this relationship was achieved is not known. Some viruses, in fact, are so lacking in hereditary capabilities that they possess only enough information to code for their own protein coat.

Whether viruses are living organisms is an academic question. In many respects they can be handled experimentally as though they were cells or unicellular organisms. Their relation to disease in both plants and animals makes them of much importance in human affairs, but for problems of heredity, mutability, immunity, biochemistry, and molecular biology in general, they are remarkably useful experimental objects. Their heredity is bound up in nucleic acid molecules, so that an understanding of their inheritance patterns is possibly transferable to higher forms; they mutate like any cell or organism and their genes are similar even if their patterns of hereditary transmission may be unique; their very simplicity of structure and their limited hereditary potential are advantages in trying to correlate structure and function; and some of them, such as the bacteriophages, possess prodigious powers of reproduction, with thousands of individual virus particles being released within a very short period of time after infection.

The origin of viruses in the evolutionary scheme of life is uncertain. Their dependence upon living cells as a culture medium for growth and reproduction would indicate that they could not have preceded cellular organisms in origin. This fact, plus their lack of usual cellular machinery, would suggest that they might be very degenerate cells or fragments of cells, having somehow rid themselves of all unnecessary features except their hereditary apparatus in the form of nucleic acids and their protective and infective apparatus in the form of protein. In any event, viruses are intriguingly packaged units of genetic information, and much of what will be described in the next chapter on cellular information was first apprehended and understood in these simple units.

Tools and techniques

Any science is mature to the extent that it has acquired a sound and acceptable theoretical base. The theory of evolution through natural selection, the cell doctrine, and the theory of the gene provide the fundamental structure of biology, but progress in theory

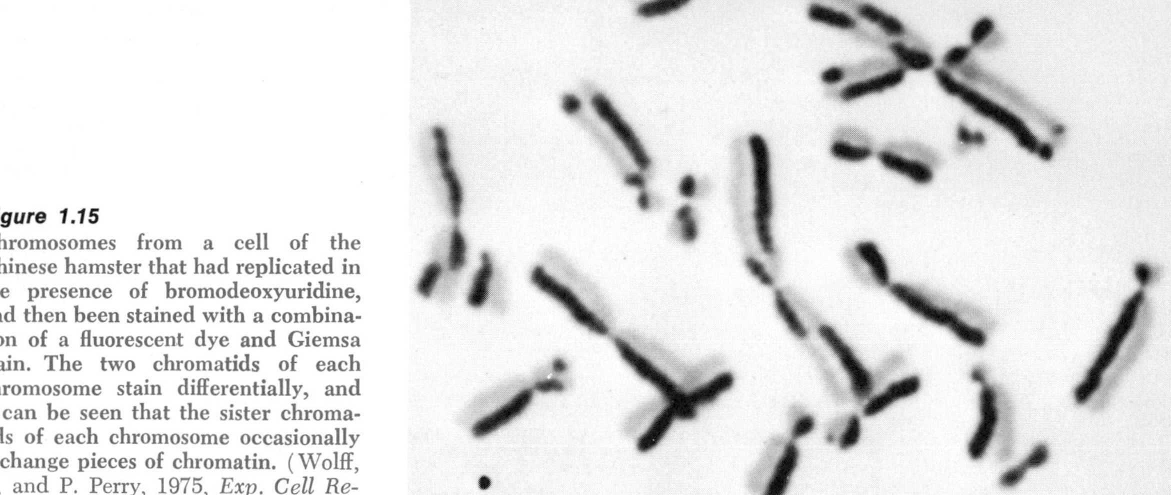

Figure 1.15
Chromosomes from a cell of the Chinese hamster that had replicated in the presence of bromodeoxyuridine, and then been stained with a combination of a fluorescent dye and Giemsa stain. The two chromatids of each chromosome stain differentially, and it can be seen that the sister chromatids of each chromosome occasionally exchange pieces of chromatin. (Wolff, S., and P. Perry, 1975, *Exp. Cell Research* 93, 23–30; courtesy of Dr. Sheldon Wolff.)

often plays leapfrog with new instrumentation, new techniques, and new questions to be answered. This situation has been especially true for cytology. Some cells may be large enough to see with the unaided eye. But to identify their internal organization, we must magnify them greatly and often use specific dyes, radioactive atoms, or fluorescent molecules to highlight selected parts of the cell (Figure 1.15).

In attempting to study in detail the objects they observe, adequate magnification is as much of a problem for the cytologist, who has to overcome very small sizes, as it is for the astronomer, who has to overcome great distances. For our purposes, the problem of magnification can be best considered in terms of resolving power, which is the ability of an optical system to reveal details of structure, or more specifically, to show as discrete entities two small points or bodies that lie close together. In observing a double star (for example, one found in the handle of the Big Dipper, seen in the Northern Hemisphere), some of you will be able to discern but a single star; others, with better resolving power, will see two separate stars. In a compound microscope, the resolution of the first lens is the critical factor. As Figure 1.16 indicates, the lens nearest the specimen being examined, called the objective lens, is the key element of a compound miscroscope, be-

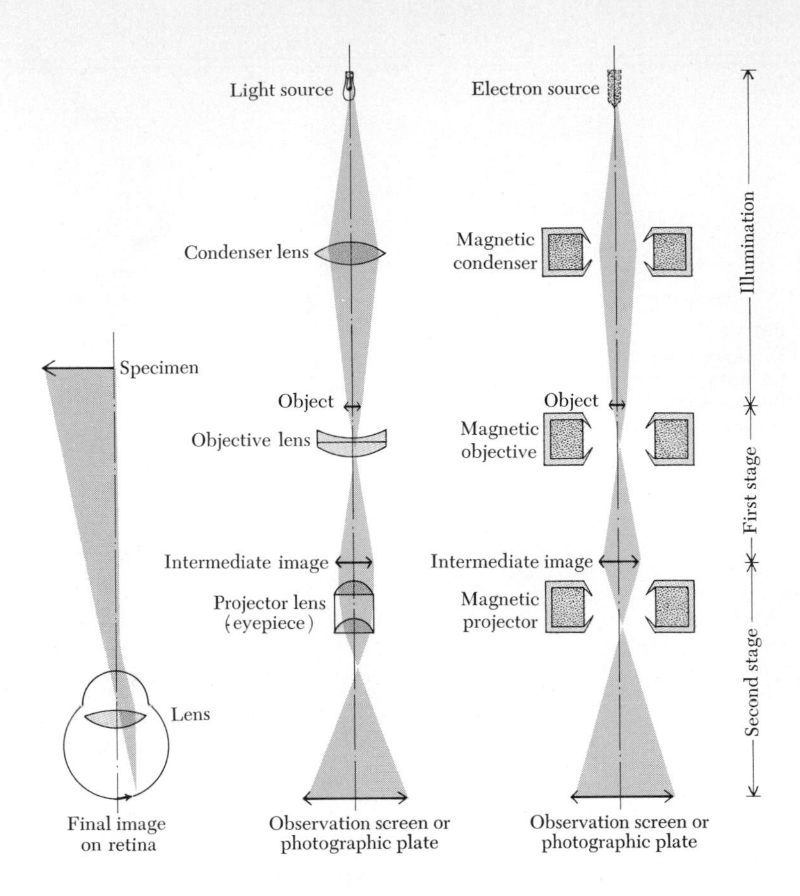

Figure 1.16
Schematic representation of the optical systems of the human eye, the light microscope, and the electron microscope.

cause the projector lens, the ocular, can magnify only what the objective has resolved.

The unaided human eye has a resolving power of about 0.1 to 0.2 mm (millimeters). Lines closer together than this will be seen as a single line, and objects that have a diameter smaller than this range will be invisible or seen only as blurred images. The human eye, however, has only slight powers of magnification; each of us, therefore, must calculate sizes mentally, and experience is probably the largest factor in our ability to judge accurately when we are using our unaided visual sense. Microscopes, of course, both resolve and magnify, but with the lenses obtainable today the ability to resolve is limited by the kind of illumination used. Objects that are closer than about one-half the wavelength of the illuminating light cannot be clearly distinguished in a light microscope. Thus, even with the most perfectly ground lenses, and with white light having an average wavelength of 5,500 Å (angstroms), an oil immersion objective cannot resolve two discrete points separated from each other by a distance less than 2,700 Å, or 0.27 μm (micrometers). Since many parts of cells have smaller dimensions, their presence and structure were undetected until a means of greater resolution was found.

The electron miscroscope provides this increased resolving power by making use of "illumination" of a different sort. High-speed electrons are employed instead of light waves. As the electrons pass through the specimens being viewed, parts of the cell absorb, permit the passage of, or scatter electrons differentially because of differences in density. An image of the specimen is thus formed on an electron-sensitive photographic plate or fluorescent screen. The human eye is not stimulated by electrons, hence the need for plates or screens. The "optical" system is similar to that in the light microscope (Figure 1.16), except that the illumination is focused by magnetic lenses instead of conventional glass lenses.

Although electrons are particles and hence possess mass, they behave like radiant energy and can be characterized by their wavelength, which varies with voltage. When accelerated through the microscope by a potential difference of 50,000 volts, they have a wavelength of about 0.05 Å. This is 1×10^{-5} that of average white light. An electron microscope can thus theoretically resolve objects that are separated by a distance of about 0.01 Å. This dimension is far less than the diameter of an atom (the hydrogen atom has a diameter of 1.06 Å), but owing to limitations in the way that magnetic lenses function, the actual resolving power of the best modern instrument is about 5 to 10 Å. At this level, individual atoms cannot be distinguished, but large molecules of biological importance are readily visible (Figures 1.17 and 1.18).

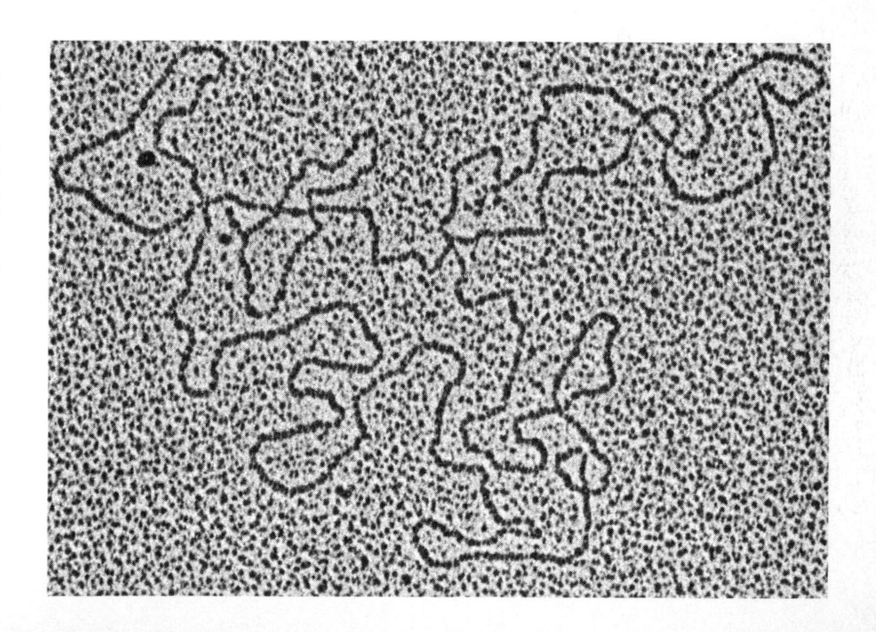

Figure 1.17
Electron micrograph of the circular chromosome of the P22 bacteriophage, which infects the bacterium, *Salmonella typhimurium.* This is basically a naked DNA molecule, released from a disrupted viral particle, spread on a protein film layered on water, and then picked up on a grid before being photographed. The background is the surface of the grid. (Courtesy of Dr. L. McHattie.)

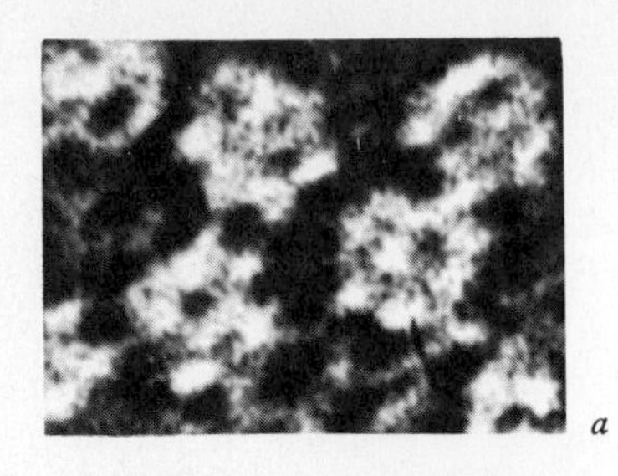

a

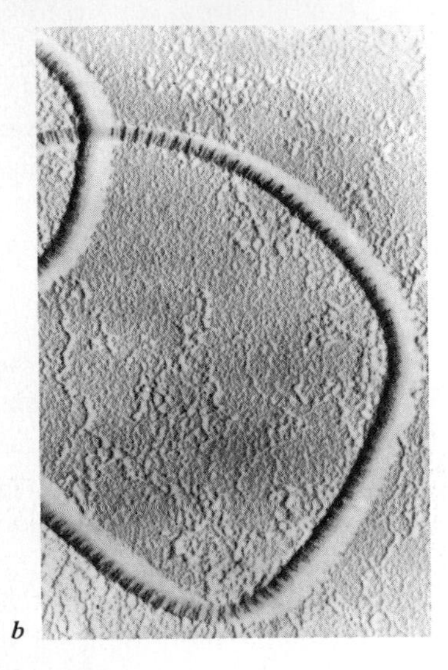

b

Figure 1.18
Electron micrographs of protein molecules: (*a*) aggregates of the carbon dioxide-fixing enzymes extracted from spinach chloroplasts; and (*b*) collagen, an animal protein found prominently in cartilage, showing its characteristic banded structure. (Courtesy of Dr. E. Moudrianakis.)

In approximate figures, then, the human eye can resolve down to 100 μm, the light microscope to 0.2 μm, and the electron microscope to 0.001 μm. Or, to put it another way, if the normal human eye has a resolving power of 1, that of the light microscope is 500, and of the electron miscroscope is 100,000. The electron microscope has thus opened a whole new domain to the cytologist by making visible a number of cellular structures that would have otherwise remained undetected.

The significance of magnification in microscopy, as distinguished from that of resolution, is that the smallest objects resolvable by the light miscroscope need to be magnified to about 2,000 to 4,000 diameters to be readily discernible visually. In the electron microscope, objects 10 Å in diameter require magnifications of 200,000 to 300,000 times.

The increasing degree of resolution made possible by advancements in microscopy is indicated in Figure 1.19, as well by the other electron micrographs in this book. How much more refined we can become in our visualization of cellular structure and organization is open to question. An electron microscope that would resolve structure at the level of 1 Å would enormously expand our field of vision, for it would enable us to visualize molecular organization directly, as well as molecular aggregates such as those in Figures 1.17 and 1.18. Reliable preservation of the fine structure of the cell seems to have been achieved through freeze-drying techniques, as well as through the more conventional means of fixation, sectioning, and staining, so this aspect of microscopy does not seem to be the bottleneck for progress. Neither is voltage a problem, for present voltages seem more than adequate. Lens construction, however, presents inherent difficulties, and it well may be that as in light microscopy, a limitation has been reached, but with lens design rather than wavelength of radiation the limiting factor.

A word needs to be said about the differences in preparation of cells for light and electron microscopy, differences dictated by the character of the radiation used. Visible light readily passes through whole cells, or those that have been sliced in a microtome at thicknesses of 5 to 10 μm. Electrons of the voltage customarily used would be totally absorbed in such thick masses, so cells must be sliced at about 100 Å thickness to be useable. For electron microscopy, an ultramicrotome, using diamond or glass knives, is needed for precision cutting, and the cells must be embedded in a plastic rather than the customary paraffin used by the light microscopist.

Regardless of the instrument used or the method of preparation of cells for examination, the parts of the cell being investigated

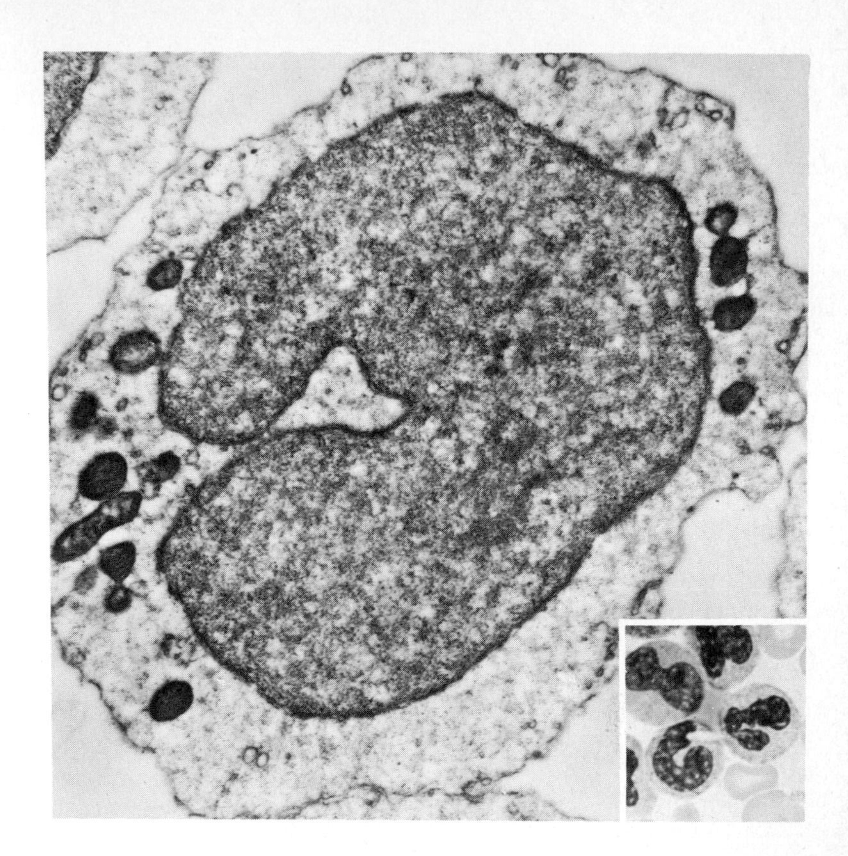

Figure 1.19
Differences in resolution: an electron micrograph of a human lymphocyte; and (inset) a group of comparable cells photographed in the light microscope.

must be clearly distinguishable from their immediate surroundings.

In the electron microscope, this contrast is possible because some structures are or can be made more electron "dense" than others, and photosensitive film is darkened to the degree that it is struck by the electrons passing through a specimen. Since the degree to which electrons are scattered is a function of the mass of the atom, and since the light atoms of organic materials—hydrogen, carbon, nitrogen, and oxygen—have little scattering power, the parts of the cell to be examined may be selectively "stained" with heavy metals to show contrast. Heavy metals such as osmium, bismuth, uranium, and manganese are customarily used. Contrast is equally difficult to achieve with the light microscope because most parts of the cell are transparent to light. To overcome this problem, the cytologist uses the proper killing agent (fixative) and stain to color selectively the parts to be examined. Literally hundreds of fixing and staining procedures are known; they are the cytologist's recipes, continually improved upon in the search for better ways to study cells. Since many molecules, because of their chemical makeup, will selectively absorb or interact with particular dyes, some of them fluorescent, the staining procedures are used not only to reveal cellular structure but also to assist in the identification and distribution of molecules that could not be detected in any other way.

One of the limitations of both light and electron microscopy is that transmitted illumination, whether visible light or electrons, reveals only a two-dimensional image of the specimen being examined. A three-dimensional idea of the cell or its parts can be constructed only from serial sections, a difficult, if not impossible, task for the electron microscopist when one realizes that it would require 1,000 to 3,000 sections from an ordinary-sized cell of 30 μm in diameter. Furthermore, it is impossible in the transmission electron microscope to visualize whole cells. The development of the scanning electron microscope (SEM) overcomes these difficulties in part, and within a certain range of magnification.

The scanning electron microscope enables the development of three-dimensional images for several reasons. It does not record the electrons passing through the specimen, but collects rather than focuses all of the secondary electrons released from the specimen by the impinging electrons. Since the impinging electrons constitute needlelike probes, providing thereby a constant depth of focus regardless of magnification, the hills and valleys of the specimen are clearly revealed instead of simply a single plane of focus (Figure 1.20; compare with Figure 1.23). In addition, the

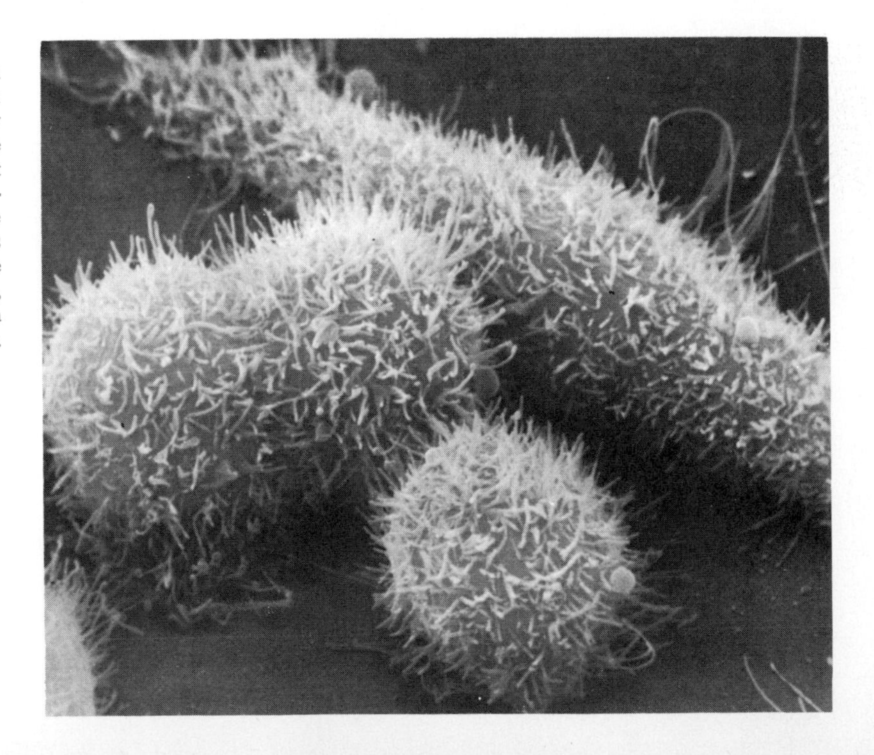

Figure 1.20
A scanning electron micrograph (SEM) of a living, malignant human (Hela) cell, cultured in the laboratory and growing on the surface of the medium. The rounded cells are about 30 μm in diameter. Since only the surfaces of objects can be viewed in the SEM, the thickness of the specimen is not of critical importance. Compare this figure with that in Figure 1.23, which is a similar cell taken in a phase microscope. (Courtesy of Dr. Keith Porter.)

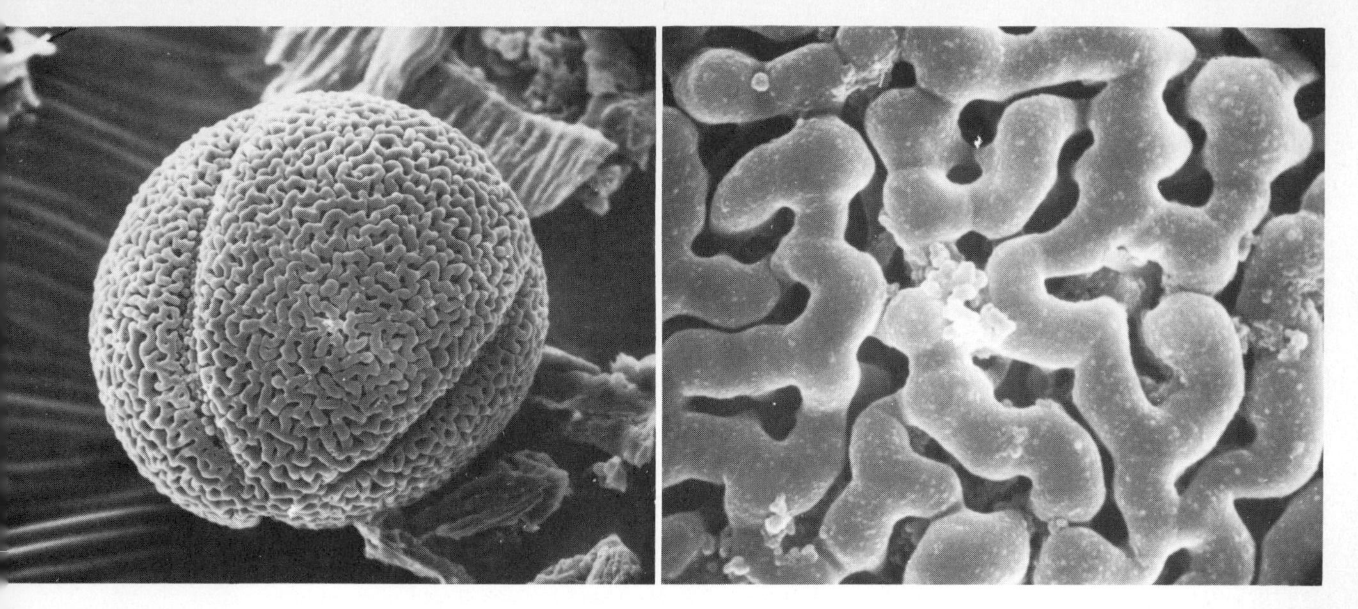

Figure 1.21
Images obtained with the scanning electron microscope. Left, pollen grain of the lotus, *Nelumbo lutea* (× 1900), and right, a higher magnification of the sculpturing of the pollen wall (× 10,000). (Courtesy of Dr. James Walker.)

scanning electron microscope possesses a continuous range of magnifications, and can yield clean images from a magnification of about 15 diameters, also possible from good hand lenses, to 20,000 diameters (Figure 1.21). The scanning electron microscope, therefore, spans the resolution and magnification ranges of both the light and transmission electron microscopes, but in a complementary way, thus adding greatly to the ability of the cytologist to study cells.

Parts of cells also can be selectively studied through the use of molecules containing radioactive atoms, particularly phosphorus 32 (^{32}P), carbon 14 (^{14}C) and tritium (^{3}H), an isotope of hydrogen. When radioactive molecules are taken up and react chemically with particular parts of the cell, their presence and location can be detected through autoradiography. That is, a thin layer of photographic emulsion is spread over the flattened cells, and as the radioactive atoms disintegrate, the rays or particles released from these atoms pass into the emulsion and cause a darkening of the emulsion much in the manner that exposure to light causes a photographic film to darken. When such cells are viewed under the microscope they may appear as in Figure 1.22. Molecules possessing radioactive atoms are, therefore, like dyes or fluorescent molecules having a high degree of specificity of attachment: they are cellular "probes" or "labels" that enable the microscopist to distinguish one kind of molecule from another, or to follow the reaction pathway of particular molecules or their parts.

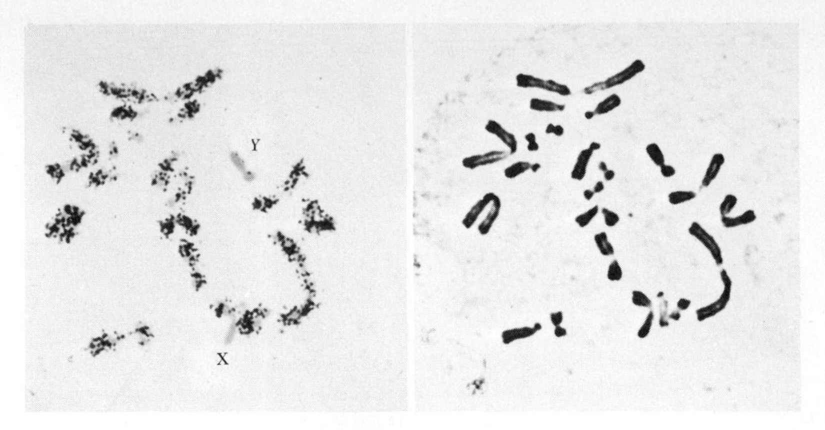

Figure 1.22
A metaphase cell of the male Chinese hamster that had gone through part of its life cycle in the presence of radioactive tritium (^3H). The tritium replaces hydrogen in thymidine, a molecule that becomes incorporated selectively in the DNA of the chromosomes. Right: the cell was stained with Feulgen and then photographed. The coverglass was then removed, and an emulsion was spread over the preparation and allowed to remain for some time before being rephotographed (left). When the tritium atoms disintegrate, they darken the emulsion as they pass through, much as a photographic film is darkened when struck by light. Notice that the X and Y chromosomes have fewer disintegrations over them than the other chromosomes; this is because they undergo replication at a different time than do the remainder of the chromosomes, and at a time when the tritiated thymidine was not available. (Courtesy of Dr. T. C. Hsu.)

A living cell, however, is always more fascinating than a dead one. To watch cells divide is to witness one of the most dramatic of biological phenomena. This cannot be done at the level of the electron microscope because of the vacuum in which the specimen is placed, but a special light microscope, called a phase-contrast microscope, permits the examination of living material. It does so by taking advantage of the fact that the higher the refractive index of a medium (or cell part) the more is light retarded in velocity as it passes through. Transmitted light from one part of a cell is, therefore, out of phase with that from another part, and the phase-contrast microscope converts these differences in phase into differences in brightness. Figure 1.23 is a photograph of a living human cancer cell taken through a phase-contrast microscope. Under the conventional light microscope, such a cell would appear almost structureless.

Another way to study living cells is to "grind" them up and examine their parts. This is done with a special mortar and pestle to burst the cells and release their contents into a solution. When

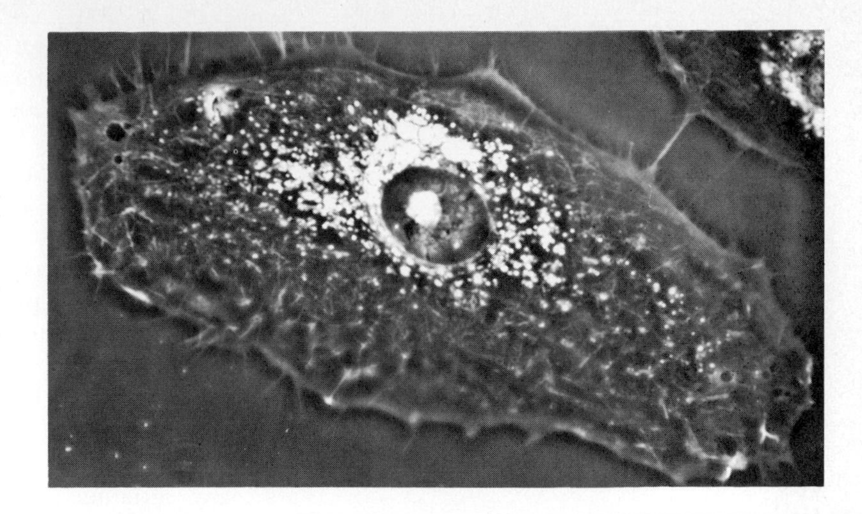

Figure 1.23
A Hela cell grown in tissue culture and photographed through a phase-contrast microscope. The nucleus with its nucleolus is visible in the center; the white mass is liquid refractile material taken in from the culture medium; the slender rods inside the cell are mitochondria; and the outer fine projection are microfibrils frequently formed by mammalian cells grown in tissue culture. Compare this cell with that in Figure 1.20. (Courtesy of Dr. George O. Guy.)

this solution is centrifuged at carefully regulated speeds, the different parts of the cell, suspended in solution, settle out and sediment according to their density; that is, the denser portions settle out at lower speeds, the less dense ones at higher speeds (Figure 1.24). Even large molecules of different molecular weights can be separated from each other, but this is done by density gradient centrifugation rather than by the differential centrifugation depicted in Figure 1.24. In density gradient centrifugation,

Broken cells in suspension
600 × g, 10 minutes
Nuclei
Supernatant
8,000 × g, 10 minutes
Mitochondria
Supernatant
25,000 × g, 10 minutes
Lysosomes
Supernatant
100,000 × g, 10 hours
Ribosomes
Soluble enzymes and nucleic acids

Figure 1.24
A flow diagram showing the fractionation of the cell into its component parts by means of differential centrifugation. If one were using plant cells, chloroplasts capable of carrying out photosynthesis can be obtained by centrifuging at 2,000 to 3,000 × g for 1.0 minute. The soluble enzymes, DNA, and the several RNAs could be separated still further by either chromatographic or electrophoretic techniques.

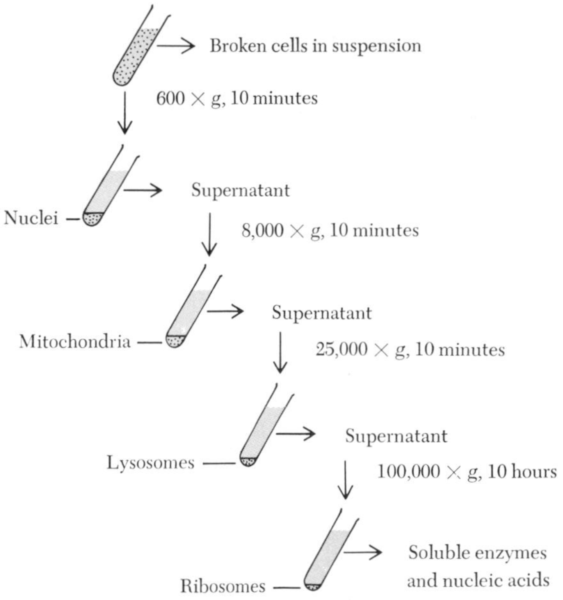

a concentrated solution of cesium chloride or sucrose is centrifuged at about 100,000 times gravity for about 20 hours. At this force the molecules of cesium chloride, for example, move toward the base of the centrifuge tube, forming a gradient of density from top to bottom. Any molecules suspended in this solution prior to centrifugation, say molecules of DNA or protein, will also move toward the bottom of the tube, and will come to rest at that place in the density gradient equal to the density of the molecules. The centrifuge tube then can be punctured at the bottom, and the solution can be collected in fractions and analyzed for content or tested for chemical activity.

Mixtures of molecules of biological importance also can be separated and individual molecular species often purified by either chromatography or electrophoresis. In chromatographic techniques—there are many variations, each of particular usefulness—solutions of molecules are poured through a vertical column of specially prepared materials and collected in fractions of equal amounts at the bottom of the column. The molecules move down the column according to size or degree of attraction to the materials in the column. In electrophoretic procedures, the mixture in question is placed in an electrical field, and the net electrical charge of the molecules causes them to move along a moistened paper or a gel. The greater the charge on the molecule the farther it will move. Individual molecular species will appear as spots or bands when properly stained, and then they can be isolated and examined for purity, chemical structure, and/or biological activity.

Centrifugation, chromatography, and electrophoresis are, of course, aids in studying cell function and chemical activity, but it should be realized that an isolated particle or molecule may or may not behave in a test tube as it does in a complex cell. Techniques for studying the intact cell are limited, particularly at the level of chemical activity and interaction, and other approaches such as those described are useful.

Living cells of plants and animals, including humans, also can be cultured with ease in much the same way as are bacteria or fungi and subjected to a wide variety of procedures to test their response and future behavior. These cells can be probed by microneedles, dissected by the techniques of microsurgery, and injected with solutions through micropipettes. Cells of different animal species—for example, rat and human cells—also can be cultured together and be caused to hybridize through cell fusion, thus permitting an examination of hybrids that could not be produced by any other means. The cytologist, therefore, has a large and varied arsenal of instruments and a whole battery of procedures for making the cell give up its secrets. Any one tool or technique, however, is usually not enough, and several methods are often employed before an answer can be found to the particular problem being investigated. Such knowledge as we have already garnered from the cell has strengthened our belief that it is the basis of life, and at the same time has made us acutely aware of how little we know of its many complexities.

Bibliography

DYSON, R. D. 1975. *Essentials of Cell Biology.* Allyn & Bacon, Inc., Boston, Mass. An introductory text for the student with a limited background in biology and chemistry.

FAWCETT, D. W. 1966. *The Cell: Its Organelles and Inclusions.* W. B. Saunders Co., Philadelphia. A volume of superb electron micrographs of many kinds of cells.

HSU, T. C. 1974. Longitudinal differentiation of chromosomes. *Annual Review of Genetics* 7, 153–176. A review of recent staining methods, including fluorescence, for the identification of particular chromosomes.

HUGHES, A. 1959. *A History of Cytology.* Abelard-Schuman, Ltd., New York. An excellent historical account of microscopical observations, the cell theory, cell division, studies of the cytoplasm, theories of inheritance, and the place of cellular theory in biology as a whole.

LEDBETTER, M. C., and K. R. PORTER. 1970. *Introduction to the Fine Structure of Plant Cells.* Springer-Verlag, New York. Superb collection of electron micrographs, coupled with brief explanatory descriptions.

LEHNINGER, A. L. 1965. *Bioenergetics.* W. A. Benjamin, Inc., New York. An elementary treatment of how energy is managed by living systems.

MCELROY, W. D. 1973. *Cell Physiology and Biochemistry.* Prentice-Hall, Inc., Englewood Cliffs, N. J. A small but complete volume with emphasis on the chemical structure and behavior of cells.

Scientific American. September, 1970. The Biosphere. The entire issue is devoted to energy and its relation to living systems.

———. September, 1971. Energy and Power. Articles by F. J. Dyson, D. M. Gates, C. M. Summers, M. Tribus, and E. C. McIrvine are pertinent to an understanding of energy systems and energy flow.

WOLFE, S. L. 1972. *Biology of the Cell.* Wadsworth Publishing Co., Belmont, Cal. A comprehensive treatment of cells with excellent illustrations.

Note: In 1974, three scientists were awarded the Nobel Prize for their efforts in dissecting the cell by various methods. Their lectures, given before the Nobel Institute, describe their findings and are published as follows:

CLAUDE, A. 1975. The coming of age of the cell. *Science* **189**, 433–435.

DEDUVE, C. 1975. Exploring cells with a centrifuge. *Science* **189**, 186–194.

PALADE, G. 1975. Intracellular aspects of the process of protein synthesis. *Science* **189**, 347–358.

The diagram in Figure 1.4, Chapter 1, indicates that the flow of energy through a living system is not haphazard or random, but is directed by an informational source within the cell. Since structure and function are intimately related, structure also must be determined by this same source of information. Hence the cell as a whole, structurally as well as functionally, is an expression of its own information; it directs and regulates the flow of energy, and provides the mechanisms for doing so.

That this is so should be expected. Each individual, plant or animal, that owes its existence to some form of sexual reproduction, began life as a single cell, the fertilized egg. Each such cell is different. The cell from an oak tree will, through growth and differentiation, produce another oak tree; a human egg eventually will produce a human being. Not only will the eggs of different species be different, so will the eggs within a species, although to a different degree. We can witness this most dramatically in identical twins. Such twins result when a mass of cells, produced from a single fertilized egg, splits in two at a very early stage of development. Each segment, consisting of a few hundred undifferentiated cells, will produce an entire individual made up of many billions of cells, yet the twins will show such remarkable physical and behavioral similarities that they are often told apart only with difficulty.

The striking similarity of identical twins forces us to recognize that their very existence, as well as that of any other individual, is an example of biological engineering of great precision. Our mass production assembly lines are scarcely more exact than these processes of life. A number of questions, therefore, arise. Where is the "blueprint" of the individual to be found? In what form is the information stored? How is the information released so that it can be acted upon? Or, to put our questions in terms of Figure 1.4, how does the information govern the flow of energy in a living system so that growth, reproduction, and individual expression are achieved?

Our answers must be sought at the cellular level, for the cell is the basic unit of organization. We now know that this information is found largely within the nuclei of eukaryotic cells and the nuclear area of prokaryotes, and it is coded in the form of genes, which collectively constitute the nucleic acids of each cell. Such a brief statement, however, compresses the results of over a century of investigation before a full understanding of hereditary information was gained.

Enzymes

Once radiant energy has been trapped and converted into chemical bond energy (Figure 1.4), all other events taking place within the cell are through the medium of chemical reactions. What a cell is and does are consequences of the chemical reactions that occur within it, and that have occurred in the past. All of these chemical reactions, and consequently all energy manipulations taking place within and between cells, are governed by enzymes; it is the enzyme complement of the cell, therefore, that, by controlling the flow of energy through the cell, provides the cell with its own uniqueness. As we shall see, enzymes are the first detectable expression of the coded information put to use by the cell, although we need to remember that the cell in its entirety is also an expression of the same source of coded information.

Enzymes are highly specific biological catalysts. Each kind of enzyme governs a particular chemical reaction, determining not so much whether or not a reaction will take place (most reactions will occur spontaneously, although very slowly, if left alone), but rather the rate at which the reaction proceeds. Like any catalyst, enzymes are not used up in the reaction events they govern. How do they accomplish this? In answering that question we must continually bear in mind that the cell is a closely regulated system, and that if it is to function properly, the chemical reactions taking place within it must be governed as to time, place, and rate.

The timing of enzymatically governed reactions is complicated by many factors, intracellular and extracellular, and they will not be discussed here. Since enzymes do not act at a distance from the site of a reaction, the place of a reaction is determined by the location of the enzyme. This is not a random affair. Enzymes as a rule do not float freely around in the cell, particularly in a eukaryotic cell, but tend to be strongly compartmentalized or localized—in or on membranes, in nuclei, mitochondria, chloroplasts, lysosomes, peroxisomes, and vacuoles, and indeed wherever membranes exist, such as the endoplasmic reticulum and the Golgi complex. This permits a spatial control of enzyme actions and tends to keep chemical pathways independent of one another within the cell. However, as pointed out earlier, the only intrinsic function of an enzyme is to govern the rate of a reaction, and it does this by lowering the *energy of activation*.

Molecules consist of two or more atoms united to each other by chemical bonds, and these bonds represent a certain amount of energy depending upon the kind of atoms and the kind of bonds uniting them. For these molecules to react, bonds must be dissolved or broken, and new ones formed, but to break a bond requires as much or more energy than that represented by the bond itself. That amount of energy, or the barrier to a reaction, is called the *energy* of *activation*. To react with each other, molecules must collide or be brought into close juxtaposition. This can occur normally but generally at a very slow rate for molecules of biological importance, but the rate of reaction can be increased by heating the system. This increases the internal energy of the molecules and also increases the likelihood of their collision, and hence of their reacting. The cell, however, functions only within a limited range of temperatures, and the reacting molecules are rarely in high concentration, thus lessening the chances of a collision. Not all collisions will lead to a reaction, but a catalyst introduced into the system will increase the reaction rate by lowering the energy of activation, thus permitting a larger number of collisions to lead to a successful reaction.

Let us assume that molecules A and B react very slowly in solution to form the products C and D. In time an equilibrium

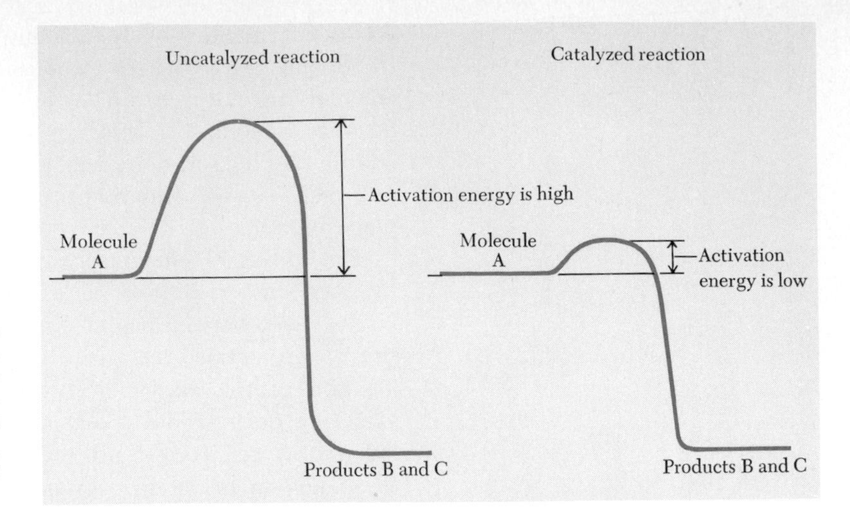

Figure 2.1
Curves showing the lowering of the activation energy by an enzymatic catalyst as contrasted with the larger amount of energy needed to accomplish the same thing in the absence of the enzyme. The enzyme makes the molecule more reactive without itself being consumed in the process.

will be reached that can be described by the formula $A + B \rightleftharpoons C + D$. Each reaction required that the energy barrier to the reaction be breached in some way for the products C and D to be formed. This is indicated diagrammatically in Figure 2.1, which also depicts how an enzyme, specific for the reaction, can increase the rate by lowering the energy of activation. If this occurs in a cell, an equilibrium is not reached because the cell is an open system and is continually exchanging energy and matter with its environment.

We can then ask how an enzyme can lower the energy of activation. This results from its structure. Structurally an enzyme is a macromolecule, a long-chain protein made up of a sequential array of amino acids. The number, kind, and sequential position of these amino acids depend upon the coded information in the cell (see p. 67), but they also determine the three-dimensional geometry of the enzyme, and its *active site* (Figure 2.2). In terms of the reacting system just described, the enzyme forms an unstable, intermediate complex with its substrate, that is, the two molecules A and B, at its active site, a complex that decomposes rapidly into C and D, and that serves as an energy bypass route to overcome the initial energy barrier. In other words, the enzyme lowers the energy of activation, thus speeding up the rate of reaction. Once the products C and D are formed, they are released from the enzyme, and the enzyme is now ready to catalyze another similar reaction.

The life of a cell is, therefore, an exquisitely coordinated and modulated series of chemical reactions occurring within an intricately compartmentalized and orderly structure. When we rea-

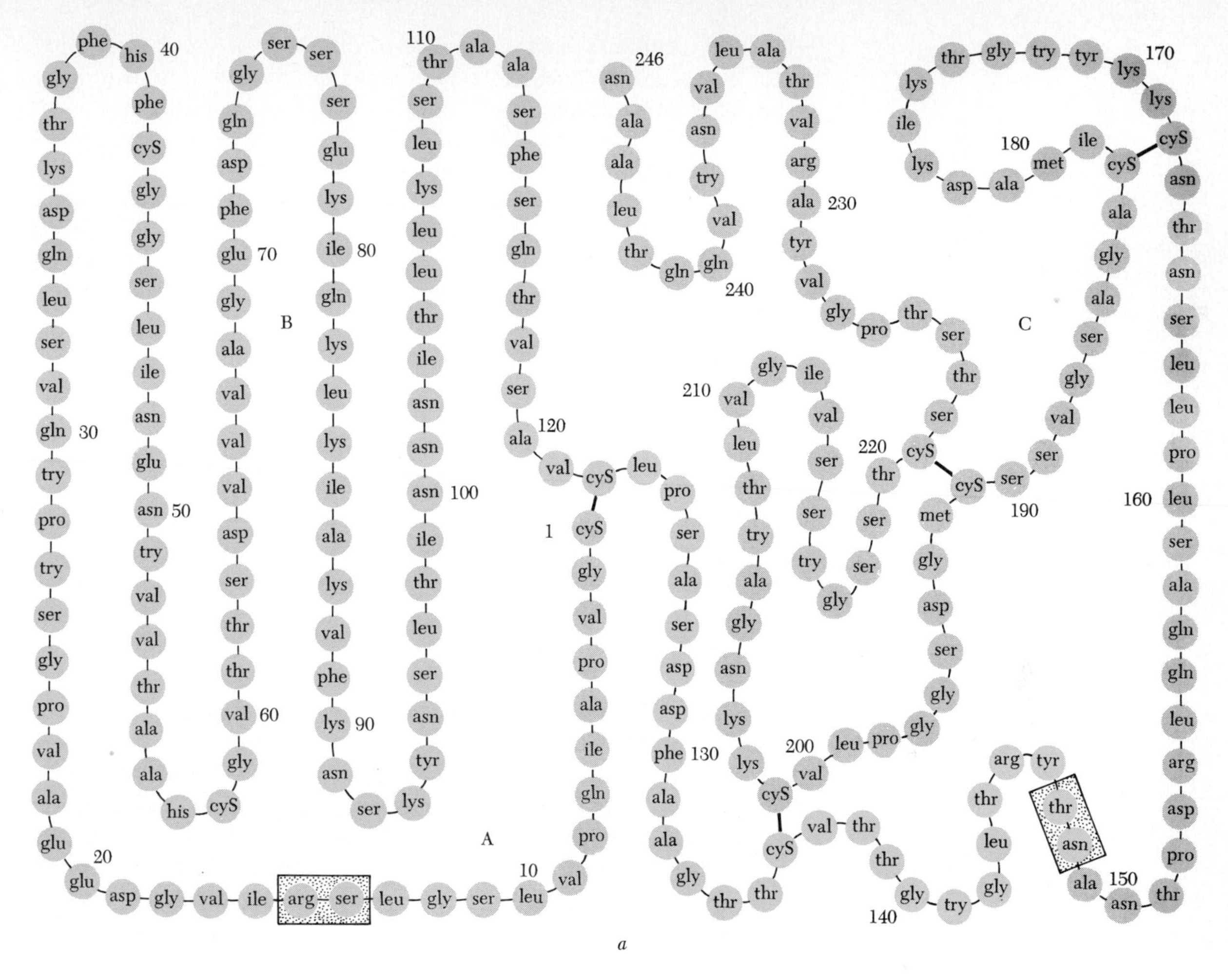

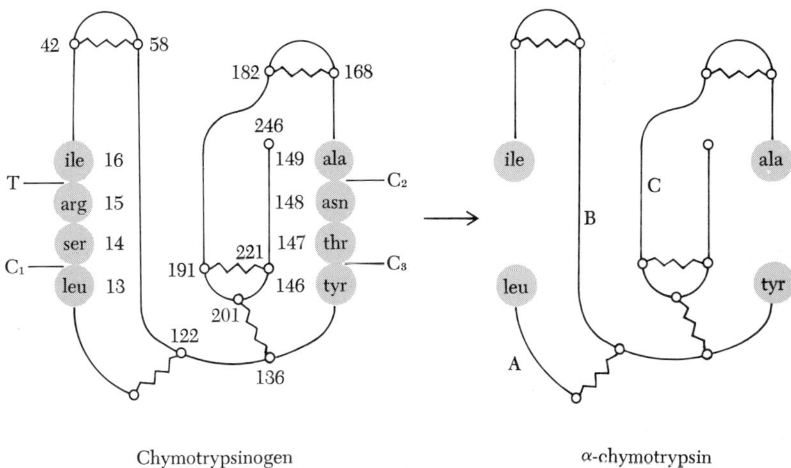

Figure 2.2
(*a*) the amino acid sequences making up the structure of bovine chymotrypsinogen A. It consists of three polypeptide chains (A, B, and C), held together with disulphide (S-S) bonds. The structure in (*a*) is an inactive form that is converted into an active enzyme by the removal of two dipeptides (in the shaded boxes). (From F. Wold, *Macromolecules: Structure and Function*, Englewood Cliffs, N. J., Prentice-Hall, Inc., 1971.)

lize that a cell may have thousands of different enzymes controlling that many different reactions, we begin to have an appreciation of the basis of our uniqueness, and of the fact that this has been achieved "by the elegant use of a small number of building blocks organized into specific macromolecules" (p. 8).

The nucleus

The membrane-bound nucleus is the most prominent feature of the eukaryotic cell (see Figures 1.19 and 1.23, Chapter 1). Schleiden and Schwann, when setting forth the cell doctrine in the 1830s, considered that it had a central role in growth and development. Their belief has been fully supported even though they had only vague notions as to what that role might be, and how the role was to be expressed in some form of cellular action. The membraneless nuclear area of the prokaryotic cell, with its tangle of fine threads, is now known to play a similar role.

Some cells, like the sieve tubes of vascular plants and the red blood cells of mammals, do not possess nuclei during the greater part of their existence, although they had nuclei when in a less differentiated state. Such cells can no longer divide, and their life span is limited. Other cells are regularly multinucleate. Some, like the cells of striated muscles or the latex vessels of higher plants, become so through cell fusion. Some, like the unicellular protozoan Paramecium, are normally binucleate, one of the nuclei serving as a source of hereditary information for the next generation, the other governing the day-to-day metabolic activities of the cell. Still other organisms, such as some fungi, are multinucleate because cross walls, dividing the mycelium into specific cells, are absent or irregularly present. The uninucleate situation, however, is typical for the vast majority of cells, and it would appear that this is the most efficient and most economical manner of partitioning living substance into manageable units. This point of view is given credence not only by the prevalence of uninucleate cells, but because for each kind of cell there is a ratio maintained between the volume of the nucleus and that of the cytoplasm. If we think of the nucleus as the control center of the cell, this would suggest that for a given kind of cell performing a given kind of work, one nucleus can "take care of" a specific volume of cytoplasm and keep it in functioning order. In terms of energy (see Figure 1.4), this must mean providing the kind of information needed to keep the flow of energy moving at the correct rate and in the proper channels. With the multitude of enzymes in the cell, energy can of course be channelled in a multitude of ways;

it is the function of some informational molecules to make some energy channels more preferred than others at any given time. How this regulatory control is exercised is not entirely clear.

The nucleus is generally a rounded body. In plant cells, however, where the center of the cell is often occupied by a large vacuole, the nucleus may be pushed against the cell wall, causing it to assume a lens shape. In some white blood cells, such as the polymorphonucleated leucocytes, and in the cells of the spinning gland of some insects and spiders, the nucleus is very much lobed (Figure 2.3). The reason for this is not clear, but it may relate to

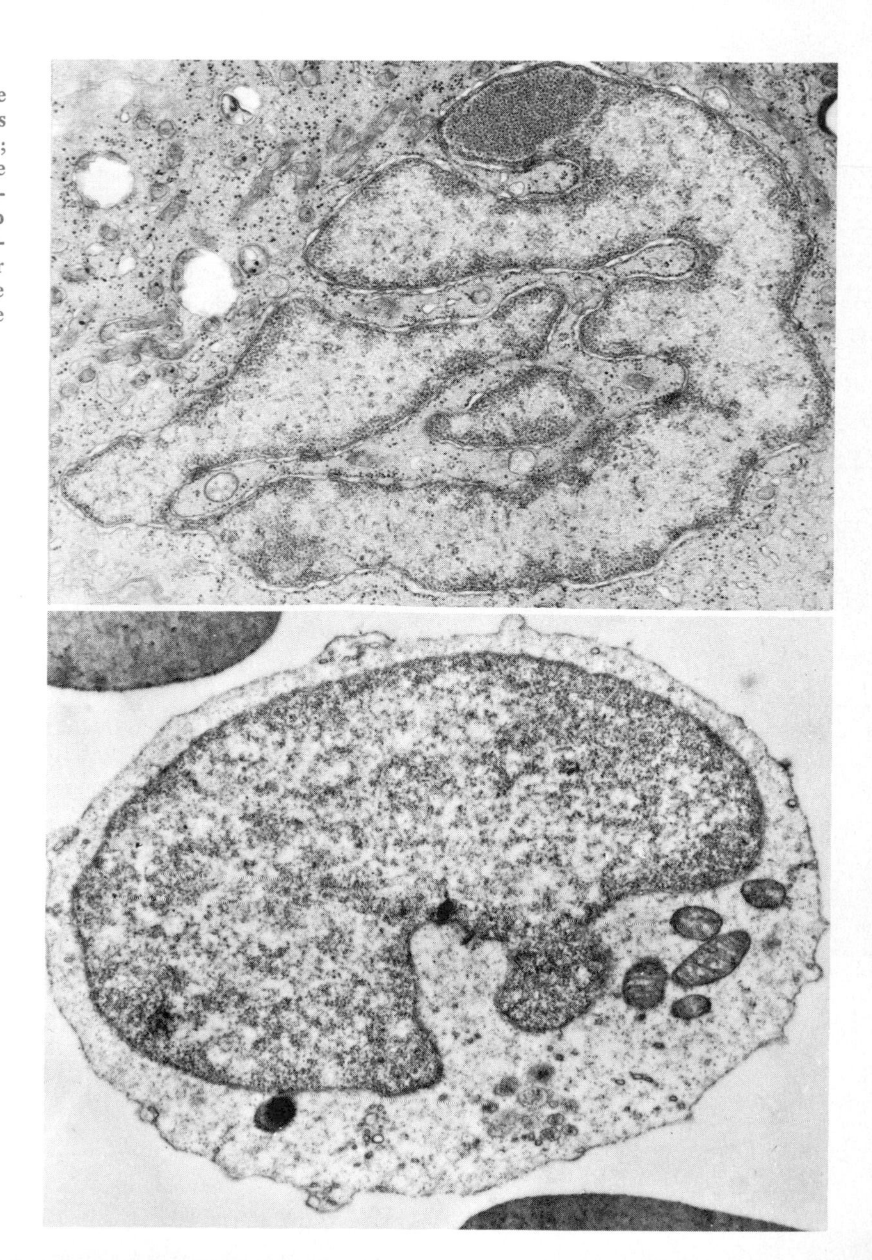

Figure 2.3
Electron micrographs of human white blood cells. Top: highly lobed nucleus of a polymorphonucleated leucocyte; concentrations of chromatin can be seen adjacent to the nuclear membrane and filling the lobe at the top of the nucleus. Bottom: small leucocyte with its nucleus lobed to a lesser extent. Both of these cells have characteristically large nuclei relative to the amount of cytoplasm.

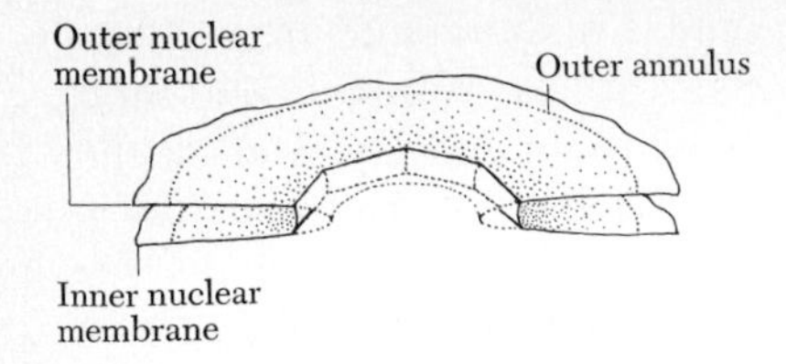

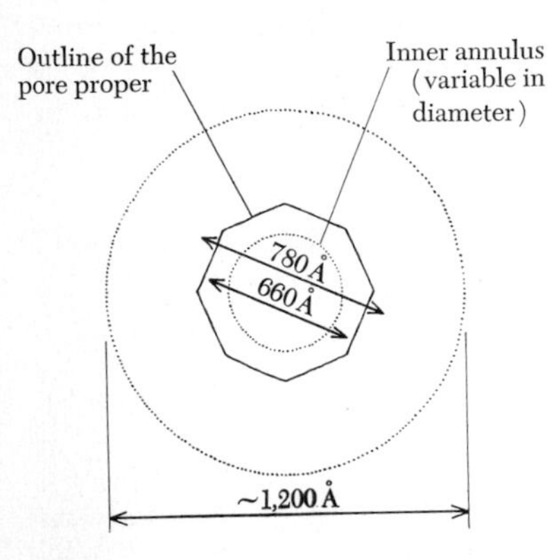

Figure 2.4
Diagram showing the character and dimensions of the pores in the nuclear envelope of an amphibian, as interpreted by Gall. Compare with the electron micrograph in Figure 2.3. (Courtesy of Dr. J. D. Gall.)

the fact that for a given volume of nucleus, a lobate form provides a much greater surface area for nuclear-cytoplasmic exchanges, possibly affecting both the rate and the amount of metabolic reactions.

The nucleus, whatever its shape, is segregated from the cytoplasm by a double membrane, the *nuclear envelope,* with the two membranes separated from each other by a *perinuclear space* of varying width. The envelope is absent only during the time of cell division, and then just for a brief period. As Figure 5.2, Chapter 5, shows, the outer membrane is often continuous with the membranes of the endoplasmic reticulum, a possible retention of an earlier relationship since the envelope, at least in part, is formed at the end of cell division by coalescing fragments of the endoplasmic reticulum. The cytoplasmic side of the nucleus is frequently coated with ribosomes, another fact that stresses the similarity and relation of the nuclear envelope to the endoplasmic reticulum (see Chapter 5).

Everything that passes between the cytoplasm and the nucleus in the eukaryotic cell must traverse the nuclear envelope. Since some fairly large molecules are known to move from nucleus to cytoplasm, and since the structure of cellular membranes of all sorts would seem to preclude such movement, a paradox appears evident. This well may be more apparent than real for it is known that ribosomes must pass from nucleus to cytoplasm, but what happens to the membrane at the time of passage is not known. However, nuclear membranes are periodically interrupted by pores, at the edges of which the inner and outer nuclear membranes appear to be continuous (Figure 2.4). In plant cells these are irregularly and rather sparsely distributed over the surface of the nucleus, but in such cells as the amphibian oocyte, the pores are numerous, somewhat octagonal in shape, and regularly arranged (Figure 2.5). Their presence would suggest an easy passageway for molecules, even ones of substantial size, but high resolution electron micrographs, particularly of animal cells, show that the pores are not simple openings in the membrane, but are often filled

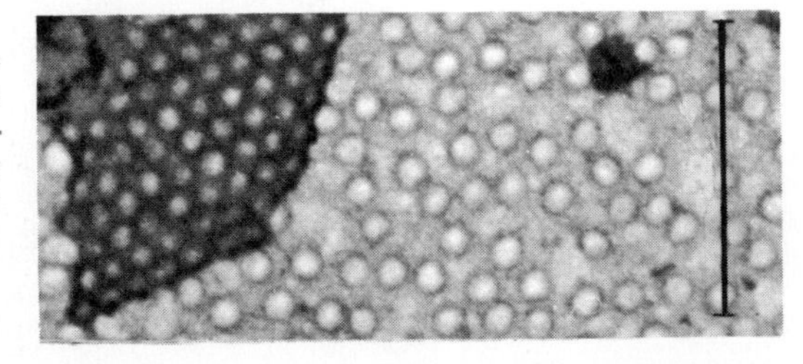

Figure 2.5
Face-on view through the electron microscope of the nuclear membrane from a frog's egg, showing the regular arrangement of the nuclear pores; the darker portion is an overlying membrane fragment. The distribution of the nuclear pores in the cells of higher plants is far more irregular, as indicated in Figure 5.2. (Courtesy of Dr. R. W. Merriam.)

with a plug of electron dense material. Differences in pH (hydrogen ion concentration) between nucleus and cytoplasm also suggest that the passage of materials is not as free as the membrane structure might suppose. The significance and the origin of the pores, therefore, remain uncertain.

The nucleus stains readily with basic dyes after acid fixation to reveal a network of fine threads, among which coarser masses of stained material are distributed, much of it adjacent to the inner nuclear membrane (Figure 2.6). The electron microscope, with its greater powers of resolution, reveals much the same network of

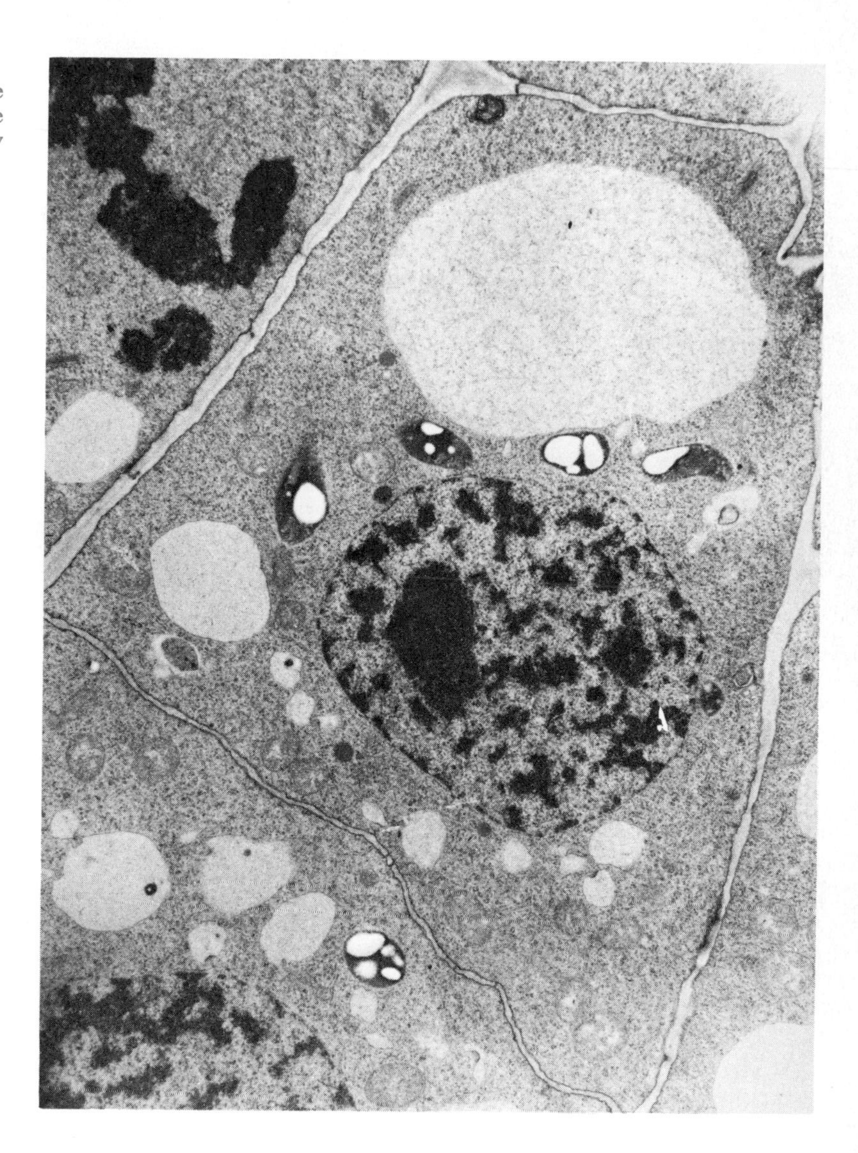

Figure 2.6
An interphase cell of a root tip of the pea, *Pisum sativum*, showing how the chromatin is often appressed closely to the nuclear envelope.

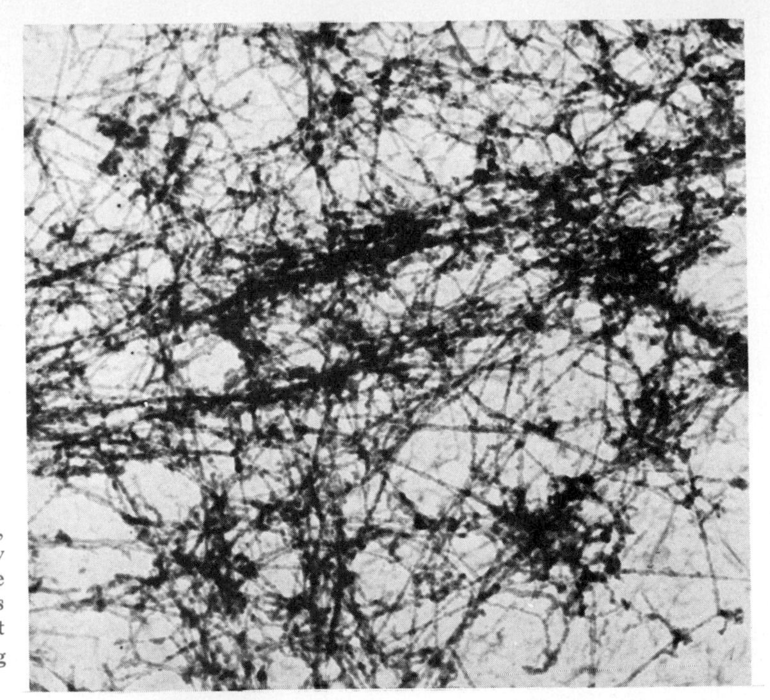

Figure 2.7
Electron micrograph of a portion of a nucleus, prepared by floating it on water, at a very early stage in the division of a cell from the testis of the milkweed bug, *Oncopeltus fasciatus*. The strands of chromatin are about 100 Å in diameter and reveal no distinguishing features. (Courtesy of Dr. S. Wolfe.)

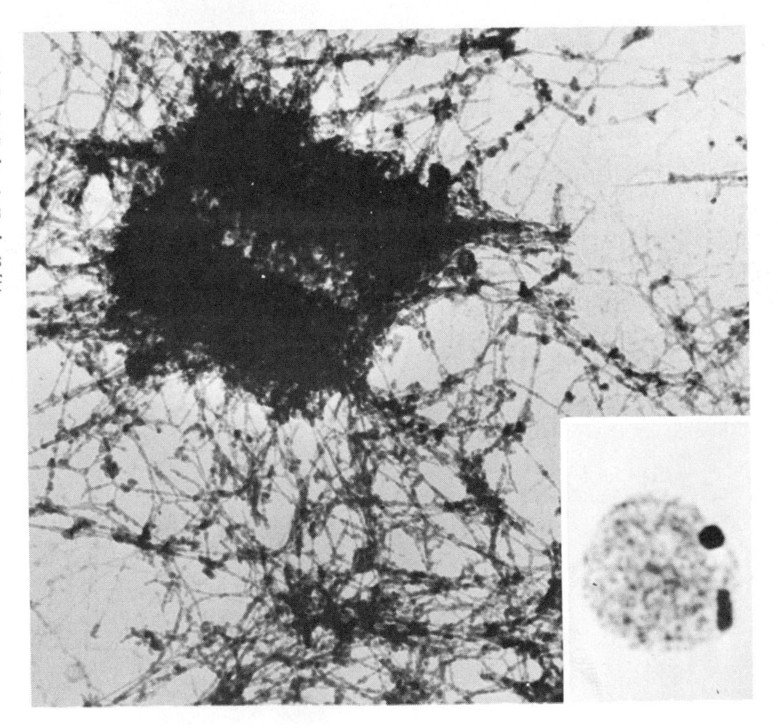

Figure 2.8
Comparable stages of a spermatocyte (a cell of the testis that will eventually produce sperm) of the milkweed bug, *Oncopeltus fasciatus*, taken at high magnification in the electron microscope and at a much lower magnification in the light microscope (inset). The contracted chromatin in both photographs consist of sex chromatin, which often undergoes contraction at an earlier stage than does the remainder of the chromatin. (Courtesy of Dr. S. Wolfe.)

fine threads (Figure 2.7), with no additional details of striking significance except that the threads are more numerous and of smaller diameter (Figure 2.8). This stainable material is the *chromatin,* with the fine threads being *euchromatin,* the coarser masses *heterochromatin.* It is the chromatin that, through condensation and contraction during cell division, will transform itself into those bodies we customarily think of as chromosomes, with the chromosome number, size, and shape being characteristic of each species (Figure 2.9). It is also the chromatin, and particularly the DNA contained within it, that is stained in a highly specific manner by the Feulgen reaction; by measuring spectroscopically the amount of Feulgen dye per nucleus, it is possible to show that the amount of DNA per nucleus is constant for each species (Table 2.1).

In addition to the chromatin, the nucleus contains one or more dense bodies, or *nucleoli* (singular, *nucleolus*). These are attached

Figure 2.9
Meiotic cell (microsporocyte) in rye, showing the attachment of the nucleolus to particular chromosomes (the centrally located pair). (Courtesy of Dr. R. A. Nilan.)

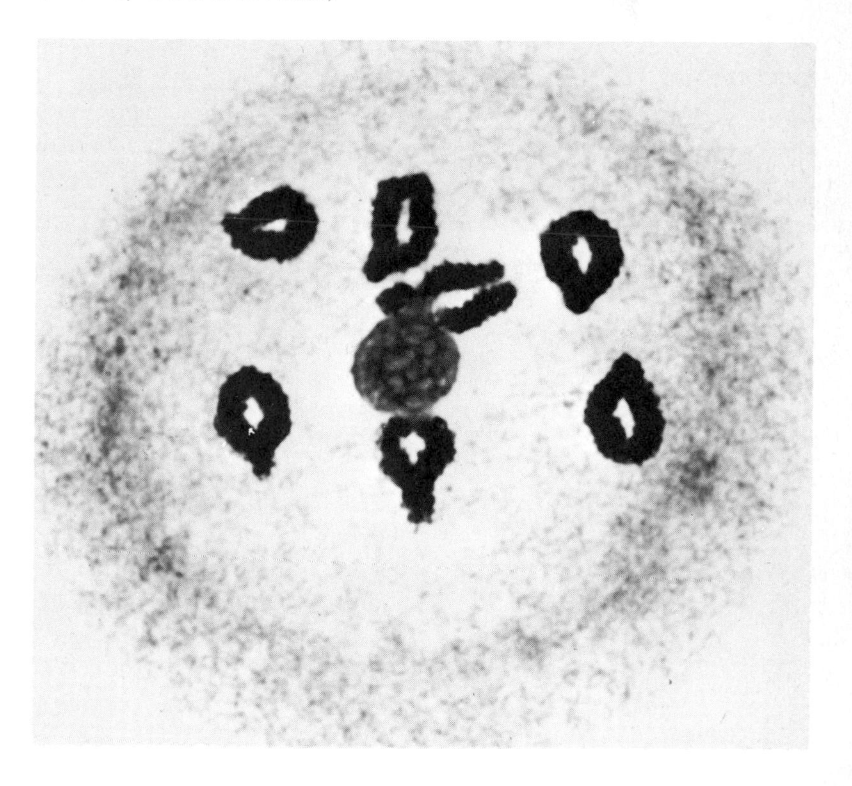

Table 2.1
Amount of DNA (in 10^{-9} milligrams) per haploid cell in a variety of organisms. The amount from somatic cells would be double the amounts given.*

Mammals		*Reptiles*		*Fish*	
Man	3.25	Snapping		Carp	1.64
Beef	2.82	turtle	2.50	Shad	0.91
Rat	3.4	Alligator	2.50	Lungfish	50.0
Dog	2.75	Water snake	2.51	*Miscellaneous*	
Mouse	3.00	Black snake	1.48	Maize	8.4
Marsupial	4.5	*Amphibians*		Drosophila	0.085
Birds		Amphiuma	84.0	Aspergillus	0.043
Chicken	1.26	Necturus	24.2	Neurospora	0.020
Duck	1.30	Frog	7.5	*E. coli*	0.0040
Goose	1.46	Toad	3.66		

* From C. P. Swanson, T. Merz, and W. J. Young, *Cytogenetics* (Englewood Cliffs, N.J.: Prentice-Hall, Inc., 1967), p. 177.

to, and formed by, special regions of particular chromosomes called *nucleolar organizer regions* (Figure 2.9, see also Figures 8.10 and 8.15, Chapter 8), which not only synthesize the nucleic acid portion of the nucleolar material but also organize it into a dense body. The character of the nucleolus varies with the type of cell and its metabolic state; it is larger and more dense in active cells, rapidly growing embryonic cells, and those engaged in protein synthesis. It disappears and reappears during the course of cell division (see Chapter 7).

Electron microscopy reveals an internal differentiation of the nucleolus in the form of a loose network of strands and granular material (Figure 2.10). As will be discussed later in this chapter, it is now known that the nucleolus, with its associated chromatin, is concerned with the formation of ribosomes, which eventually will accumulate in the cytoplasm to become organized as part of the protein-synthesizing machinery of the cell. This is demonstrated by the fact that radioactive uridine, a nucleoside that becomes incorporated as a base into the RNA of the ribosomes, is first concentrated in the nucleolus, after which it passes to the cytoplasm. The fact also that the nucleolus does not stain with Feulgen dye is indication that the nucleolus and the chromatin differ in the kind of nucleic acid present in each structure.

Nucleus as a control center A number of observations support the idea of the nucleus as not only an integral part of the eukaryotic cell, but also as its center of control. The whole science of genetics,

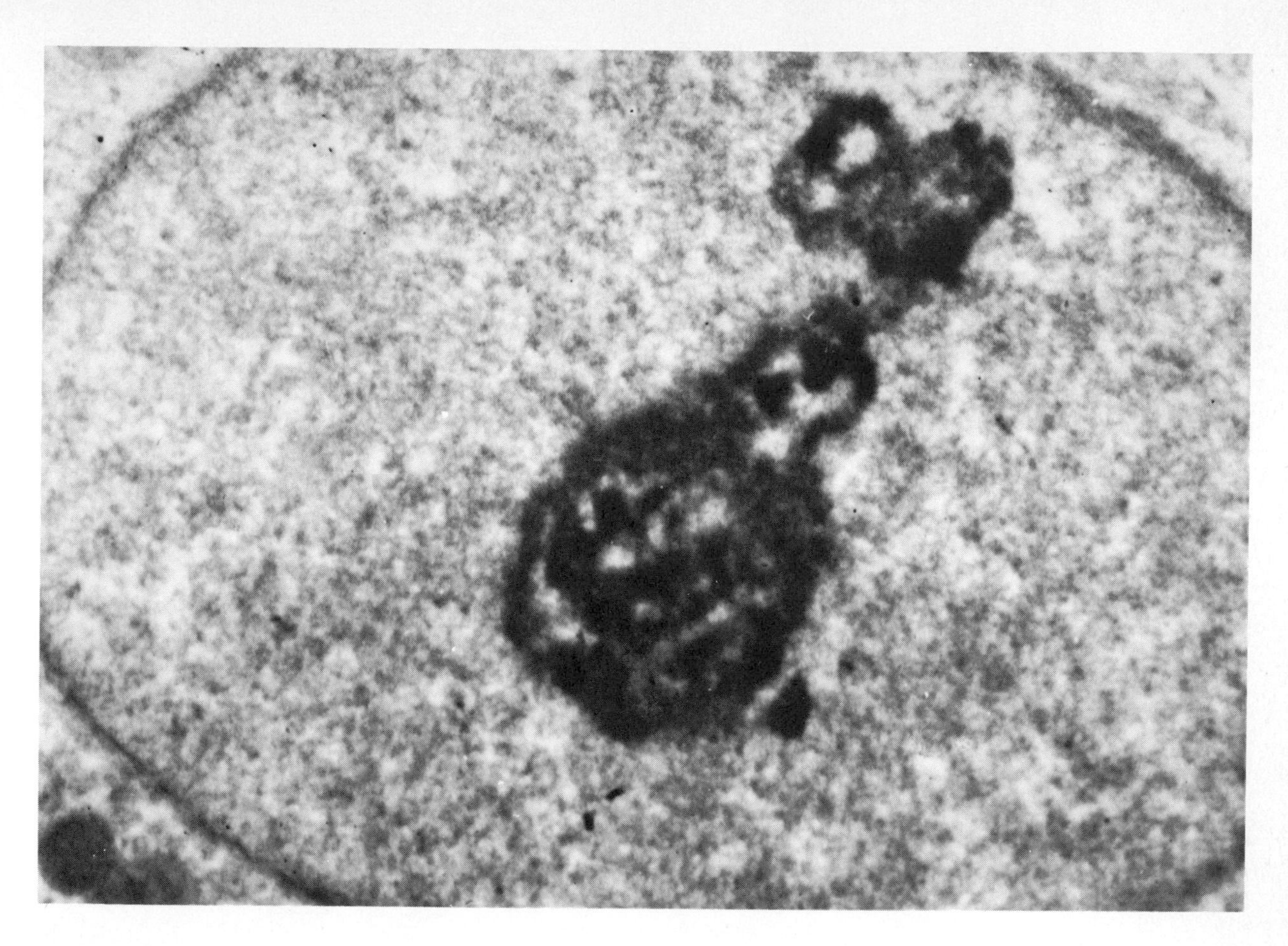

Figure 2.10
The nucleus of a mouse fibrocyte, showing the fibrous nature of a portion of the nucleolus. The dark strands are the nucleolonema; although not evident in this illustration, there is often a distinct clear area in the nucleolus called the pars amorpha. The granular material consists of ribosomes or of other RNA-containing particles. The nucleolus is subject to many changes in the cell and can assume many forms; this is often true in malignant or diseased cells. (Courtesy of Dr. E. Borsyko.)

in fact, is based on this assumption, with the chromosomes the vehicles of hereditary transmission, and the DNA within the chromosomes the molecular basis of heredity.

The mammalian red blood cell lacks a nucleus as it passes from its origin in the bone marrow and enters the bloodstream. It is a cell with a limited, albeit crucial, respiratory role—that of moving oxygen from the lungs to tissues, and carbon dioxide from the tissues to the lungs. It is also limited in growth and in length of life (about 120 days), and it is incapable of further division and of continuous repairs. From these observations, coupled with the almost universal occurrence of a single nucleus per cell, the equal contribution of the sperm to the zygote even though it is little more than a nucleus, and the fact that in cell division it is principally the nucleus that goes through an exact partitioning of its chromatin content, we can point to the nucleus as a control center of the cell.

Three other experiments reinforce this concept. If the nucleus of a fertilized frog egg is removed, a not particularly difficult operation, and the nucleus of a cell of a frog of a different species is substituted, a tadpole and eventually a frog will result, and it will be of the character of the frog that donated the substitute nucleus. The cytoplasm of the first frog is of little consequence after the very early stages of embryogenesis. It is also possible to remove the nucleus of an amoeba by pinching off a piece of the cytoplasm containing the nucleus. The remaining cytoplasm is not damaged, and the enucleated fragment can metabolize for days or even weeks afterward. The cell will eventually run down, however, and will be unable to rejuvenate itself unless another nucleus is put back into the cytoplasm. A cell without a nucleus is, therefore, a cell with a limited future and no posterity, and the nucleus is the necessary organelle that provides information and/or parts to keep the cytoplasm functioning properly for an indefinite period.

The influence of the nucleus in controlling the morphology of an organism is made strikingly evident by the work done on the green alga *Acetabularia*, a single-celled organism of large dimensions—5 to 10 cm (centimeters) in height—found in warm marine waters, and fastened to a substrate of sand or rocks by a rhizoid (Figure 2.11).

Figure 2.11
Acetabularia mediterranea, single-celled green alga of warm marine waters, which has been used extensively for experimental studies.

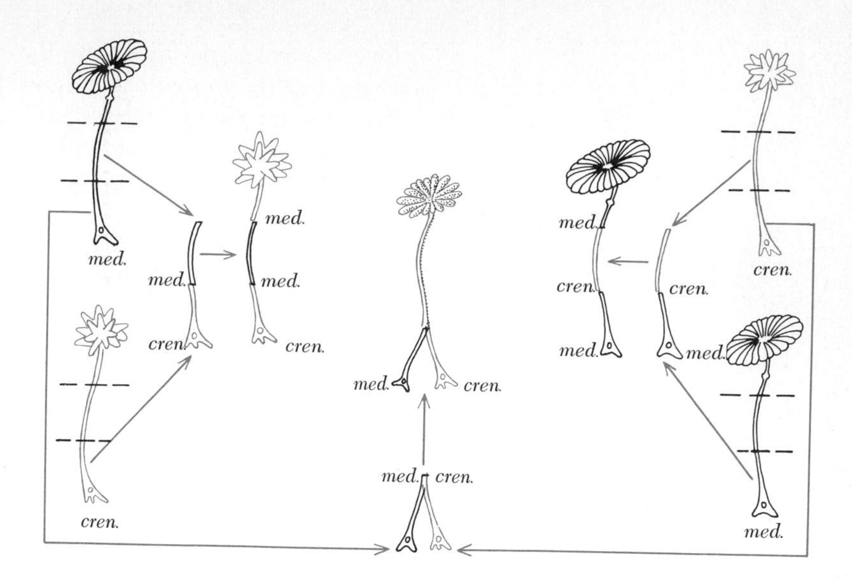

Figure 2.12
Influence of the nucleus on development in Acetabularia. Stalk segments of *A. mediterranea* grafted onto nucleus-containing rhizoids of *A. crenulata,* and vice versa produce caps characteristic of the species contributing the nucleus. When two nucleus-containing rhizoids are grafted together, the cap consists of loose rays, as in *A. crenulata,* but their points are more rounded, as in *A. mediterranea.*

A. *mediterranea* and A. *crenulata* are two species that differ as to the morphology of their caps. If a cap is cut off, it will be regenerated (Figure 2.12). It is also possible to graft a piece of stalk of one species onto the decapitated, nucleus-containing rhizoid of the other species. When the cap forms, it will be somewhat comparable to the species that donated the stalk, but if it is removed and allowed to regenerate, it will be characteristic of the species that contributed the nucleus; the stalk cytoplasm will have exhausted its influence and that of the nucleus will become evident. The nucleus, therefore, causes the cell to do as it dictates; the next problem is to determine how this jurisdiction is exercised.

Figure 2.13
The haploid set of chromosomes of the wakerobin, *Trillium erectum.* The length of each chromosome and the position of the centromeres permits the identification of individual chromosomes. (Courtesy of Dr. A. H. Sparrow.)

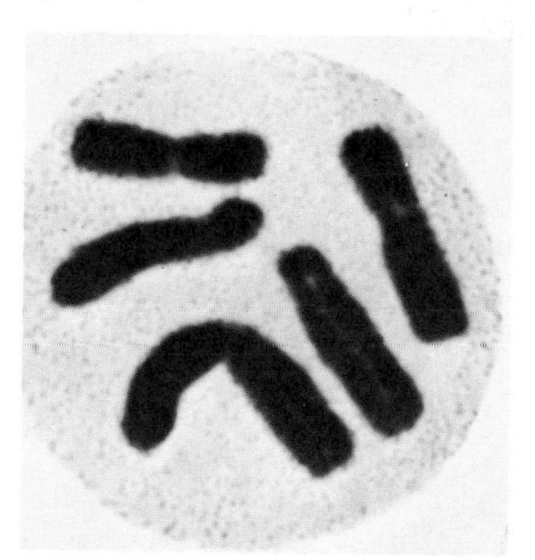

The chemistry of chromatin

As has been pointed out, cellular control must be exercised through chemical means. Furthermore, since a human cell does what a human cell is supposed to do, and the cell of an orchid does what it is supposed to do, the information somehow must be contained within the nucleus, must be capable of being decoded or "read," and must be acted upon in such a manner that the instructions from the nucleus have been carried out.

Attention was early directed at the chromosomes in the nucleus as the site of coded information. Chromosomes are constant in number, shape, and size within an individual and, generally, within a species (Figure 2.13). When chromosomes are changed, spon-

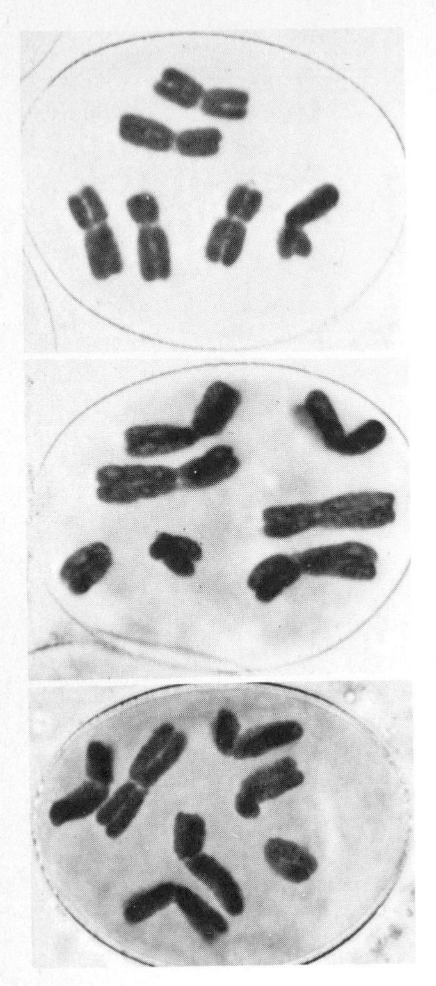

Figure 2.14
Chromosomes in the microspores of *Tradescantia paludosa,* the spiderwort. Those at the top are normal; one chromosome in each of the other two cells has been broken by x rays. The two fragments without centromeres will be unable to move properly during cell division and will be lost. The large size of these fragments would, in all probability, be lethal because of the amount of chromatin (genetic material) involved, but lesser losses can act as mutations to bring about phenotypic changes in the individual possessing such aberrant chromosomes.

taneously or experimentally, the change can often bring about a change in the individual (Figure 2.14), and when genes that control the expression of specific characters in an individual are followed from one generation to the next, they follow the same pathways and show the same inheritance patterns as do the chromosomes.

Chromosomes, or chromatin, can be isolated quite readily from both prokaryotic and eukaryotic cells. When analyzed chemically, chromatin from prokaryotes is naked DNA, that is, it is not associated with other kinds of molecules. Eukaryotic chromatin, however, is far more complex, consisting of four species of molecules—DNA; RNA; a basic protein of low molecular weight called *histone,* rich in the amino acids lysine and arginine, and existing in at least five different forms; and an acidic, nonhistone protein component. During isolation procedures, the DNA and histone are generally associated as a nucleoprotein complex, with the ratio of DNA to histone very close to 1:1, even though the amount of DNA per nucleus varies with the particular species (Table 2.1). The RNA and acidic protein vary in amount quite widely from one kind of cell to another even within an individual, being greater in the more actively metabolizing cells.

A number of experiments have demonstrated that the coded and controlling information in the cell is bound up in the molecules of DNA. It took a long time to prove this for several reasons. It was long thought that the DNA molecules, which are of very high molecular weight, were of monotonous regularity, being made up of units of four bases (adenine, guanine, cytosine, and thymine), regularly repeated. Such a molecule, it was believed, could not conceivably possess the internal diversity required of a molecule of inheritance. By this time hundreds and hundreds of genes, from many organisms and governing the expression of every imaginable kind of character, had been discovered and tested, and it was argued that only a protein consisting of many amino acids of 20 different kinds, and linked together in every possible combination and permutation, could provide the necessary diversity.

A series of experiments on bacteria pointed research in the right direction and eventually settled the issue. Frederick Griffiths, a British bacteriologist, was carrying out a series of studies on pneumococcus, a bacterium responsible for respiratory illness in mice and guinea pigs. One strain of the bacterium, called Type 1, was highly virulent, causing death when injected into an experimental animal. Type I bacteria, when plated out on culture dishes, formed shiny, regularly spherical colonies. Another strain, Type II, was avirulent; it did not produce the disease when injected into organisms, and it differed further in producing dull, irregularly shaped colonies when cultured Both the degree of virulence and the shape and character of the colonies depended upon the presence or absence of a polysaccharide capsule external to the cell membrane; Type I possessed the capsule, Type II lacked it.

In one experiment, Griffiths injected heat-killed Type I bacteria into a group of animals. This, by itself, would not induce symptoms of the disease, but he then injected a dose of avirulent (Type II) bacteria into the same animals. One would normally assume that the second injection would be equally ineffective in causing death, but the doubly injected animals died. The bacteria isolated from these dead animals were Type I, virulent and capsulated (Figure 2.15).

Two hypotheses can be advanced to explain these observations. Either the heat-killed Type I cells somehow had been reactivated by the presence of the avirulent Type II cells, or the Type II cells had received something from the dead Type I cells that converted them into virulent cells. The conversion was permanent, not simply a temporary phenomenon. The latter explanation proved to be correct, because the contents of burst Type I cells added to a Type II culture in a test tube brought about a similar transformation of some Type II cells. The rate of transformation was low but easily detectable. This experiment also eliminated the idea that the mice or guinea pigs might have had something to do with the experimental results. *Bacterial transformation,* as the phenomenon came to be known, was a one-way transfer of heritable characteristics from one cell to another.

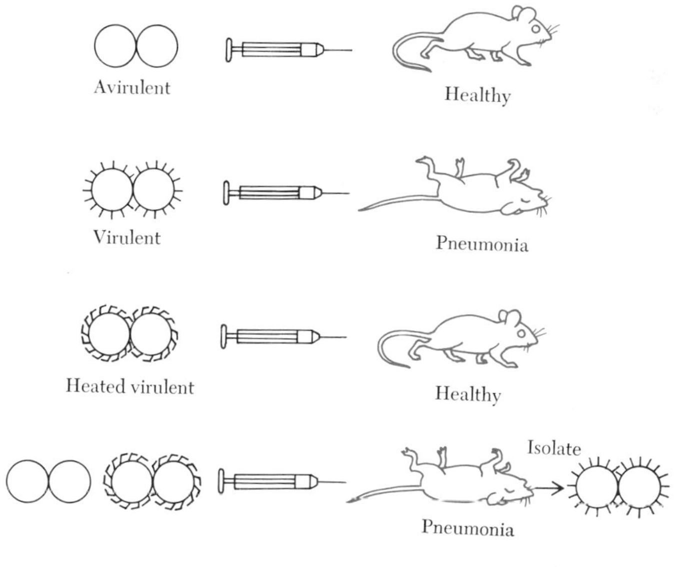

Figure 2.15
Griffiths' experiment showed that heat-killed virulent bacteria could contribute "something" to avirulent strains and cause them to be changed to a virulent form. It was eventually showed that this "something" was DNA.

Griffiths carried out his experiments in 1928, but it was not until 1944 that the contents of these bacteria had been fractionated and tested for transforming ability, and DNA was identified as the transforming principle. It was further shown that not only virulence but also any other heritable character could be transferred in a similar way; therefore, if genes are responsible for heritable characters, then genes must be made of DNA. The central problem then became one of determining the structure of the DNA molecule to understand how heritable information could be stored in, and released from, a chemical structure, and subsequently be given expression by way of the chemical reactions taking place within the cell.

Chemical analysis had revealed that DNA was a polymer of high molecular weight, that is, a macromolecule of repeated units. Each of these units proved to be composed of three smaller molecules linked together to form a nucleotide: these included a five-carbon sugar (deoxyribose), phosphoric acid, and a nitrogen-containing moiety generally referred to as a base. These bases are of four kinds (Figure 2.16): two are pyrimidines (thymine and cytosine) and two are purines (adenine and guanine). At one time it was believed that the DNAs from all organisms were similar, with the four bases existing in a ratio of 1:1:1:1, but clearly this notion had to be revised if the genes, of which there were endless kinds, were composed of DNA. More critical analyses of DNAs from a variety of organisms showed that the base ratios of different species were indeed different (Table 2.2), and that the differences could not be blamed on faulty isolation procedures. In addition, it was also

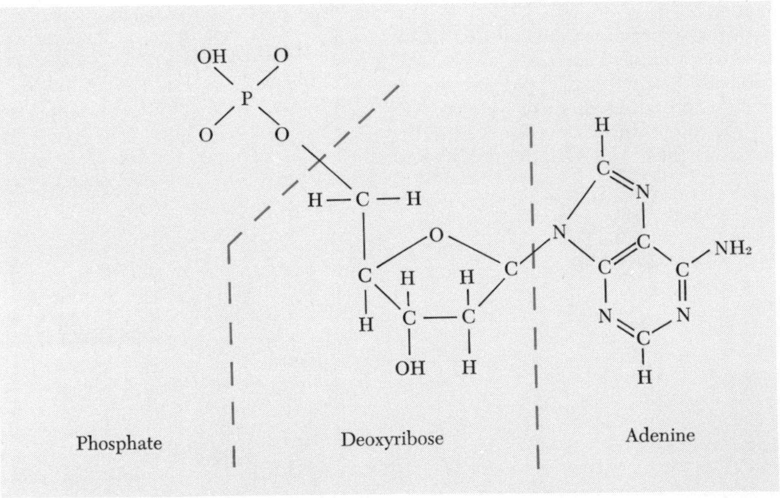

Figure 2.16
The linkage of the three kinds of molecules to form a nucleotide, in this instance, deoxyadenylate. Linkage to form a chain of nucleotides will be through an oxygen in the phosphate portion and the three carbon in the sugar portion. Linkage to form a double helix will be by hydrogen bonds as indicated in Figure 2.17.

Table 2.2
Base Ratios of Several Well-Known Organisms *

	Adenine	*Thymine*	*Guanine*	*Cytosine*	*Ratio of* $\frac{A+T}{C+G}$
Man	29.2	29.4	21.0	20.4	1.53
Sheep	28.0	28.6	22.3	21.1	1.38
Calf	28.0	27.8	20.9	21.4	1.36
Salmon	29.7	29.1	20.8	20.4	1.43
Wheat	27.3	27.1	22.7	22.8	1.19
Yeast (fungus)	31.3	32.9	18.7	18.1	1.19
Virus, vaccinia	29.5	29.9	20.6	20.0	1.46
Bacteria					
Staphylococcus	31.0	33.9	17.5	17.6	1.85
Pseudomonas	16.2	16.4	33.7	33.7	0.48
Colon bacterium	25.6	25.5	25.0	24.9	1.00
Clostridium bacterium	36.9	36.3	14.0	12.8	2.70
Pneumococcus	29.8	31.6	20.5	18.0	1.88

* Values are arbitrary but accurate as ratios.

demonstrated that in DNAs from most sources the purines equalled the pyrimidines in molar amounts, that the amount of adenine (A) was equal to that of thymine (T), and that cytosine (C) was equal to that of guanine (G). Thus, $A + G = T + C$, and $A = T$ and $C = G$, but A or T are not necessarily equal to C or G. There was no indication that the bases were arranged in any particular order, suggesting that ATGTTCACG might very well have a different genetic meaning from ACACTTGTG, even though the number and kinds of bases are similar. In a sense this would be no different from the words *team, mate,* and *meat;* the letters are the same but the meanings are not.

The deoxyribose sugar was attached to a particular carbon atom of each base, and successive sugars in the macromolecule were connected by phosphate bonds derived from the phosphoric acid (Figures 2.17 and 2.18). Remembering that $A = T$ and that $C = G$ in a sample of DNA, it was argued that the bases are arranged in pairs, and that the molecule as a whole is in the form of a double helix (Figure 2.19). Even before base pairing had been proposed, however, physical measurements had shown that the DNA molecule consisted of two chains, which were formed by sugar and phosphate residues linked alternately. The question remaining was how the bases fitted in with this structure. Model building, based on stereochemical considerations, not only supported the concept of a double helix but also demonstrated how the stability of the

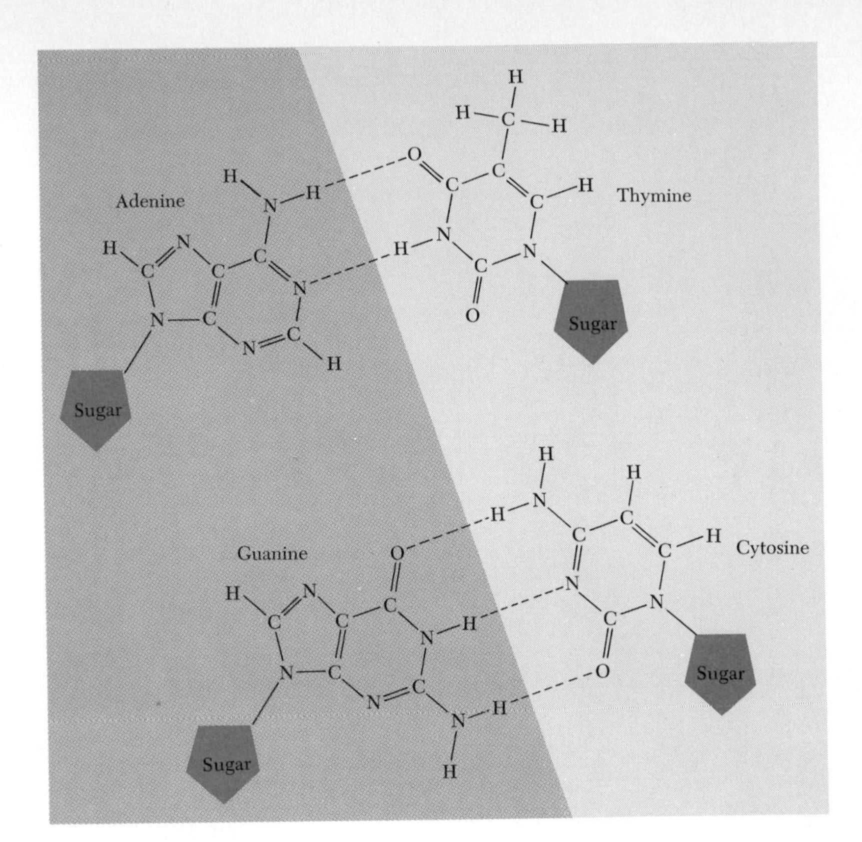

Figure 2.17
Chemical configurations of the four bases found in the DNA molecule, arranged as base pairs. Thymine and cytosine are pyrimidines; adenine and guanine are purines.

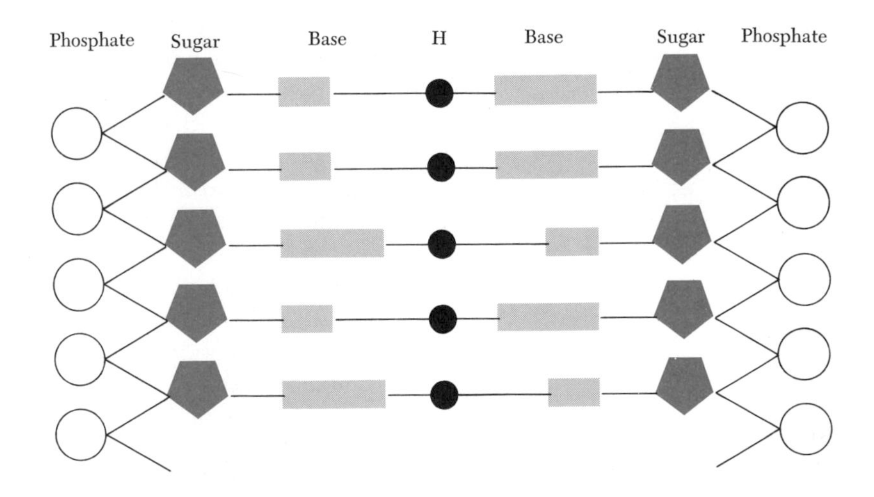

Figure 2.18
Schematic and flattened arrangement of phosphates, sugars, and bases to form the DNA molecule, with the right- and left-hand portions of the molecule held together by hydrogen bonds. This arrangement, twisted into a helix by molecular forces, gives the configuration seen in Figure 2.19.

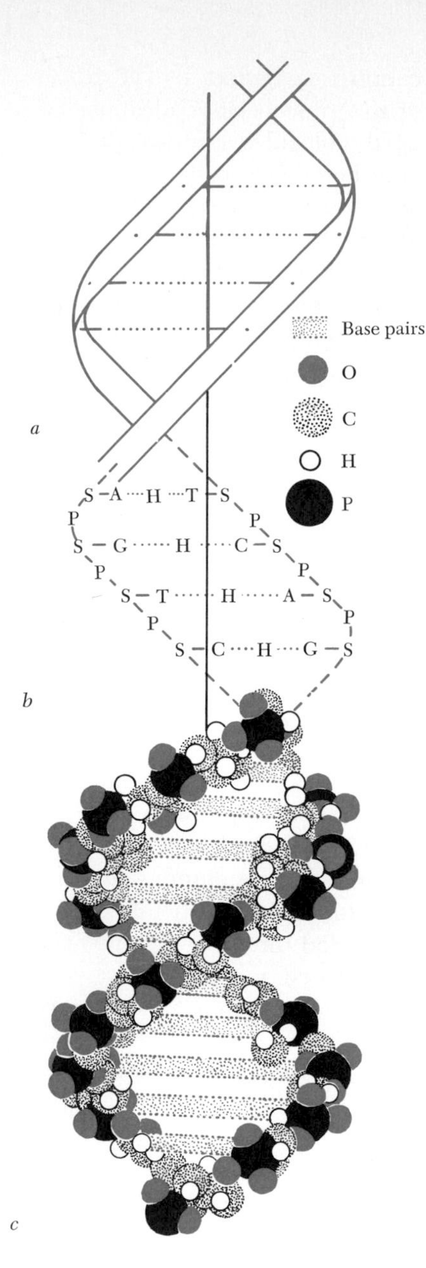

Figure 2.19
The DNA helix, with three different ways of representing the molecular arrangement. (*a*) General picture of the double helix, with the phosphate-sugar combinations making up the outside spirals and the base pairs the cross bars. (*b*) A somewhat more detailed representation: phosphate (P), sugar (S), adenine (A), thymine (T), guanine (G), cytosine (C), and hydrogen (H). (*c*) Detailed structure, showing how the space is filled with atoms: carbon (C), oxygen (O), hydrogen (H), phosphorus (P), and the base pairs.

DNA molecule was achieved and explained the significance of the base ratios. An enormous amount of critical experimental evidence has confirmed this structure beyond any reasonable doubt. The molecule of DNA consists, therefore, of two helically entwined molecular strands, each having a sugar-phosphate backbone and with the bases projecting inward from the sugar moiety. The bases are joined between the two strands by hydrogen bonds, thus giving stability to the molecule. The pairing, however, is strictly complementary in that A is always paired with T and C with G. The two strands of the molecule are, as a consequence, complementary with each other throughout their entire length. We find, as a result, that the only base pairs that are possible are A–T, T–A, C–G, and G–C. Any other base pairs are noncomplementary, and, because of stereochemical differences, cannot be fitted into the helical structure.

In addition to accounting for the base ratios set forth in Table 2.2, the nature of the double helix explains other phenomena. It reveals how the hereditary substance of the cell can be so stable in its metabolic characteristics and its amount per cell; it permits us to understand how the DNAs of a wide variety of organisms can be so similar physically and chemically, and yet so different in an hereditary sense; it offers an explanation of how a sequence of base pairs can be the source of coded information in the cell; and it provides a mechanism for the exact distribution of genetic information from one cell to another.

It was known, for example, that when a cell divides, the two daughter cells are genetically identical. The mother cell must, therefore, have replicated, in a highly precise manner, its hereditary and controlling substance prior to cell division so that during division each daughter cell would be, qualitatively and quantitatively, similar in genetic content. Figure 2.20 indicates how this is accomplished. The double helix separates into its two constituent polynucleotide strands by breaking the hydrogen bonds that hold the bases together in pairs. As each strand unwinds, the metabolic machinery of the cell forms a complementary copy alongside,

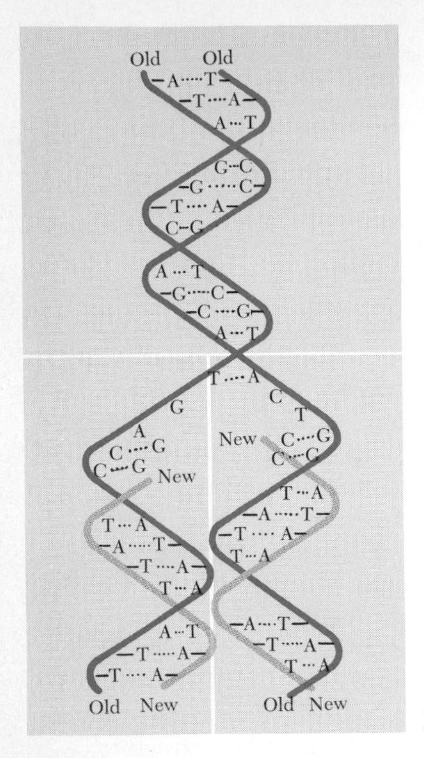

Figure 2.20
Replication of the DNA molecule, which takes place during interphase. The old helix unwinds (center) and the two new helices are formed.

so that when the process of replication is complete two identical helices are formed. Confirmation of this process of replication has been proven through the use of tagged molecules, and is explained in Chapter 7. By this mechanism, the hereditary content of all of an organism's cells remains constant, and human cells form only human cells, elephant cells form only elephant cells.

The amount of DNA per cell varies with different species (Table 2.1). The DNA of the common colon bacterium, *Escherichia coli*, is a single macromolecule 1,100 microns long and contains approximately 3.4 million nucleotide pairs. Each nucleotide pair occupies 3.4 Å of the double helix, and one complete turn of the helix consists of 10 nucleotide pairs, 34 Å in length. The *E. coli* DNA, therefore, has about 340,000 turns in its double helix, each of which must be undone when the DNA replicates. The feat can be accomplished by the bacterium in about 20 minutes under normal growing conditions. As organisms increase their morphological, physiological, and behavioral complexity, they obviously require more coded information in the form of more DNA. This general increase is indicated in Table 2.1, but there are also some discrepancies suggesting that there is no fixed relation between DNA amount, structural and functional complexity, and evolutionary position.

The amount of DNA per cell raises another problem that has long interested biologists. How is the DNA packed into a nucleus? In the prokaryotes there appears to be only one chromosome per cell—it consists of naked DNA and is found in the membraneless nuclear area as a long, tenuous, and tangled thread or fiber. When we consider such units as the mammalian cell, we are dealing with DNA that is well over a meter in length when fully extended, contains billions of nucleotide pairs, is divided into a specific number of chromosomes characteristic of each species, is contracted into units that can maneuver within the cell readily during cell division (Figure 2.21), and is packaged within a nucleus only 10 μm in diameter. As Figure 2.22 shows, the chromatin appears to be a series of "beads" on a string, and analysis suggests that the beads consist of about 200 to 210 nucleotide pairs compacted together with four of the five known histones. The interbead regions are about 140 Å (or 42 nucleotide pairs) long and are associated somehow with the fifth histone. This topic will be examined later at some length (see page 69), but it would appear that the beads are portions of the DNA that cannot be decoded while in a compacted form, and that both the histones and the acidic proteins of the nucleus are concerned with the inhibition or release of information from the DNA.

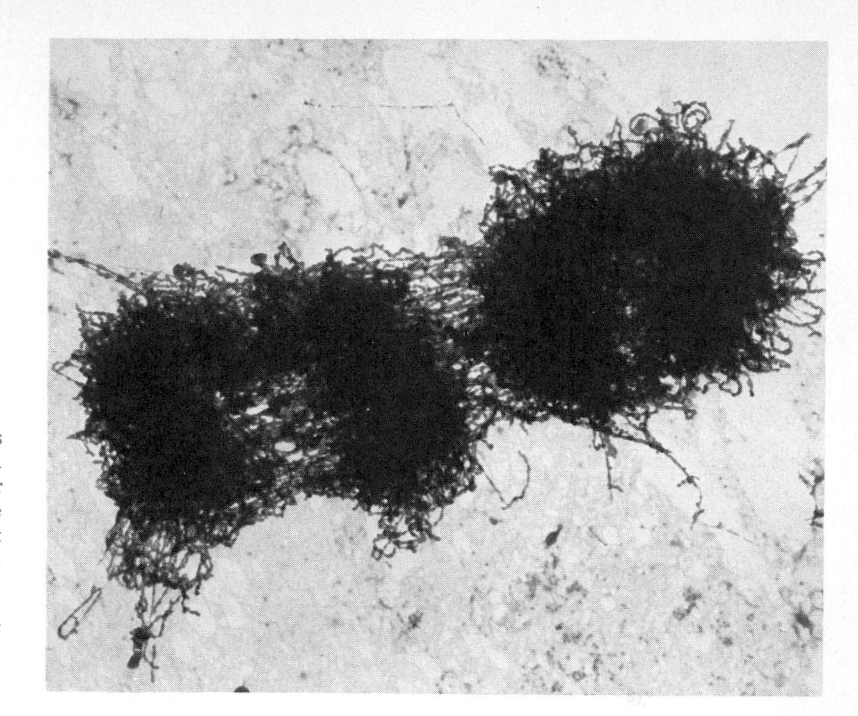

Figure 2.21
A pair of homologous chromosomes from the meiotic cell of the milkweed bug, *Oncopeltus fasciatus.* The manner of packaging the DNA in these chromosomes seems quite different from the more orderly arranged coils seen in the chromosomes in Figure 8.22, but it is quite obvious that these "balls of yarn" contain strands of DNA of very substantial length. (Courtesy of Dr. S. Wolfe.)

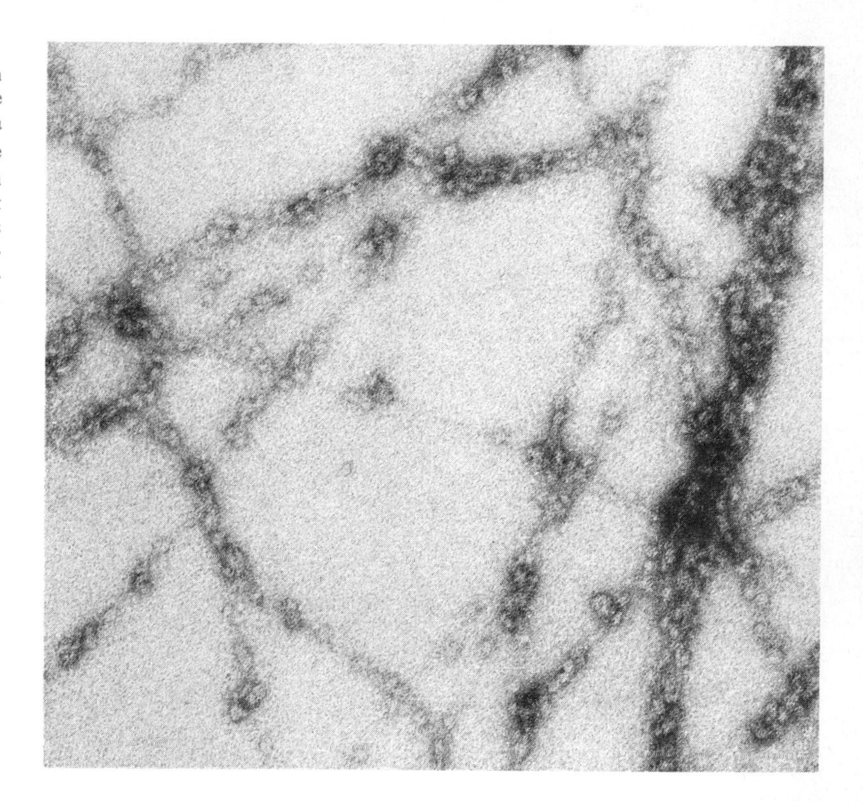

Figure 2.22
Isolated chromatin from chicken erythrocytes showing the beaded nature of the strands. Each strand has a diameter of 75 Å. Four of the five known histones are associated with the beads, and it is assumed but not proved that the fifth histone is coupled with the DNA of the interbead regions. Some investigators consider the interbead areas to be artifacts of preparation. (Courtesy of Dr. C. Woodcock.)

The genetic code

Recognizing that the DNA in each cell is the critical molecule of inheritance, and that the structure depicted in Figures 2.19 and 2.20 has been verified, it then follows that the source of information needed by the cell to carry out its functions and build its necessary structures in an orderly and predictable manner must be found, in some sort of coded form, in the sequence of nucleotide pairs in each molecule.

It was demonstrated in 1941 that some genes influence the expression of heritable traits through the medium of enzymes. Thus, if a product P, which is necessary for the expression of a particular trait, is formed by a series of chemical reactions,

$$A \rightarrow B \rightarrow C \rightarrow D \rightarrow\rightarrow P$$

each step in the series is controlled by an enzyme. If the enzyme controlling the reaction $A \rightarrow B$ is missing or rendered nonfunctional by a defective gene, the remaining steps in the series cannot occur because the product B similarly would be missing. Figure 2.23 illustrates the series of reactions and the genetic circumstances leading to the formation of the amino acid arginine.

Without exception, enzymes are proteins. Therefore, the uniqueness of every cell, individual, or species is based on the uniqueness of its proteins, which either control the chemical reactions taking place or enter into the structure of every cellular organelle, and particularly of its membranes. DNA, as the crucial molecule of inheritance, must then in some way determine the specificity of the proteins of the cell. The specificity of proteins

Figure 2.23
The biochemical reactions required for the formation of arginine from glucose. The reactions in the conversion of ornithine to arginine are unique to arginine formation, while the earlier steps are common to the formation of other substances. The enzymes governing the reactions indicated are themselves under the control of certain genes, which have been identified. If the genes mutate and produce defective enzymes, the steps leading to the next reaction are blocked.

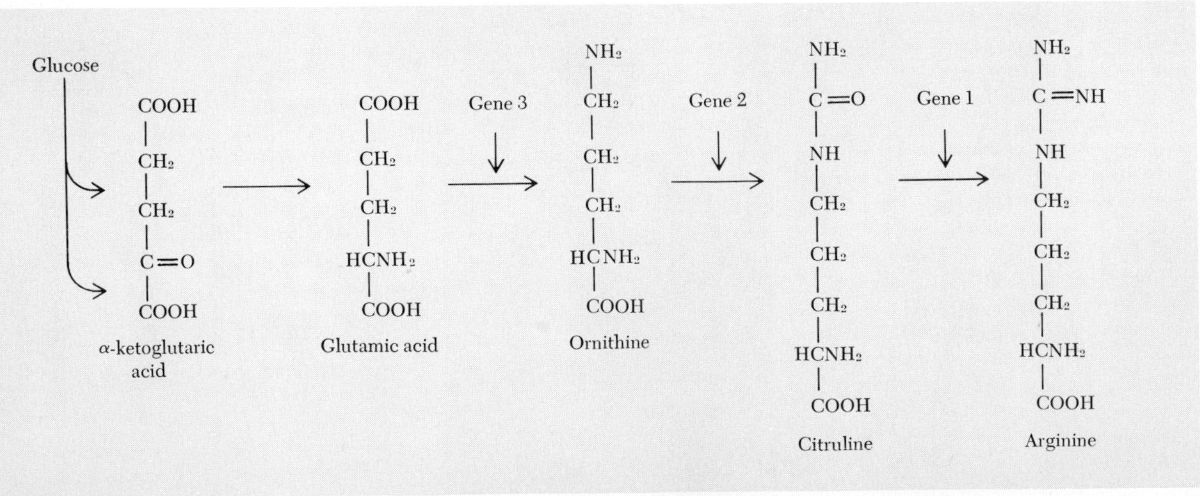

Glucose → α-ketoglutaric acid: $COOH-CH_2-CH_2-C{=}O-COOH$

→ Glutamic acid: $COOH-CH_2-CH_2-HCNH_2-COOH$

Gene 3 ↓ → Ornithine: $NH_2-CH_2-CH_2-CH_2-HCNH_2-COOH$

Gene 2 ↓ → Citruline: $NH_2-C{=}O-NH-CH_2-CH_2-CH_2-HCNH_2-COOH$

Gene 1 ↓ → Arginine: $NH_2-C{=}NH-NH-CH_2-CH_2-CH_2-HCNH_2-COOH$

must, in turn, somehow reside in the sequence of its units—linked amino acids—that previously had been determined by the sequence of nucleotides in the DNA. It is no accident that both DNA and proteins are macromolecules made up of repeated units.

Figure 2.2 illustrates the amino acid sequence of an enzyme, bovine chymotrypsinogen, concerned with digestion in the small intestine of cattle. It will be immediately evident that there are far more kinds of amino acids (there are 20 different ones) that enter into the structure of proteins than there are nucleotide bases (4). Quite obviously then, if DNA, in any kind of direct manner, determines the sequence of amino acids in the protein, one nucleotide in DNA cannot specify the kind and position of any single amino acid in a protein; such a *singlet* code would take care of only four amino acids. A *doublet* code is similarly inadequate, since adjacent nucleotides, taken in pairs and making use of all possible combinations, could specify only 16 (4×4) amino acids (Table 2.3). A *triplet* code, however, is more than adequate, since 64 ($4 \times 4 \times 4$) triplets are possible. By means of artificial RNAs

Table 2.3

The possible numbers of code words based on singlet, doublet, and triplet codes are shown here. We now believe that the triplet code is correct.

Singlet code (4 words)	*Doublet code (16 words)*				*Triplet code (64 words)*			
					AAA	AAG	AAC	AAU
					AGA	AGG	AGC	AGU
					ACA	ACG	ACC	ACU
					AUA	AUG	AUC	AUU
					GAA	GAG	GAC	GAU
					GGA	GGG	GGC	GGU
A	AA	AG	AC	AU	GCA	GCG	GCC	GCU
G	GA	GG	GC	GU	GUA	GUG	GUC	GUU
C	CA	CG	CC	CU	CAA	CAG	CAC	CAU
U	UA	UG	UC	UU	CGA	CGG	CGC	CGU
					CCA	CCG	CCC	CCU
					CUA	CUG	CUC	CUU
					UAA	UAG	UAC	UAU
					UGA	UGG	UGC	UGU
					UCA	UCG	UCC	UCU
					UUA	UUG	UUC	UUU

of known composition, the triplet codes have been determined in the laboratory, and have been found universally applicable to all tested species. If all of the triplet codes are functional, it would mean that most amino acids can be specified by more than one code. Table 2.4 indicates that this is indeed the case.

It will be readily evident that the codes in Table 2.4 are indicated in terms of RNA and not of DNA. The reason for this practice will become clear when it is realized that the message that will be read directly by the cell is encoded in an RNA molecule and is not directly taken from the DNA of the chromosome. A piece of DNA is, however, the basic source of the coded message "read" by the cell, and this message is eventually translated into a particular kind of protein, structural or enzymatic. The enzyme depicted in Figure 2.2 has 246 amino acids in its makeup; the

Table 2.4

The genetic code that has been found to exist in all organisms. UAA (ochre), UAG (amber), and UGA (umber) are chain-terminating codons; the translating mechanism is unable to read these codons, and the protein-forming action ceases. AUG plays two roles, depending upon where it is located in a message: it is a chain-initiating codon standing for n-formylmethionine in *E. coli*, but not in higher organisms; if found in the middle of a message it stands for methionine in all organisms.

First letter	Second letter: U	Second letter: C	Second letter: A	Second letter: G	Third letter
U	UUU, UUC – Phe; UUA, UUG – Leu	UCU, UCC, UCA, UCG – Ser	UAU, UAC – Tyr; UAA OCHRE; UAG AMBER	UGU, UGC – Cys; UGA UMBER; UGG Tryp	U C A G
C	CUU, CUC, CUA, CUG – Leu	CCU, CCC, CCA, CCG – Pro	CAU, CAC – His; CAA, CAG – GluN	CGU, CGC, CGA, CGG – Arg	U C A G
A	AUU, AUC, AUA – Ileu; AUG Met	ACU, ACC, ACA, ACG – Thr	AAU, AAC – AspN; AAA, AAG – Lys	AGU, AGC – Ser; AGA, AGG – Arg	U C A G
G	GUU, GUC, GUA, GUG – Val	GCU, GCC, GCA, GCG – Ala	GAU, GAC – Asp; GAA, GAG – Glu	GGU, GGC, GGA, GGG – Gly	U C A G

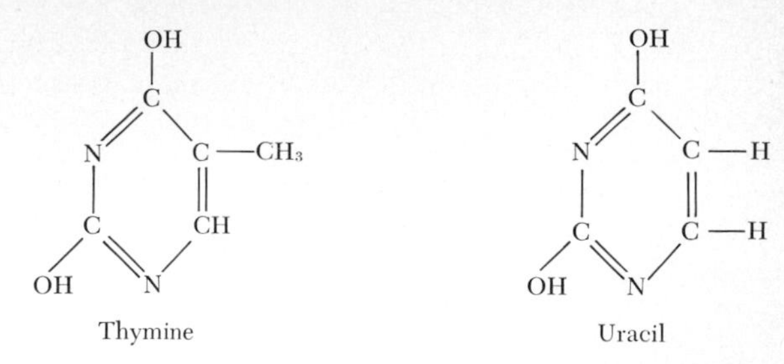

Figure 2.24
The molecular configurations of thymine, found in DNA, and uracil, which replaces it in RNA.

gene or genes that determine its structure must, therefore, in aggregate be at least 738 nucleotides long.

DNA, however, at least in eukaryotic cells, is in a membrane-bound nucleus, while protein synthesis is carried on in the cytoplasm. Somehow a message must be sent from nucleus to cytoplasm, where it is to be translated into protein formation. The only logical means by which this can be done is a chemical one. A series of brilliant experiments has shown that DNA does not act directly, but rather through the medium of several kinds of RNAs. Although RNA is complementary to one of the polynucleotide strands of the double helix from which it originates, it differs from DNA in that the thymine in DNA is replaced in RNA by a very similar molecule, uracil (Figure 2.24). In addition, RNA is single-stranded rather than being in the form of a double helix.

DNA, therefore, is a versatile molecule (Figure 2.25): it can make more of itself through a process of *replication,* which occurs once every cell division, and, through a process called *transcription,* it can make the three kinds of RNAs that perform different functions in the sequence of cytoplasmic events which leads to protein formation. *Messenger* RNA (mRNA) gets its name from the fact

Figure 2.25
Diagram to indicate the relation of the several RNAs to DNA, and the roles played by the several RNAs in protein synthesis. The number of tRNAs required for the formation of any protein will depend upon the number of different amino acids contained within the protein.

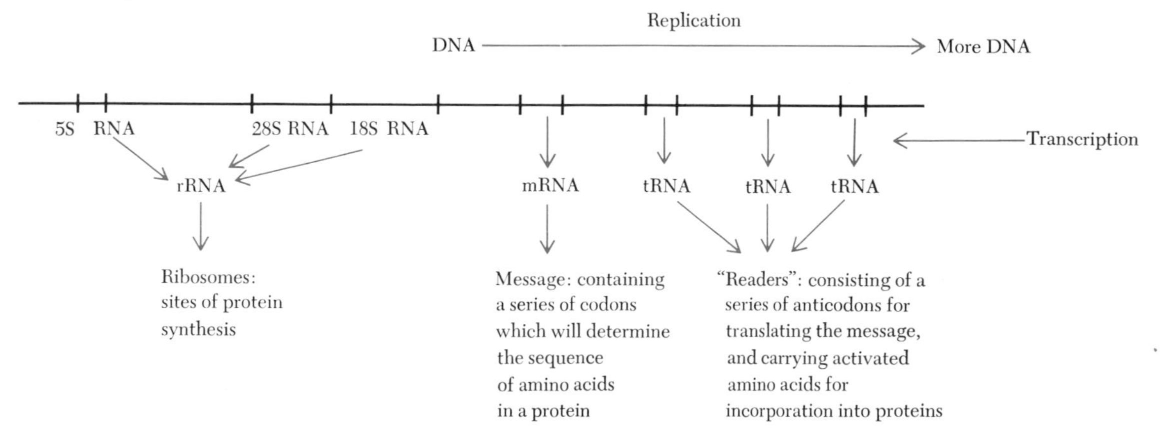

that it is the message that is to be read. Derived from DNA, the message is in the form of triplet *codons* arranged sequentially. This provides the rationale for putting the codons in RNA units (Table 2.4) rather than in terms of DNA. Since the code is a triplet one, any single mRNA will have at least three times as many nucleotides in the message as there will be amino acids in the protein to be formed. The *transfer* RNAs (tRNAs), of which there are many different ones, are cloverleaf-shaped molecules (Figure 2.26). They contain unusual nucleotides in their makeup, a free end terminating in the nucleotides -CCA, and a set of three nucleotides at the tip of the middle leaflet, which constitutes an *anticodon*. These anticodons can pair, in complementary fashion, with any three nucleotides forming a triplet codon in the mRNA—hence the name "anticodon." The tRNAs are, therefore, the "readers" of the message encoded in the mRNAs. The unusual nucleotides in the tRNAs are formed from normal nucleotides after the tRNA is released from the DNA. The shape of the tRNAs results from the fact that internal complementary pairing of nucleo-

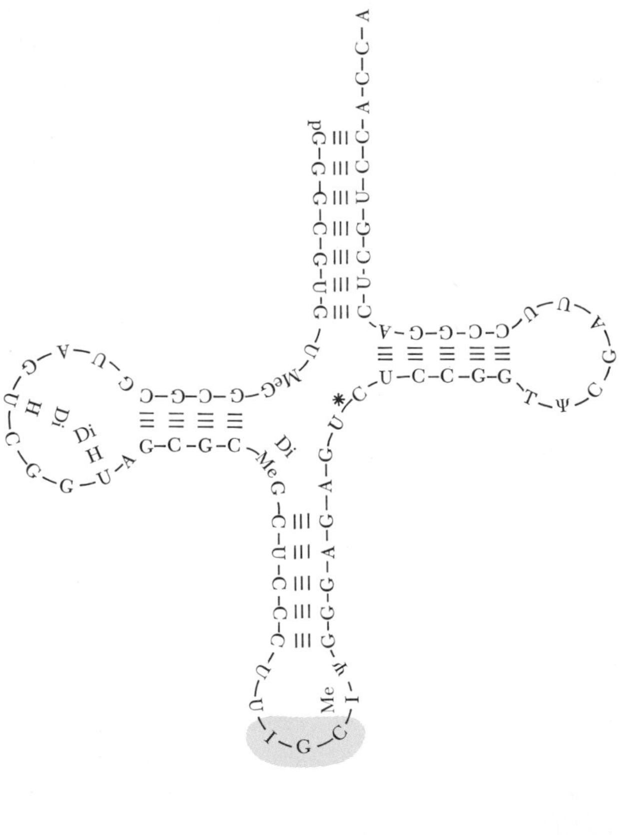

Figure 2.26
The cloverleaf model of an alanine tRNA molecule from yeast, the first of the tRNAs to be analyzed. This is not the only secondary form the molecule can take, but is the most probable one. The unusual nucleotides in the tRNAs are formed after the tRNAs are transcribed, and their symbols and names are as follows: I, inosine; Ψ, pseudouridine; T, ribothymidine; MeI, 1-methylinosine; MeG, 1-methylguanosine; DiMeG, N-2-dimethylguanosine; and DiHU, 5,6-dihydroxyuridine. The anticodon, I-G-C, is found in the base of the middle loop (boxed). (After R. W. Holley et al., *Science* 147, 1462, 1965.)

tides, most of them of the C–G pair, causes the molecule to loop back on itself in several places. The anticodon, being complementary to the triplet codons in mRNA, has theoretically 64 different forms, but fewer than 61 tRNAs are known (it seems unlikely or unnecessary that tRNAs would exist for the termination codons—Table 2.4). The -CCA tip of the tRNA is the place where activated amino acids can be terminally attached, the particular amino acid being determined in some unknown fashion by the nature of the anticodon, and put in place by a specific activating enzyme for each of the 20 amino acids.

The ribosomal RNAs (rRNAs) are structural elements, and in eukaryotic cells are of three kinds: 28S, 18S, and 5S, with S being a sedimentation constant. The numerical values are measures of size relationships, the larger values sedimenting most rapidly during centrifugation. The 28S and 18S rRNAs are made in the nucleolar organizer region of the chromosome (in heterochromatin) (Figures 2.27 and 2.28), while the 5S rRNA is made elsewhere in the chromosome complement. These are combined with proteins in the nucleolus, and they will eventually unite in the cytoplasm to form 80S ribosomes having the shape depicted in Figure 2.29. The prokaryotic ribosomes are somewhat smaller and always lie free in the cytoplasm of the cell; those in eukaryotic cells also may lie

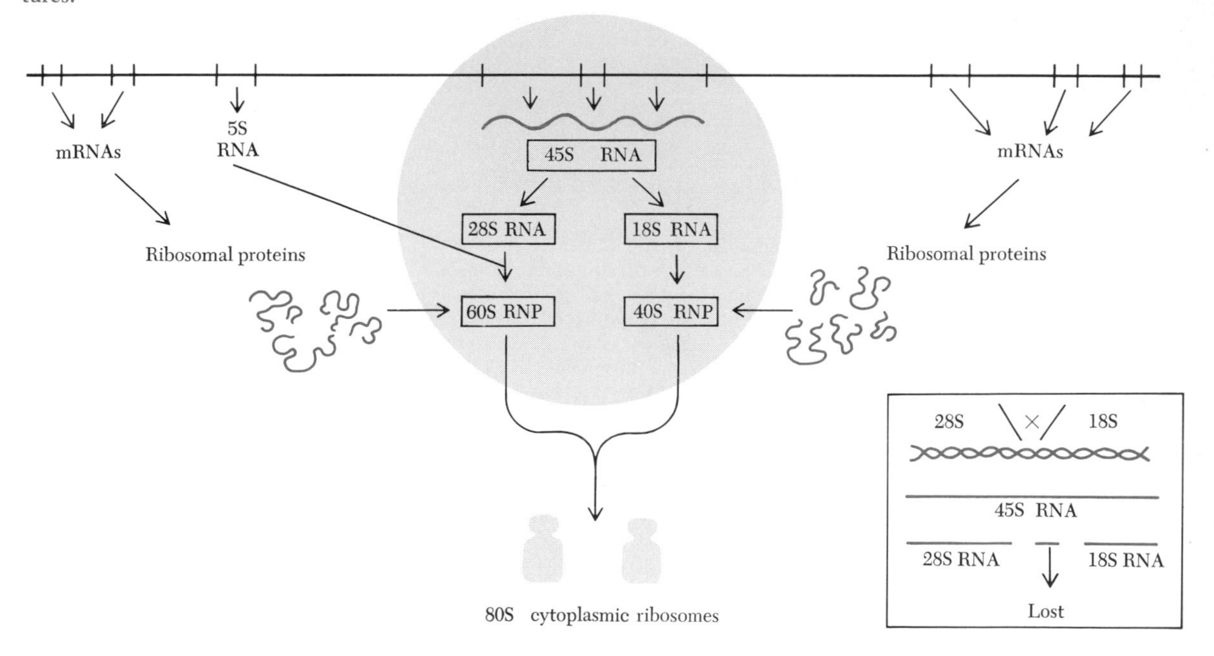

Figure 2.27
Formation of ribosomes in eukaryotic organisms. Units of 45S rRNA are transcribed in the nucleolar organizing region of the chromosome, are strongly methylated on being released from the DNA, and are then dissociated into 28S and 18S rRNA units, with a loss of some RNA derived from the spacer (×) DNA (see inset). The 5S rRNA is made from DNA outside of the nucleolar region and transported into the nucleolus. The 28S (plus the 5S) and the 18S rRNAs are then coupled with a large number of individual proteins to form 60S and 40S RNP (ribonucleoprotein) units, which pass rapidly from the nucleolus into the cytoplasm where they are joined to form ribosomes. The ribosomes are made up of 60% RNA and 40% protein, plus Mg^{++} ions, which aid in binding the units together. The diagram indicates that the ribosomal proteins are translated from mRNAs formed elsewhere on the chromosome; this may or may not be so since some investigators suggest that the rRNAs may be translated to form ribosomal proteins being incorporated into the ribosomal structures.

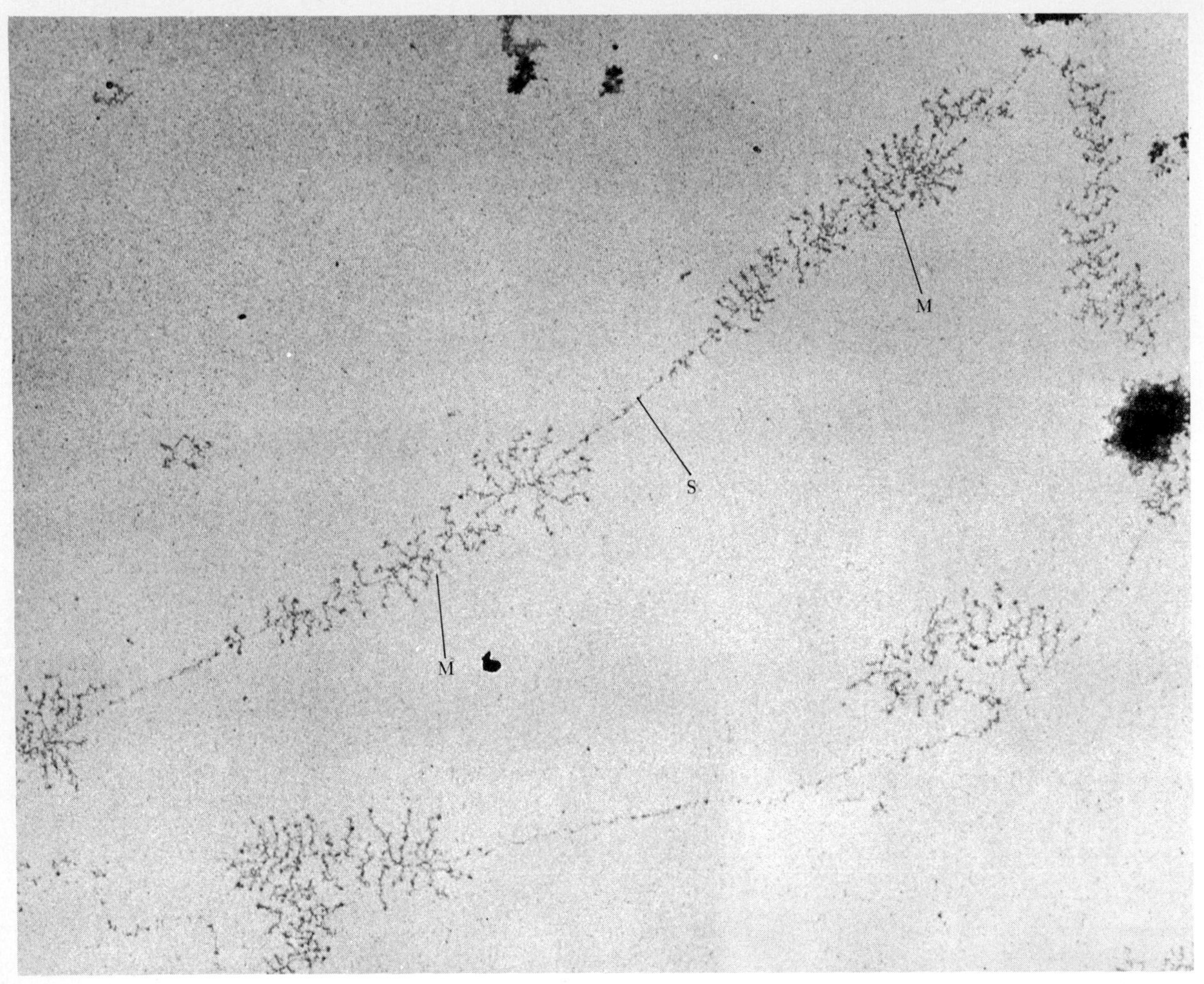

Figure 2.28
Genes in the nucleolar region of the oocyte of the newt, *Triturus viridescens*, transcribing the ribosomal precursor RNA (probably the 45S unit), which will give rise to the 28S and 18S units, which will enter into the structure of the mature ribosome. The long axis is DNA, possibly existing as a DNA-histone complex. Each matrix (M) unit is 2.5 μm long, and successive matrices are separated from each other by a spacer DNA (S). Each matrix consists of a series of RNA fibrils, which increase in length as one goes from one end of the matrix to the other; these are released when they reach the proper length and are then processed in the nucleolus for incorporation into the ribosomes. Each matrix is, therefore, engaged in the simultaneous transcription of many rRNA molecules; the shortest have just begun to be formed, the longest are about ready to be released. (Courtesy of Dr. C. Woodcock.)

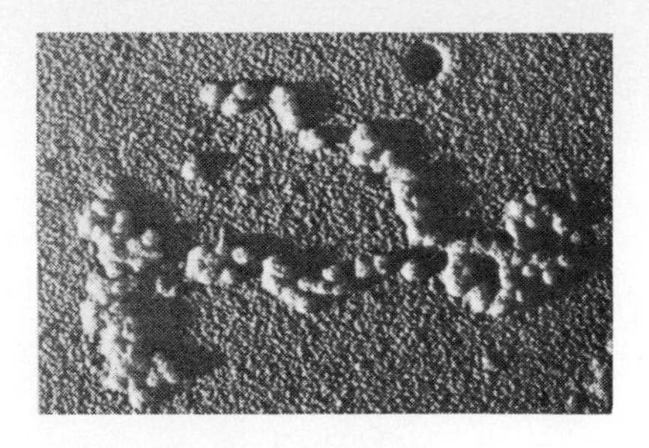

Figure 2.29
Electron micrograph of a cluster of ribosomes, held together as polyribosomes or polysomes by a strand of mRNA. The ribosomes move along the mRNA and are believed to have two receptive sites for tRNAs, as indicated in Figure 2.30.

free in the cytoplasm, or they may be attached to the membranes of the endoplasmic reticulum.

Despite its small size, about 250 Å in diameter, the ribosome is a complicated structure. Based on knowledge of the 70S ribosomes of the bacterium *E. coli,* the smaller (30S) portion contains a 16S rRNA molecule (molecular weight = 600,000), together with about 20 different proteins. The larger (50S) portion contains the 23S and 5S rRNAs (molecular weight = 1,200,000 and 40,000 respectively), together with 40 different proteins. In both portions, the different proteins are each thought to be represented by a single molecule, but this remains to be verified. The eukaryotic ribosome has larger pieces of rRNA and a larger number of proteins, but is similar in all other respects to those of the prokaryotes. It is of interest that the rRNAs and the proteins of a particular species, if added together in solution, will spontaneously form ribosomes, but the rRNAs of one species will not combine in this fashion with the ribosomal proteins of another species.

The relationship of DNA and the several RNAs in protein synthesis is indicated in Figure 2.25. Only one of the two polynucleotide strands of the DNA can be transcribed, and the double helix must open up for this to take place. This is accomplished by an enzyme, RNA polymerase, which binds to the double-stranded DNA, causing it to unwind and open up, and at the same time causing the formation of an RNA that grows in length as the polymerase moves along the DNA. Thus, if the portion to be transcribed is as follows:

T—A—G—C—C—T—T—G—A—C—T—G—C—A—C

A—T—C—G—G—A—A—C—T—G—A—C—G—T—G

and only the bottom half of the helix is transcribed, the mRNA will be

U—A—G—C—C—U—U—G—A—C—U—G—C—A—G

The mRNA will then be released from the DNA and pass from the nucleus to the cytoplasm where it will become attached to one or more ribosomes. Here the message in the mRNA will be "read" by the appropriate tRNAs (Figure 2.30). As this is done, the amino acids attached to the -CCA end of the tRNAs will become linked together to form a growing protein chain. When the tRNAs reach a termination codon (UAA, UAG, or UGA), the process will cease, and the protein will be released for use in the cell either as

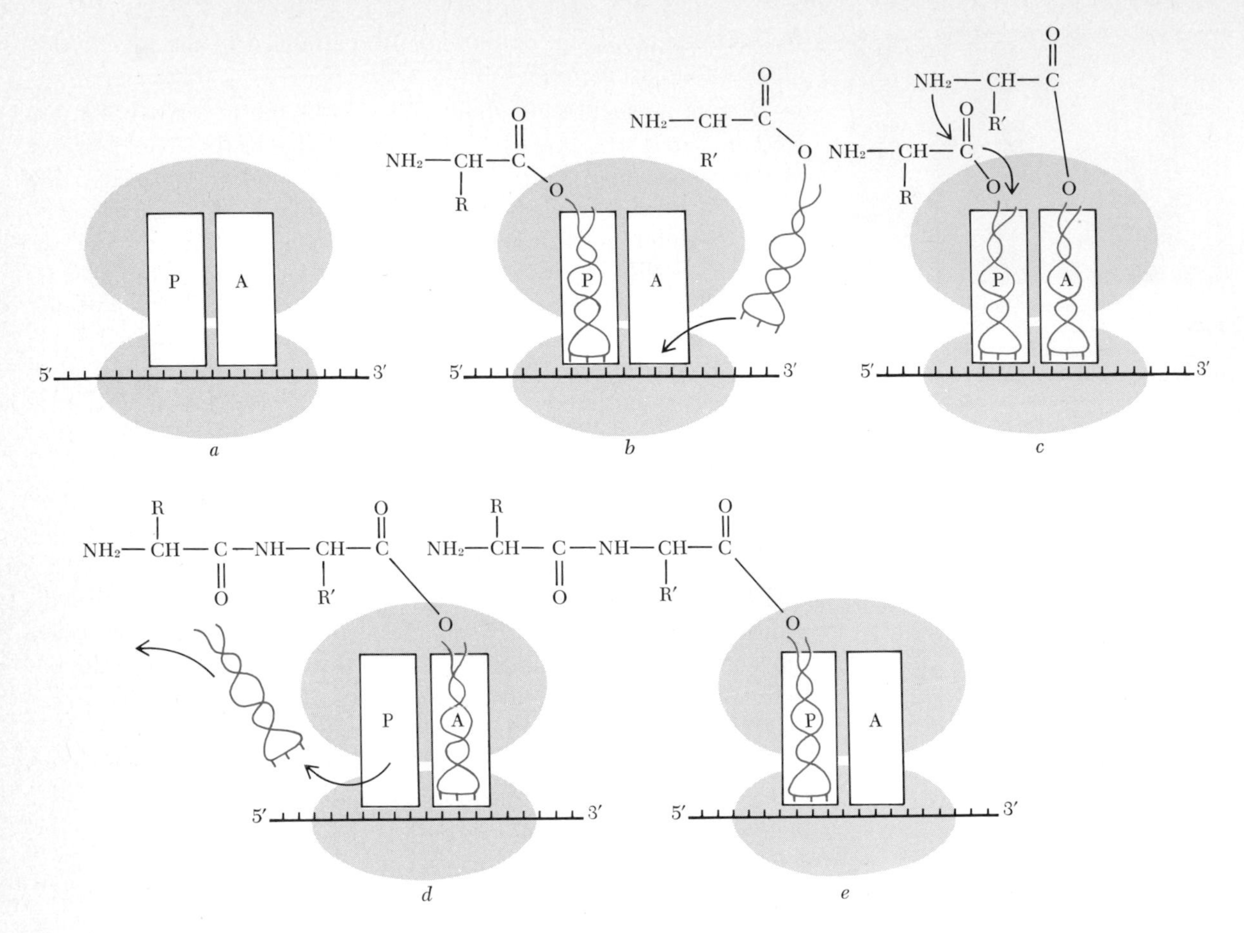

Figure 2.30
The ribosome cycle of protein synthesis. (*a*) Representation of the ribosome with its two sites of activity and the relation of the mRNA to these two sites. The P, or peptidyl, site is where the amino acids are joined through peptide bond formation to form a growing polypeptide, and the A, or aminoacyl, site is where the incoming activated tRNAs become attached. (*b*) The first activated tRNA occupies the P site, which also determines the match between the codon in the mRNA and the anticodon in the tRNA. (*c*) Both tRNAs in place and with the amino acids in the process of being linked by a peptide bond. (*d*) As the peptide bond is formed, the first tRNA is released, and the second tRNA is ready to move from the A to the P site, leaving the A site available to receive another activated tRNA. (*e*) When a termination codon is reached in the mRNA, the two halves of the ribosome fall apart and the mRNA is released. The reading of the mRNA is from the 5-prime (5′), or the phosphate, end of the molecule to the 3′, or the exposed hydroxyl, end of the terminal ribose sugar. The mRNA, therefore, has a polarity, and the polypeptide being formed reflects this polarity since it grows from its amino, or -NH_2, end toward its carboxyl, or -COOH, end. There is good indication that the two portions of the ribosome are distinct from each other and readily disassociate. The mRNA is clearly attached to the smaller, or 30S, unit of the ribosome, and it is believed that this attachment occurs in the nucleolus, with the 30S unit, plus its attached mRNA, moving through the nuclear membrane into the cytoplasm. Although this figure depicts the ribosome as a single unit of two parts, the tRNAs are believed to be brought into position attached to a 50S units, and when they have fulfilled their mission in leaving an amino acid for attachment to the growing polypeptide chain, they are believed to leave, still attached to a 50S unit. The 30S unit plus the mRNA is, therefore, the constant feature of the ribosome, and the 50S unit plus the activated tRNA moving into position, dropping its amino acid, and departing in an unactivated state. The cell, as a consequence, must have a pool of 50S units, with attached tRNAs, which can move in and out of position as the proteins are being formed.

an enzyme or as a structural molecule in membranes or other organelles (Figure 2.31).

The entire process of RNA and protein formation is, like all cellular reactions, governed by specific enzymes for each step taken, and is possible only when all of the necessary nucleotides, amino acids, and energy sources are available. Despite the complexity of the whole process, the triplet code and the process of protein formation are universally applicable to all organisms from simplest to most complex. It appears that chloroplasts and mitochondria, which also possess heritable DNA, produce RNAs and ribosomes of their own, with the ribosomes being more comparable in size and complexity to those of prokaryotes than to those in the cytoplasm of eukaryotic cells. The mRNAs differ according to the message contained within them and consequently have their own kind of specificity, but the tRNAs and ribosomes of both prokaryotes and eukaryotes are very similar in function, and in in vitro experiments may be used with mRNAs from other organisms in the synthesis of proteins. The role played by the ribosomes

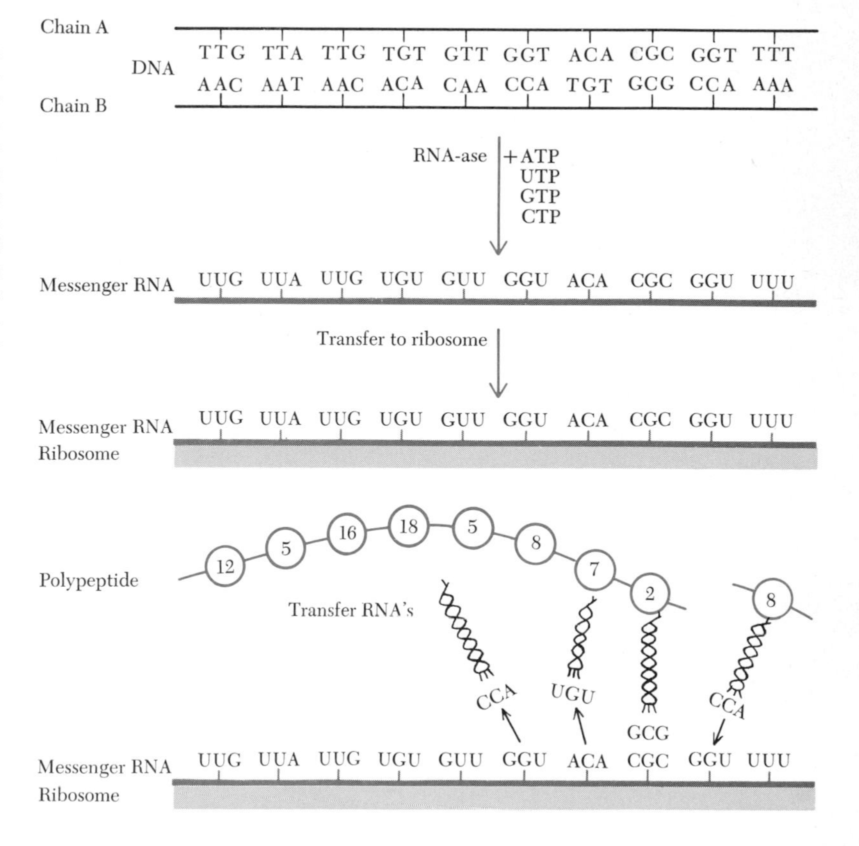

Figure 2.31
The sequence of events whereby the information coded in the DNA molecule is transcribed into messenger RNA (mRNA) and then, by translation, into the formation of a particular protein. It is assumed here that it is the Chain B of the DNA that is being "read." The mRNA contains the code to be read, while the transfer RNAs (tRNAs) possess an anticode, which matches the code by base-pairing. The tRNA at the right carries an activated amino acid that has not yet been inserted into the growing polypeptide.

as the site of protein synthesis is essential, and it is believed that the process occurs as depicted in Figures 2.30 and 2.31.

Control of transcription and translation

A molecule such as a unit of mRNA or an enzyme is like a correctly formulated sentence: it has a beginning, an end, an orderly sequence of internal units, and, when grasped as a totality, a meaning within a given context. The descriptions of the processes of transcription and translation have dealt primarily with the sequences and meanings of RNAs and proteins, that is, how a code in DNA is ultimately translated into a polypeptide by means of the several kinds of RNAs. What is not included in the descriptions is an understanding of how the process is initiated and, when the molecular sentence has been synthesized, how the process is terminated. In addition there is the problem of what particular segment of DNA is to be transcribed, as well as when and how the decision is made. Some aspects of these complex events are known, but much remains to be clarified.

Some differences in transcriptional events exist between prokaryotes and eukaryotes, differences that, at least in part, are related to the nakedness of the DNA in one form and its complex nucleoprotein nature in the other. All RNAs, however, are transcribed from double-stranded DNA by an RNA polymerase. In the prokaryotic *E. coli,* and based on in vitro studies, this enzyme must be coupled with an additional protein, called *sigma,* before it can recognize chemically the proper initiating point for transcription, and make the correct selection of which one of the two polynucleotide strands of the helix is to be transcribed. Without sigma, the polymerase begins to transcribe at any point along the molecule; this, of course, would result in a jumble of meaningless and generally incomplete molecular sentences. The presence of sigma is not sufficient by itself to identify the initiating point, and from the analysis of a number of mRNAs, it appears that the first nucleotide is either A or G, suggesting that the polymerase, coupled with sigma, first recognizes either T or C. Whether this holds similarly for all RNAs is not known.

Termination of the RNA chain must be similarly precise, but although the nature of the termination point is not known, it is determined in part by the presence of a terminating protein, *rho,* which couples with the RNA polymerase. The sigma and rho proteins, therefore, provide the polymerase with a high degree of specificity for chain initiation and termination, although in ways still to be clarified.

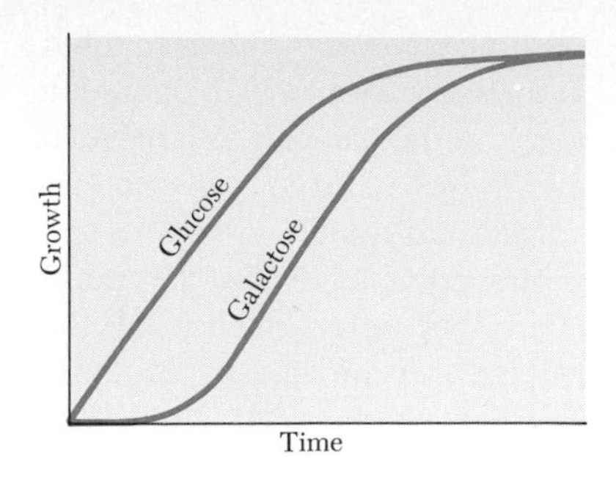

Figure 2.32
The growth of cells on glucose and galactose. The enzymes handling glucose are normal, or constitutive, elements of the cells, and glucose can be utilized without delay. The enzymes utilizing galactose, however, are not normally present and must be induced and synthesized before that sugar can serve the purposes of growth.

The naked DNA of prokaryotes appears competent to transcribe at all times. If the above discussion casts some light on how transcription begins and ends, it has not answered the question of how the cell knows when to activate a particular segment of DNA. One thing is clear, however; factors external to the coded messages in DNA must have some regulatory influence as to which genes are to be transcribed and which are to remain in an inactive state. For example, in *E. coli,* glucose is a normal energy source, and the enzymes involved in its utilization are present in the cell at all times and independent of whether glucose is present or absent. As growth increases as a result of an increase in the amount of glucose in the medium, so too does the amount of the enzyme required to handle it. If, however, glucose is replaced by galactose, a sugar not normally used as an energy source, the cell can utilize it only after a lag period (Figure 2.32). Galactose utilization requires several enzymes. The cells do not normally possess these in detectable quantities, but galactose acts as an inducer, causing the genes coding for the galactose-related enzymes to be activated for transcription. The lag period is indicative of the fact that the cell requires a period of time to get the galactose system functioning, whereas the glucose system was already functioning and ready to perform. If galactose is removed from the culture medium, the enzymes involved disappear; those concerned with glucose utilization do not.

Eukaryotic cells exhibit a number of transcription differences from those found in prokaryotes. Proteins comparable to sigma and rho have not been found coupled to eukaryotic RNA polymerase, but eukaryotic DNA is complexed with histones and nonhistone proteins. Four of the five known histones compact the DNA into beads (Figure 2.22), in which state it is inactive or nontranscribable. The role of the fifth histone in the interbead area is not known, but it is believed that the acidic proteins remove the histones in such a manner as to make the DNA transcribably competent. This obviously must be done in a highly selective manner in particular cells, for not all genes are continuously active. Again, the chemical basis of such selectivity is not known, although the physiological state of the cell, brought on by differentiation or by circumstances external to the cell, influences the degree of selectivity.

Another difference relates to the manner of producing functional RNAs ready for the process of translation. In eukaryotes the rRNAs are processed in the nucleolus before appearing in the ribosomes, and the tRNAs have their unusual bases inserted after they have passed from the nucleus into the cytoplasm. Changes also take place in the mRNAs. The initial RNA transcripts are very much longer than would be calculated from the size of a single gene, and it well may be that noninformational DNA is being transcribed along with that containing a coded message. The nonessential portion, however, is degraded before it leaves the nucleus, after which a second process adds additional adenines to one end of the mRNAs. There is no code for this poly-A portion, the addition being carried out in the nucleus, but after release of the mRNA from the DNA; it is not translated into protein in the cytoplasm, and its function remains unknown.

Our understanding of the process of translation is somewhat firmer than that of transcription. As Table 2.2 indicates, termination of the translational process occurs when one or another of the three triplet codes—UAA, UAG, or UGA—are encountered in the reading of the mRNA message. No normal tRNA can interact in complementary fashion with these triplets to insert an amino acid into a growing polypeptide chain. This fact provides a basis for understanding the origin and nature of some mutations. For example, if in an mRNA coding for a crucial enzyme, the triplet code -AAA- were to mutate to -UAA-, lysine would fail to be inserted into the growing polypeptide, and the process would cease. An abbreviated protein would result, and very likely a nonfunctional one.

The initiation of protein synthesis from mRNA is as highly specific as is its termination. In *E. coli* there is an "initiator" tRNA, which complexes in complementary fashion with the triplet code -AUG-. This sets the reading frame of the message, with other nucleotides being read off in groups of three. In addition, as Table 2.2 would suggest, a methionine is inserted at the beginning of every protein. It is, however, a particular kind of methionine (Figure 2.33), and it can only be inserted at the beginning and not in the middle or at the end of a message. Should the triplet AUG

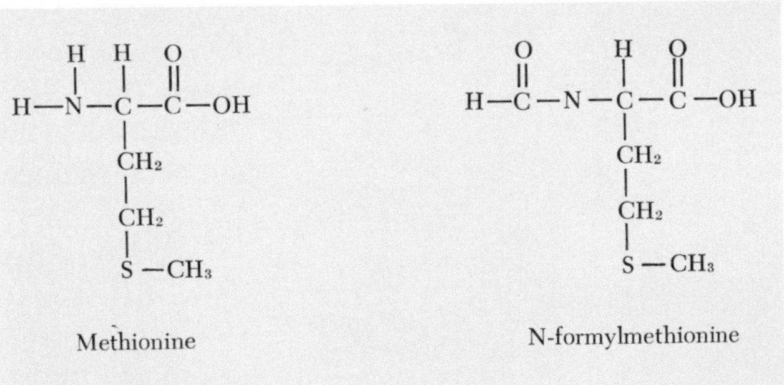

Figure 2.33
The structures of methionine (left) and N-formylmethionine (right). Both are coded for by the AUG codon, but the former can only be inserted internally into a growing polypeptide chain, while the latter functions as an initiating amino acid, being inserted in the N-terminal position.

Methionine

N-formylmethionine

appear elsewhere in the message, a normal methionine would be inserted.

This would suggest that all proteins should commence with a methionine at their starting point. Some proteins do possess methionine in this position, but not all. In functional cells, an enzymatic mechanism clips off the terminal methionine before it becomes operative. Thus in an in vitro system, a known protein begins with N-formyl-MET-ALA-SER-ASN-PHE-THR- . . . . , but the same protein isolated from cells begins ALA-SER-ASN-PHE-THR-. . . . . The above system, characteristic of *E. coli,* does not exist in plant viruses or in higher plants and animals. However, other specific initiating codons, tRNAs, and N-terminal-blocked amino acids seem to be present and to function in similar ways.

Bibliography

BARRY, J. M. 1964. *Molecular Biology: Genes and the Chemical Control of Living Cells.* Prentice-Hall, Inc., Englewood Cliffs, N. J. An elementary but complete description of genes and their modes of action.

BARRY, J. M., and E. M. BARRY. 1969. *An Introduction to the Structure of Biological Molecules.* Prentice-Hall, Inc., Englewood Cliffs, N. J. An extensive and detailed treatment of the major molecules of biological importance: nucleic acids, proteins (enzymes), carbohydrates, lipids, and their modes of formation, interaction, and action.

BEADLE, G., and M. BEADLE. 1966. *The Language of Life.* Doubleday & Co., Inc., Garden City, N. Y. An introduction to the physical basis of inheritance, simply but accurately presented.

BONNER, D. M., and S. E. MILLS. 1964. *Heredity* (2nd ed.). Prentice-Hall, Inc., Englewood Cliffs, N. J. A brief but excellent account of the hereditary roles played by DNA, RNA, and proteins, with an emphasis on microorganisms.

DUPRAW, E. J. 1970. *DNA and Chromosomes.* Holt, Rinehart & Winston, Inc., New York. A comprehensive and detailed treatment of the molecular and chromosomal aspects of cell biology; a controversial but stimulating book.

HARTMAN, P. E., and S. R. SUSKIND. 1969. *Gene Action* (2nd ed.). Prentice-Hall, Inc., Englewood Cliffs, N. J. An advanced treatment of the biochemical features of modern genetics and gene action. A difficult book, but excellent for those willing to devote the time necessary to understand the concepts and experimental details.

LEWIN, B. M. 1970. *The Molecular Basis of Gene Expression.* Wiley-Interscience, New York. Attention is devoted to the molecular mechanisms by which genes gain cellular and organismal expression through proteins.

Stahl, F. W. 1969. *The Mechanics of Inheritance* (2nd ed.). Prentice-Hall, Inc., Englewood Cliffs, N. J. A companion volume to the Hartman-Suskind book; technically difficult, but well worth the effort.

Watson, J. D. 1975. *Molecular Biology of the Gene* (3rd ed.). W. A. Benjamin, Inc., Menlo Park, Cal. An excellent treatment of molecular genetics by one of the discovers of the structure of DNA; a classic volume.

Wolfe, S. L. 1972. *Biology of the Cell.* Wadsworth Publishing Co., Inc., Belmont, Cal. Chapters 9 through 12 deal extensively with replication, transcription, translation, and the roles of the several macromolecules in chemical activities.

For cells to maintain themselves as autonomous units and to continue to function, it is vitally important that some control be exercised over the exchange of materials between the cell and its surroundings. The differences between the internal chemical composition of a cell and that of its external environment represent a degree of order, or nonrandomness, that can be maintained only by the presence of a barrier to free movement of material into and out of cells. Furthermore, since cells must also assimilate matter from their environment to function, to grow, and to reproduce, they must be able selectively to allow certain molecules or ions to enter across this barrier, often against concentration gradients, while restricting or excluding others.

All exchange of material between the cell and its environment must occur across the outer boundary of the cell, and it is in the structural organization of this *cell surface* that energy must be expended, both to maintain the barrier and to surmount it. It is also at the cell surface that information is first received, both from the immediate environment or from other cells, the latter type of communication being of fundamental importance to the integration of function characteristic of multicellular organisms.

The plasma membrane

The existence at the surface of cells of a specialized, selectively permeable film, the plasma membrane, was inferred long before it could be demonstrated directly by the powerful resolution of the electron microscope. This inference was based on observations that when cell surfaces are torn by a very fine needlepoint the contents spill out, that certain compounds can enter and spread throughout the interior of the cell only after the surface has been punctured, and that certain molecules can penetrate intact cells more readily than others. Indeed it was from physiological studies of the permeability of cells to various compounds that the first indications of the chemical composition and structure of the plasma membrane were derived. It is generally true (although there are important exceptions) that ionic, or electrically charged, compounds penetrate more slowly than nonionic compounds, that small molecules penetrate more rapidly than large molecules, and that the more soluble a compound is in lipid the more readily it can enter the cell. These generalizations lead to the idea of a membrane carrying a positive electrical charge that would impede passage of positive ions, containing pores through which small molecules can readily pass, and consisting at least in part of a lipid, or fatty, film into and out of which fat-soluble compounds can move. It was also shown by analysis of red blood cell "ghosts" (cell membranes freed of the inner contents of the cell) that these membranes contained twice the amount of lipid necessary to form a monomolecular film equivalent to the surface area of the original cell, suggesting that the lipids were arranged in a bimolecular layer within the membrane.

This bimolecular lipid layer depends on the nature of phospholipids, the major lipid components of most membranes. Phospholipids are asymmetrical molecules in that one end of the molecule carries an electrical charge and is hydrophilic while the other end is neutral and hydrophobic. Thus the most probable arrangement of phospholipids in the membrane is one in which the polar ends of the molecules extend toward the aqueous environments of the outer and inner surfaces of the membrane, while the apolar ends lie together in the interior of the membrane (Figure 3.1).

The presence of proteins in or at the surface of the plasma membrane has been inferred from measurements of surface tension and

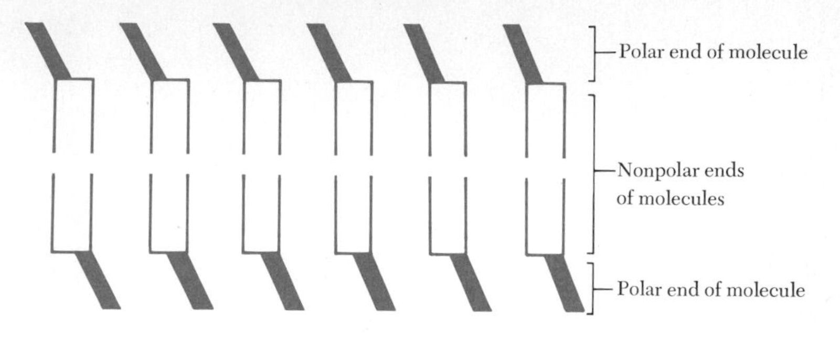

Figure 3.1
Diagrammatic representation of the arrangement of lipid molecules in a bilayer. The polar ends of the molecules are oriented toward the outer faces.

elasticity of the cell surface, these measurements being closer to those expected of protein than of lipid. As we shall see, it is in the protein moiety of the membrane that much of the specificity of membrane function and behavior resides, although other components play an important role in certain types of cells.

Since a bimolecular lipid layer with its associated proteins is well below the limits of resolution of the light microscope, direct visualization of the plasma membrane was not possible until the development of the electron microscope and the various associated techniques of preparing material for examination. It is clear from electron micrographs that the cell surface is organized in a specific manner, usually appearing in thinly sectioned fixed cells as a tripartite structure consisting of two parallel dark lines separated by a central clear area (Figure 3.2). Similar tripartite organization also can be seen in sections of artificial membranes formed by mixing phospholipids and proteins in water. Although the dimensions of the three layers can vary from membrane to membrane, they are remarkably close to those expected of a bimolecular lipid layer separating two layers of protein. Thus the central clear area was thought to represent the hydrophobic ends of the two lipid layers, and the dark lines to represent the polar ends in association with proteins, at both the interior and exterior faces of the membrane.

Figure 3.2
Electron micrograph of plasma membranes at the junction of three cells, showing trilaminar appearance. (Courtesy of J. D. Robertson.)

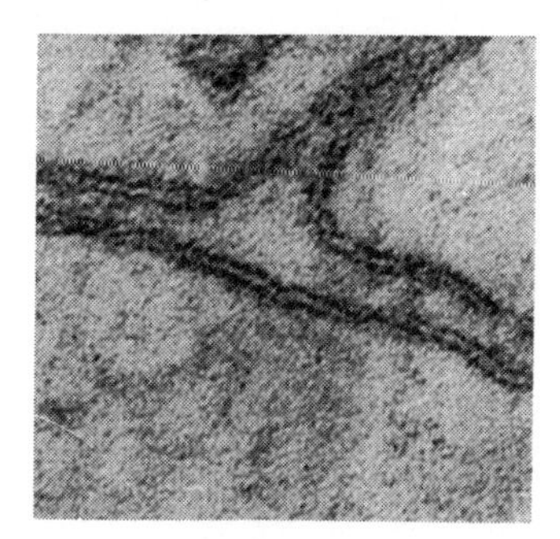

As well as revealing a specific organization of the cell surface, electron microscopy demonstrated the existence of a very highly structured complex framework of membrane systems within the cell. These will be discussed in more detail in a later chapter, but it should be pointed out here that the appearance of these internal membranes suggested that they shared with the plasma membrane a similar fundamental type of organization, based on a lipid bilayer with proteins on the outer surfaces.

Although this "unit membrane" model may be partially correct, it is now clear that membrane structure is neither as simple nor as uniform as was formerly believed. Many membranes, including plasma membranes, have been shown to display a particulate

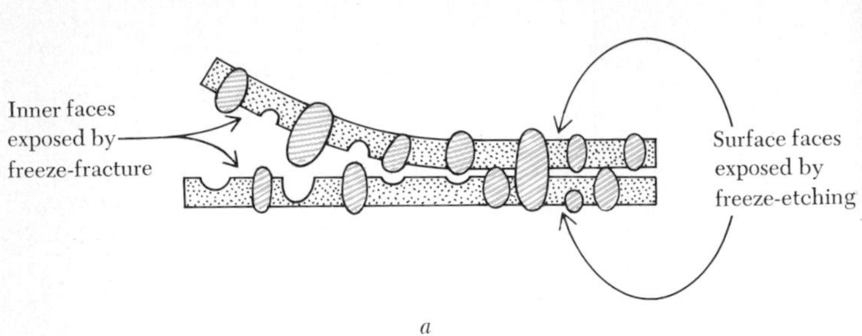

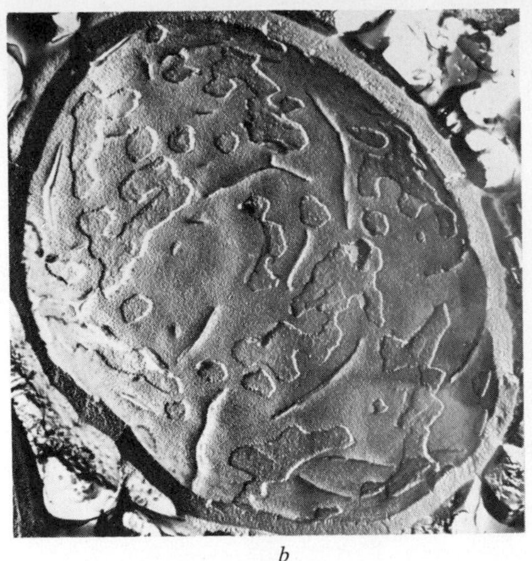

Figure 3.3
(*a*) **Diagrammatic representation of how fracture faces and surface faces of membranes are exposed by freeze-fracturing and freeze-etching.** (*b*) **Electron micrograph of faces of yeast plasma membrane exposed by freeze-fracture and freeze-etching. Note the arrays of particles in the membrane.** (Courtesy of Dr. S. C. Holt.)

appearance in the electron microscope. This is especially striking following preparation of membranes by the techniques of freeze-fracturing and freeze-etching. Both of these techniques involve freezing the specimen rapidly and then sectioning it. In freeze-fracturing, the frozen membranes are cleaved along the plane of weakness represented by the hydrophobic groups at the center of the lipid bilayer (Figure 3.3*a*). Thus the internal organization of the membrane can be examined. In freeze-etching, the ice crystals formed are allowed to sublimate, and the true membrane surfaces can be exposed (Figure 3.3*a*). Particles of various sizes are observed on both the interior and exterior faces (Figures 3.3*b* and 3.4), and these are thought to represent globular proteins, or aggregates of proteins, embedded in or at the surface of the lipid matrix of the membrane.

By using chemical reagents that can label the proteins of either one or both surfaces of the membrane, it is possible to show not only that some proteins extend all the way through the lipid bilayer but also that the membrane is asymmetric in that other proteins are restricted to either the inner or outer surface. Such asymmetry is not surprising, since the specific interactions occurring between the cytoplasm and the membrane must be different from those that occur between the membrane and the outside environment of the cell. For example, there is evidence that spectrin, one of the proteins confined to the inner surface of the red blood cell membrane, is involved in determining the distribution of some of the other protein components of that membrane.

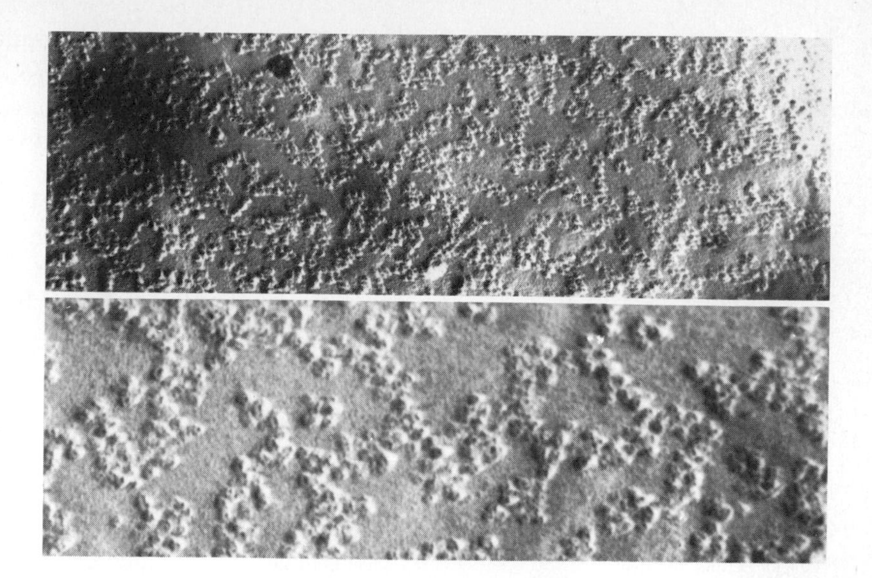

Figure 3.4
Fracture face of red blood cell plasma membranes, shown at two different magnifications. (Courtesy of Dr. S. C. Holt.)

Membranes from various cells and organisms can also differ widely in both relative amounts and chemical composition of their constituent proteins and lipids, this variation providing a basis for the wide range of physiological activities displayed by different membranes. The various structural and functional properties of membranes, influencing such diverse activities as control of transport, cell-to-cell recognition, immunological response, and even cell movement, must themselves depend on which enzyme systems comprise the protein complement of the membrane and on how these proteins are arranged in relation to each other and to the various lipids also present.

It is clear that fixed structure and composition are not invariable characteristics of all membranes. It also should be stressed that membranes are by no means static and that there can be considerable freedom of movement within the membrane of its constituent molecules. The ability of membranes to reform rapidly over exposed regions of cytoplasm, to fuse with one another, and to expand and contract during cell movement or changes in cell shape suggests the fluid and dynamic nature of such membranes. Both the lipids and the proteins of plasma membranes can be shown to be mobile to varying degrees within the membrane. This has been beautifully demonstrated by the use of cell-fusion heterokaryons, which are formed by the fusion in culture of two cells of different species. Species-specific antigens, compounds located in the plasma membranes, are initially segregated following fusion of mouse and human cells. However, within less than an hour after fusion, the mouse and human antigens are completely intermingled on the surface of the heterokaryon, indicating that lateral diffusion of these membrane components has occurred. Such movement must be controlled in some way if the functional integrity of the membrane is to be maintained; interactions between the inner face of the membrane and the cytoplasm immediately adjacent to it are believed to play a role in determining how the membrane components move and how they are distributed. The role of spectrin in the red blood cell membrane has already been mentioned, and other components of the membrane-associated cytoplasm, such as the microtubules and microfilaments (see Chapter 5) found in this region, may be involved.

Figure 3.5 illustrates a current model of basic membrane organization, including the lipid bilayer with which different proteins are associated in different ways. This fluid mosaic model of the membrane accounts for many of the properties of cell membranes.

We must also recognize, however, that the surface properties of cells may be influenced not only by the composition and activity of the underlying cytoplasmic layer, but also by the presence of additional molecules associated in various ways with the exterior surface. Some plasma membrane proteins and lipids, particularly in animal cells, are bound to oligosaccharides, and are known as *glycoproteins* and *glycolipids*. These surface compounds, which show in the electron microscope as a fuzzy layer at the cell surface (see Figure 3.11), appear to be involved in the recognition of, and the response of the cell to, factors in the environment. For example, some of the antigens that determine blood type are membrane glycoproteins and glycolipids, and the specific clumping reactions of blood cells when exposed to blood sera of different kinds depend on which antigens are present. Cells from the same tissue also exhibit the ability to "recognize" one another when in liquid suspension. For example, if heart and kidney cells from a chick embryo are dis-

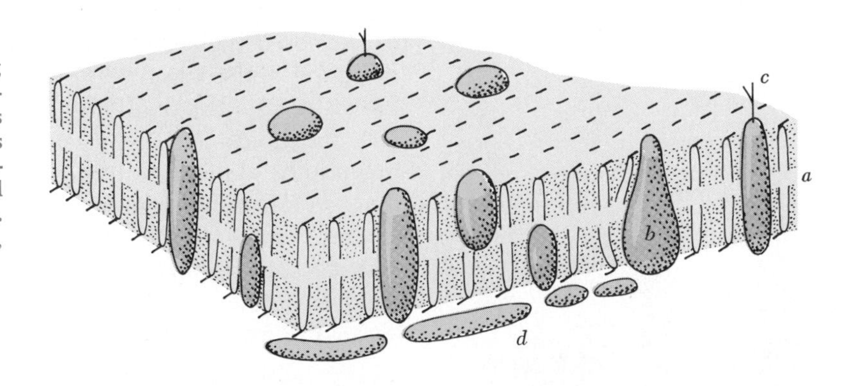

Figure 3.5
Model of membrane structure showing proteins (*b*) embedded in phospholipid bilayer (*a*). Carbohydrates associated with membrane proteins are on the outer surface of the membrane (*c*), and less tightly associated proteins (*d*) are at the inner surface. (After Branton, Singer, Capaldi, Nicolson.)

sociated to form a mixed single-cell suspension, and then allowed to remain undisturbed in culture, heart cells seek out and aggregate with heart cells, kidney cells with kidney cells. Such tissue-specific recognition between cells is a surface phenomenon that also may be mediated by the glycoproteins or glycolipids of the membrane. Glycoproteins also have been shown to be the receptors at the cell surface for *mitogens*, compounds that trigger cell proliferation in certain lymphocytes; the binding of the mitogen to the surface leads to changes in the permeability and transport properties of the membrane, which can in turn affect metabolic processes inside the cell. Both plant and animal cells are known to respond in specific ways to a range of hormones produced elsewhere in the body. The ability to recognize such chemical signals resides at the cell surface, and the initial response involves some change at the membrane itself.

Movement across membranes As discussed at the beginning of this chapter, the main role of the plasma membrane is to regulate exchange of materials between the cell and its environment. We must now consider some of the mechanisms believed to be involved in transport of matter through the plasma membrane, as well as the forces responsible for such movement.

Although passive diffusion, which results in the net movement of substances down a concentration gradient, is impeded by the plasma membrane acting as a barrier, many substances do diffuse rapidly across the membrane. As we have already seen, water and other small molecules cross the barrier readily, suggesting the presence of pores in the membrane. Although pores have not been seen directly in the electron microscope, it has been calculated that they must be very small and can occupy only a very small portion of the surface area of the membrane. Furthermore, it is possible that such pores are not permanent structural entities but represent transient and localized rearrangements in the molecular architecture of a dynamic membrane.

Another mechanism that can account for rapid movement of solutes to which membranes are otherwise relatively impermeable is *facilitated* diffusion. The kinetics of entry of various sugars, amino acids, and nucleosides strongly suggest that enzymes are involved that accelerate the rate of diffusion through the membrane. Such enzyme systems presumably constitute part of the membrane itself and may act as "carriers" of the penetrating molecules. It should be pointed out that facilitated diffusion does not imply movement against a gradient, and, therefore, is not directly dependent on expenditure of energy.

In contrast to diffusion, *active* transport mechanisms are also involved in movement, especially of ions, across the membrane. Membranes maintain not only differences in concentrations of substances between the inside and outside of the cell, but also differences in electrical potential. Active transport, therefore, requires that energy be expended to do the work involved in "uphill" movement, that is, against an *electrochemical gradient.* The reactions involved in active transport must be coupled to reactions at the membrane in which free energy is made available. Enzymes capable of catalyzing such *exergonic* reactions, which provide the energy necessary to drive the "pumping" of compounds into or out of cells, are also believed to be integral components of the membrane.

Let us consider a case of active transport, that of sodium ions (Na^+) across membranes. For most cells the concentration of Na^+ inside the cell is much lower than that outside the cell. Potassium (K^+), on the other hand, is found in high concentrations inside the cell compared to the surroundings. Thus there is a tendency for Na^+ to move into the cell from the outside and for K^+ to move out. Although membranes are usually fairly resistant to passive diffusion of ions, some leakage would take place, tending to equalize concentrations inside and out. The differences in ionic concentrations are maintained, however, by a pumping system at the membrane, which pumps Na^+ out of, and K^+ into, the cell. Energy is required for the pumping of the ions against the concentration gradients, since low temperatures or inhibitors of respiration (see Chapter 4) allow equilibrium to be reached. The necessary energy comes from the standard energy-yielding reaction of cells, that is, the hydrolysis of ATP (see Chapter 4), and the enzyme involved, ATP-ase, is in the membrane itself.

The tendency for sodium to move into the cell, that is, in the energetically favorable direction, has been utilized for the transport of various sugars and amino acids into cells. For example, the transport of glucose is dependent on Na^+, and is coupled to entry of Na^+; that is, the energetically favorable movement of Na^+ into the cell is coupled to the energetically unfavorable entry of glucose. The active transport of Na^+ out of the cell by the energy-dependent pump maintains the Na^+ concentration gradient necessary for the movement of both Na^+ and glucose into the cell.

Some of the postulated mechanisms of transport are illustrated in Figure 3.6. The carriers in the membrane have not been identified, but are thought to be proteins within the membrane structure.

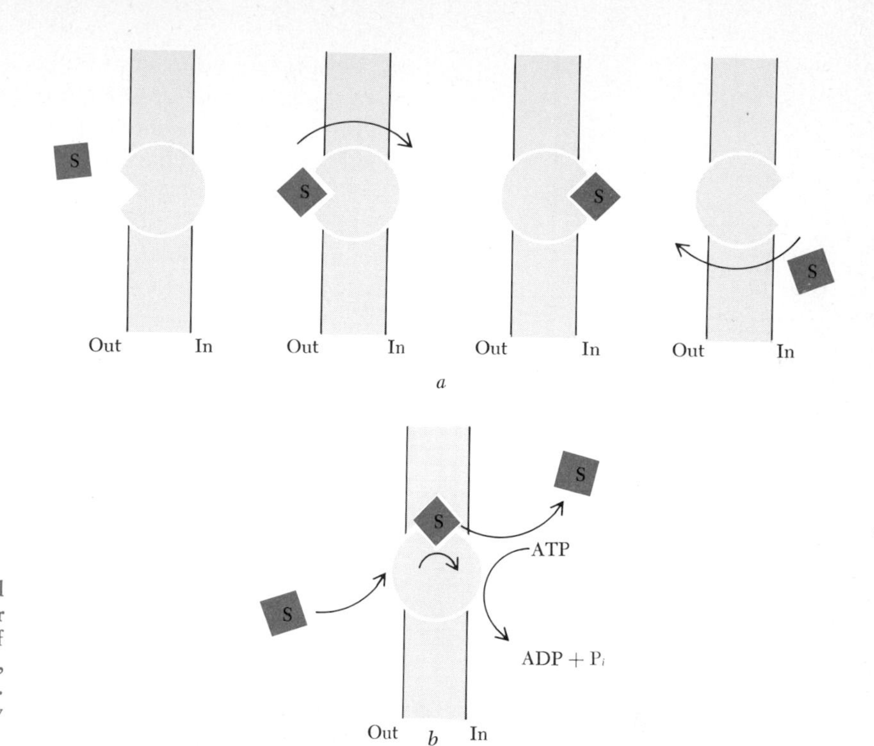

Figure 3.6
Model of how a substance *S* is carried across a membrane. (*a*) A carrier binds to the substance on one side of the membrane, transports it across, and releases it on the other side. (*b*) In active transport, the necessary energy is supplied by ATP.

Many cells can also take in material from the environment by one or both of two additional processes, *pinocytosis* and *phagocytosis*. The former name is derived from the Greek words for "drink" and "cell," and the process is literally a drinking phenomenon. The flexible plasma membrane forms a channel to get liquids into the cell, and then pinches off pockets that are incorporated into the cytoplasm to be digested (Figure 3.7). By this device, large molecules and various ions incapable of passing through the

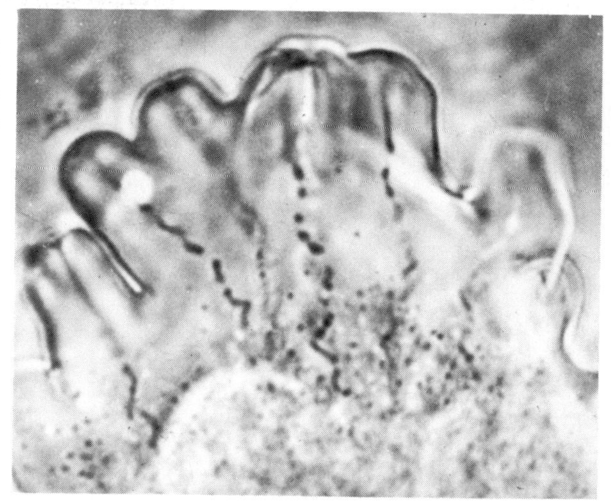

Figure 3.7
Photograph of the edge of a living amoeba, showing the pinocytotic channels (the dark lines converging toward the center of the cell). Liquids flow into the cell through these channels, to be pinched off as membrane-enclosed droplets; these eventually dissolve in the interior of the cell. (Courtesy of Dr. David Prescott.)

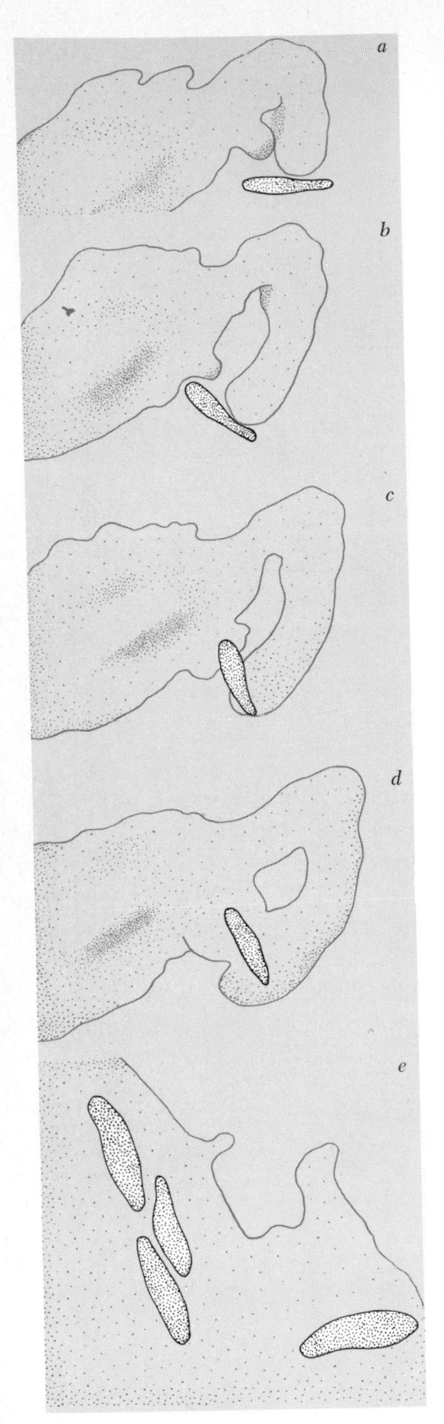

membrane can be taken up by cells. As well as being an important means of nutrient assimilation in single-celled organisms, pinocytosis occurs frequently in absorptive cells, such as those lining blood capillaries. These particular cells can incorporate material from the blood into pinocytotic vesicles, transport this material to the opposite side of the cell, and discharge it by a process that is essentially the reverse of pinocytosis, fusion of the vesicle membrane with the plasma membrane. In phagocytosis (from the Greek, *phagein,* "to eat"), arms of cytoplasm engulf droplets of liquid containing solid material, such as bacteria, and draw these materials into the cytoplasm, where the digestive enzymes break down the engulfed material into usable fragments (Figure 3.8). Phagocytosis is also an important part of the body's reaction to infection. Specialized white blood cells are attracted to invading bacteria, and after engulfing them, release enzymes that destroy the intruders. Such aspects of membrane behavior as pinocytosis and phagocytosis once again illustrate the dynamic nature of the plasma membrane, involving as they do membrane reorganization, movement, fusion, replacement, and turnover.

Modifications of plasma membrane We must also consider some of the modifications displayed by specialized regions of the cell surface membrane of various cells. One such class of modifications consists of the several types of intercellular junctions, which are regions of very close association between plasma membranes of adjacent cells and which permit these cells to interact in specific ways. These include *tight junctions,* which prevent movement through the intercellular spaces between one side of a cell layer and the other, *desmosomes* (Figure 3.9), which are involved in cellular adhesion, and *gap junctions,* which permit relatively free movement of ions and small molecules between adjacent cells. A different type of intercellular association is represented by plasmodesmata (Figure 3.10); these narrow channels run through the thick walls surrounding plant cells and are bounded by the plasma membrane, which therefore can be continuous between adjacent cells. Although neither the internal composition of plasmodesmata

Figure 3.8
Phagocytosis as observed in an amoeba. The arm of cytoplasm, coming in contact with a paramecium, surrounds it and then draws it into the cytoplasm where it can be digested. (*a–d*) The process of enveloping. (*e*) A portion of the amoeba containing several phagocytized paramecia.

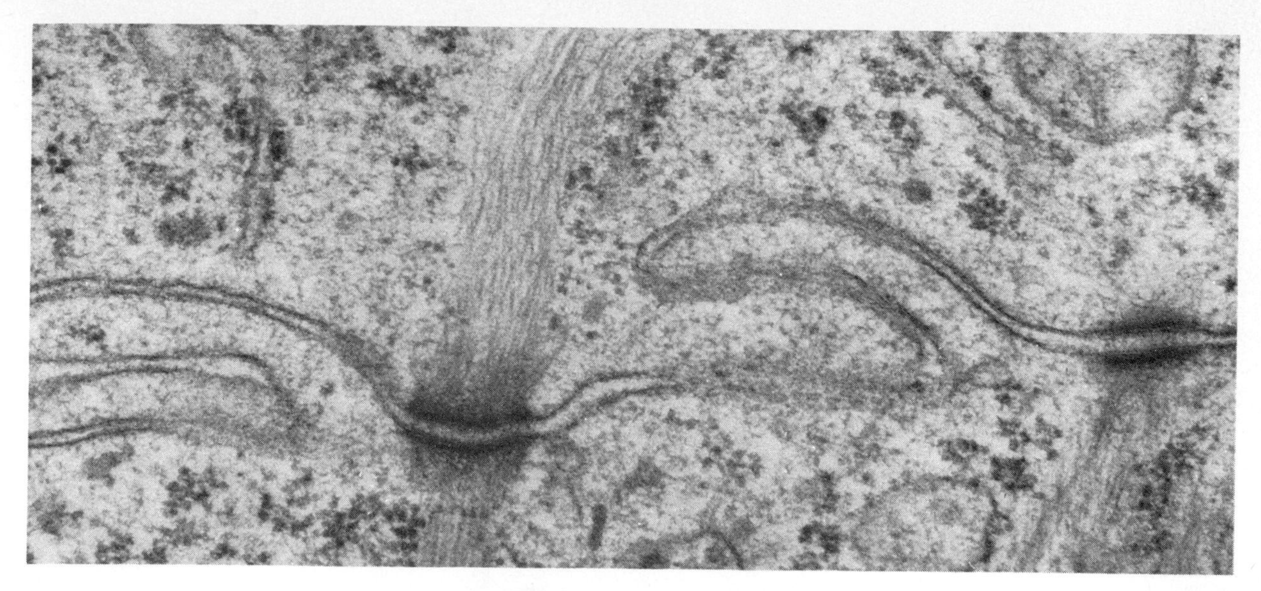

Figure 3.9
The convoluted membranes of two adjacent columnar epithelial cells, showing two prominent desmosomes. Microfilaments extend from the desmosomes into the cytoplasm. (Courtesy of Dr. C. Philpott.)

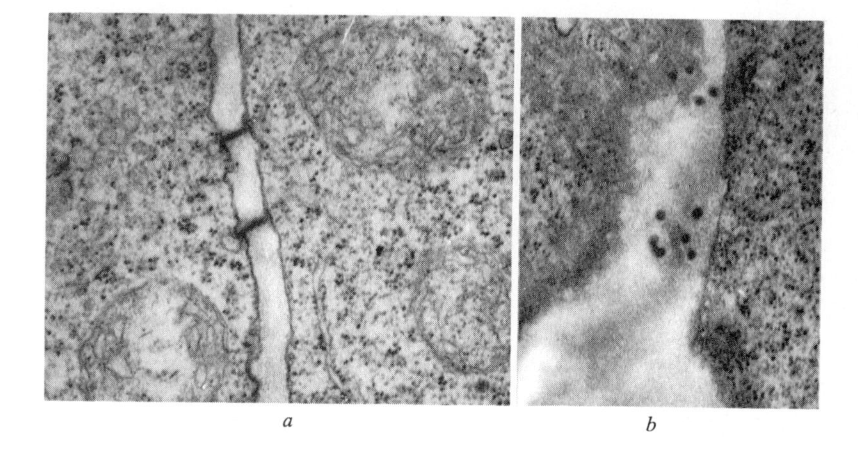

Figure 3.10
Plasmodesmata in plant cell walls. (*a*) Two plasmodesmata between adjacent cells. (*b*) Oblique section through wall showing clusters of plasmodesmata in cross-section. (Courtesy of W. McDaniel.)

nor their function are understood, these structures may represent cytoplasmic continuity and, therefore, permit direct communication between adjacent cells.

Highly convoluted regions of the membrane are also characteristic of many cell types. For example, columnar epithelium cells,

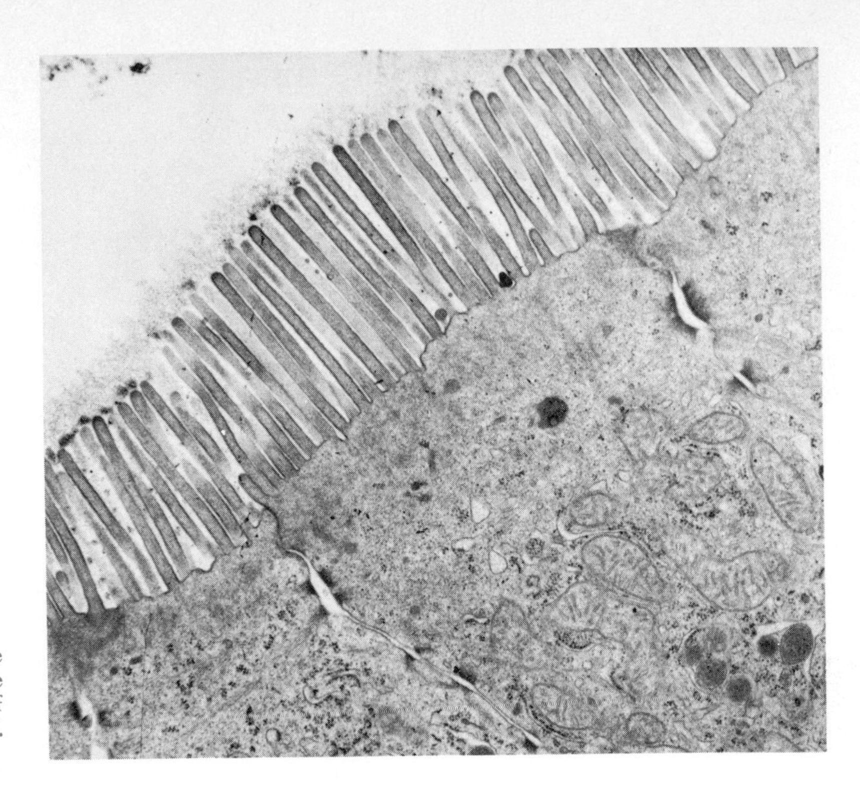

Figure 3.11
Microvilli projecting from the surface of a columnar epithelial cell. The glycocalyx is apparent at the tips of the microvilli. (Courtesy of W. McDaniel.)

which line the small intestine and are active in the absorption of digested food, form many projections, or *microvilli* (Figure 3.11) at their upper surface. These greatly increase the surface area of the cell, and hence provide a tremendous absorption area. Each such cell would have several thousand microvilli, while a square millimeter of the intestine would have as many as 200,000,000. Another modification of the cell membrane can be seen in the light-receptor cells (rods and cones) of the vertebrate eye (Figure 3.12). The outer segment of these cells is made up of a series of flattened discs, from 500 to 1,000 in some cells, stacked one on top of the other like coins. The discs are derived from foldings of the cell membrane, but they tend to break free from the membrane and appear as free-floating structures, at least in sections prepared for electron microscopy (Figure 3.13). The significance of these discs is that they represent the light-receptor surfaces of the eye. Embedded in the membranes of the rod cells are particles of rhodopsin, the only protein component of the disc membranes. These rhodopsin molecules change their molecular configuration and their orientation within the membrane in response to differences in light

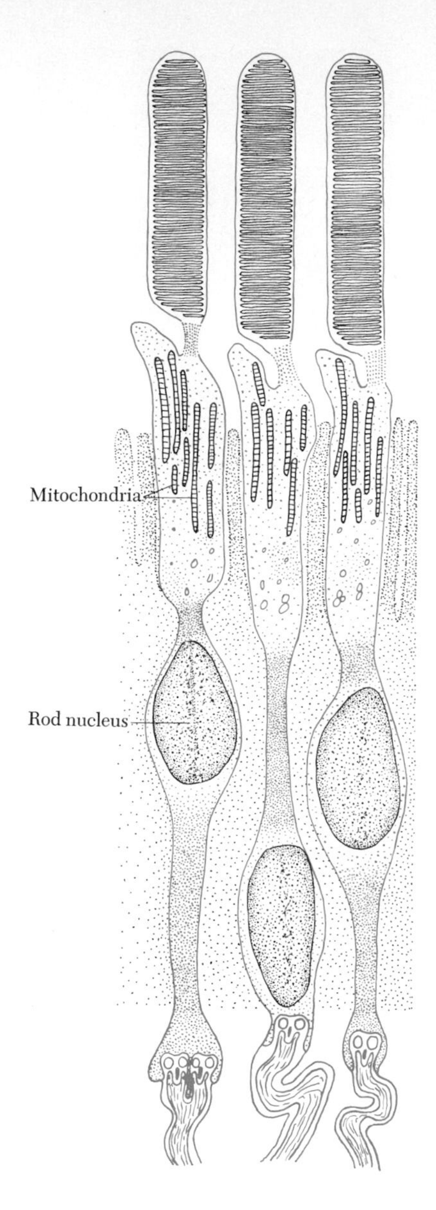

Figure 3.12
Schematic representation of a rod (light-receptor) cell in the retina of the guinea pig. The discs at the top of the cell are folded and refolded to provide many layers of membranes, each of which contains light-sensitive pigments on its surface. The mitochondria are concentrated just below the light-sensitive area; the rod nucleus is also identified. At the base, each cell has an intimate connection with a nerve fiber. (F. S. Sjöstrand, *International Review of Cytology*, 5. New York: Academic Press, Inc., 1956).

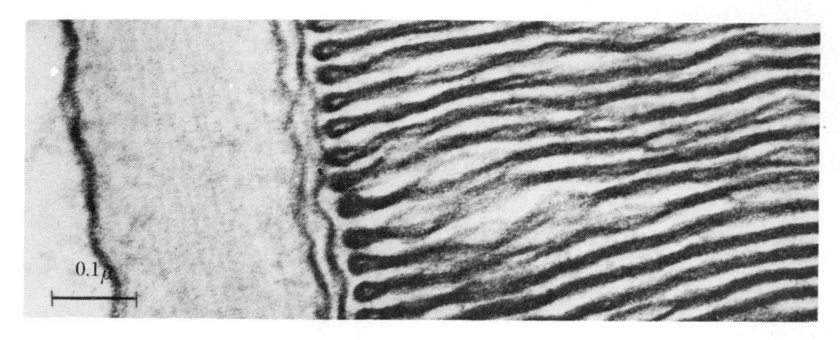

Figure 3.13
A small portion of a retinal cell, with the folded membranes where the light-sensitive pigment is located. That the membranes originate from the cell membrane is not shown here. (Courtesy of Dr. J. D. Robertson.)

intensity, and these changes are somehow communicated to the nerve fibers associated with the rod cell and transmitted to the brain. These extensive infoldings of the plasma membrane are responsible for an enormous increase in the light-receptive surface area of the retina cells.

The relation between a nerve cell and its associated Schwann, or satellite, cells presents another example of membrane flexibility. Here the entire plasma membrane of one cell forms an elaborate structural system around a portion of another cell. Figure 3.14 depicts a nerve fiber, or axon—the extended process of a nerve—with the associated Schwann cell wrapped around it. The development of the spiral proceeds as in Figure 3.15. The cytoplasm of the Schwann cell is largely squeezed to the outside, leaving the axon surrounded by a multilayered membranous system and isolated from its surroundings. The myelin sheath, as this layered structure is called, is thought to assist in the transmission of nerve impulses by insulating the nerve cell. It is unlikely that much enzyme activity is associated with these membranes, and in contrast to other membranes, few globular protein subunits can be detected at the surface.

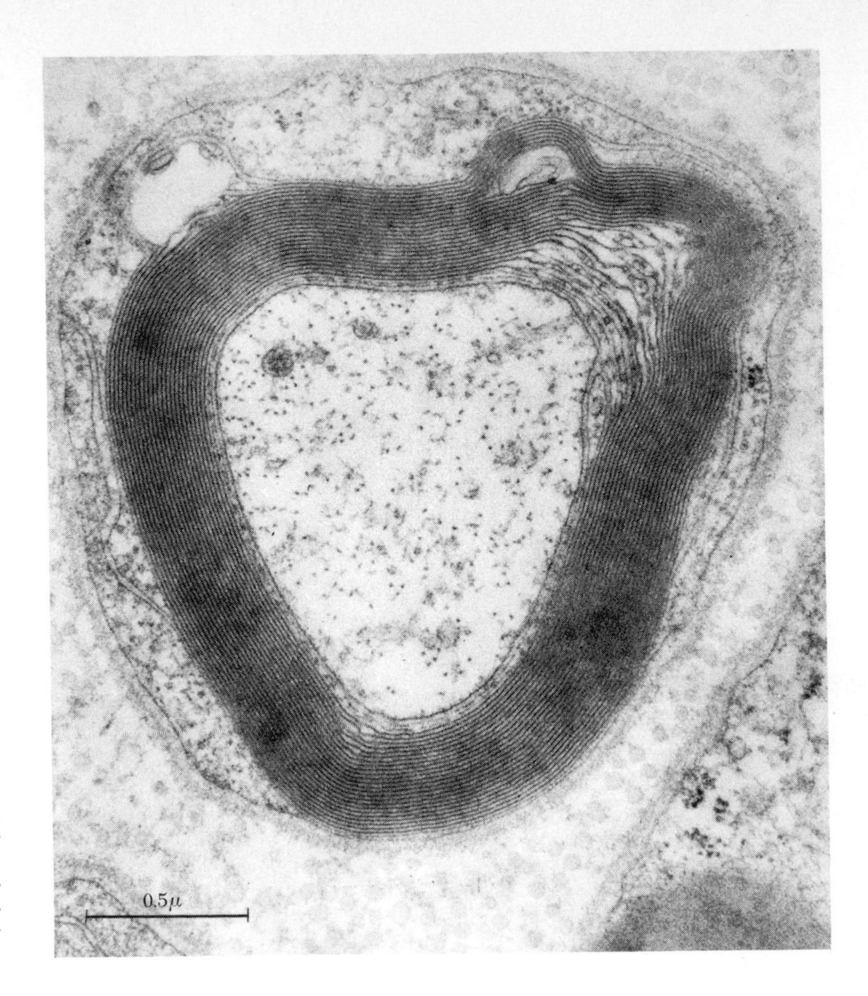

Figure 3.14
Electron micrograph cross section through a nerve fiber, with the membranes of the Schwann cell wrapped around the central axon. The cytoplasm of the Schwann cell can be seen outside the membranes and between the disrupted layers of membranes at the upper right. (Courtesy of Dr. L. G. Elfin.)

Figure 3.15
Schematic representation of the progressive envelopment of an axon by the membranes of the Schwann cell, as described by Dr. Betty B. Geren. Such an axon is said to be myelinated.

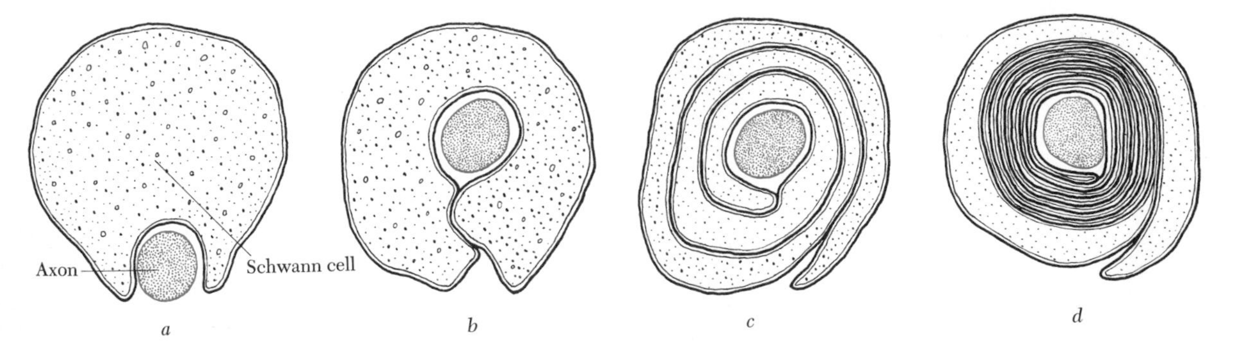

Extramembrane components of cells

Although the plasma membrane is generally considered to be the outer living limit of the cell, most cells also have extracellular components associated with the membrane. To state categorically that any part of the cell, either inside or on the immediate outside, is living or nonliving is to presume to define life, and, as everyone knows, doing so has its pitfalls. Nevertheless, extracellular materials can contribute significantly to the surface properties and the functioning of the cells they enclose. We can see such outer boundaries most readily in plant cells, many of which possess heavy walls, but animal cells and many unicellular organisms also exhibit a number of somewhat comparable external substances, some of which are visible only at the level of electron microscopy, others being readily discernible with the light microscope. This cell coat, or *glycocalyx,* varies from cell to cell in thickness, chemical composition, and in the nature of its association with the plasma membrane. We have already mentioned one type of glycocalyx in our discussion of membrane glycoproteins and glycolipids, and its possible role in cell recognition. In addition, many other functions, varied and often multiple in nature, are performed by extracellular substances: *water retention* in the case of the slimy secretions of many algae (agar is a commercially useful product derived from some algae); *protection* as provided by the tough, chitinous covering of insects; *support* as from the cellulose walls of plants, and from the collagen in cartilage and bone; *rigidity* and *hardness* as from the mineralized regions of bone, the dentine and enamel of teeth, the siliceous shells of diatoms, and chitin; *elasticity* as from the elastin fibers of skin or artery walls; *locomotion* as in the pellicles of *Euglena* and *Tetrahymena;* and *adhesiveness* as from the middle lamellae of plant cells and the hyaluronic acid and chondroitin sulfate of animal cells. The slime secreted by amoebae and other aquatic organisms can provide adhesion as well as act as a lubricating material for gliding along surfaces, while the outer character of bacterial cells determines, among other things, their immunological and virulent properties. The adhesiveness of cell surfaces is particularly critical, because without it cells would fall apart after division, and multicellularity would be impossible. The permeability of cells is generally not affected by the presence of extracellular substances except in some plant cells where waterproof lipidlike substances such as waxes are present.

The subject of extracellular substances, therefore, is large and varied, but here we shall restrict our discussion to those commonly found in higher plants and animals.

Plant cell walls Plant cells are characteristically enclosed by *cell walls,* which serve the mechanical functions of providing support for, and conferring rigidity on, the plant body. Related to this is the importance of the wall in preventing movement of water into the cell to the point where the protoplast would simply burst open. The tendency of water to enter the cell causes it to swell until pressure is exerted on the cell wall. It is this hydrostatic pressure that maintains the *turgidity* of plant parts. Furthermore, cell enlargement as a result of water moving into the cell is possible only as long as the cell wall remains sufficiently extensible.

The primary wall, the first layer to be laid down, consists of a framework of microfibrils (Figure 3.16) made up of the polysaccharide cellulose, embedded in a matrix of several other polysaccharides and glycoproteins. Between two adjacent primary walls is the *middle lamella,* which consists mainly of another polysaccharide, pectin, and which acts as an intercellular cement binding cells together. While a cell is enlarging, the primary wall is thin, elastic, and capable of great extension; although some thickening of wall can occur during elongation, in general this happens after the cell has reached its maximum size. After this time a *secondary wall* may be laid down between the primary wall and the plasma membrane. The secondary wall may be thick or thin and of varying degrees of hardness or color. It is the part of the cell that gives various woods and plant fibers (cotton, flax, hemp) their particular character, and from which is derived the cellulose used in

Figure 3.16
Fibers of cellulose as formed in the wall of an algal cell. Each fiber would be composed of many small fibrils grouped as in a rope or cable (×16,700).

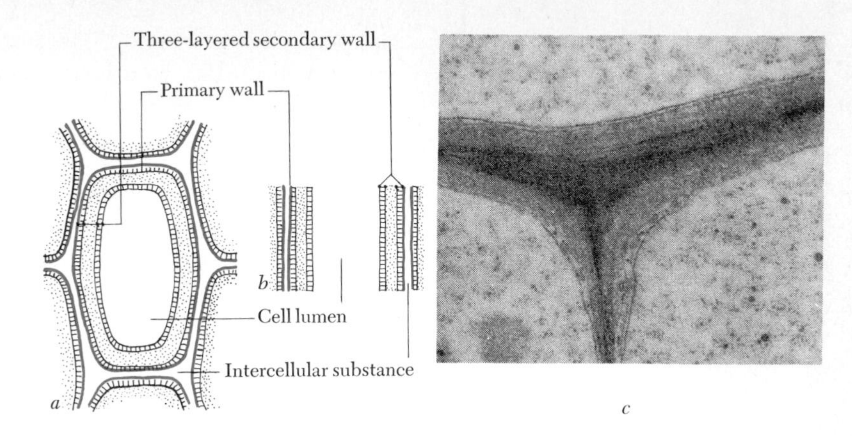

Figure 3.17
Typical wall structure of matured and lignified plant cells. (*a*) Cross section, showing arrangement of the various layers and the complex structure of the secondary wall. (*b*) Longitudinal section through a similar cell. (*c*) Electron micrograph of the cell walls of three adjacent cells: the darker middle area is the middle lamella; the lighter portions, the primary wall. (Reprinted with permission from K. Esau, *Plant Anatomy*. New York: John Wiley & Sons, Inc., 1953.)

the manufacture of rayon, nitrocellulose, cellophane, and certain plastics. The organization of the mature cell wall is illustrated in Figure 3.17.

Let us examine the growth of the cotton fiber to illustrate the principles of cell wall formation. The mature fiber, or lint as it is called, may be 0.5 to 1.5 inches long. Located in the outermost layer of cells, or epidermis, of the seed coat (Figure 3.18), each lint cell is attached to neighboring cells by a middle lamella and possesses a thin primary wall. On the day of flowering and after fertilization of the egg has taken place, the cell begins to elongate, a process that takes 13 to 20 days and terminates when the cell is 1,000 to 3,000

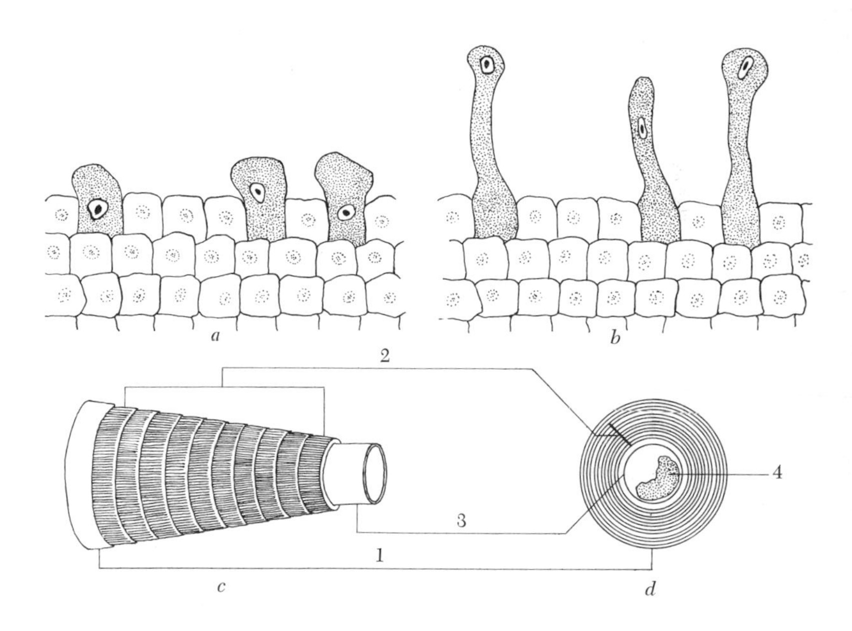

Figure 3.18
Growth and structure of the cotton fiber. (*a*) Outer layer of cells of young cotton seed, showing the beginning enlargement of the fibers on the day of flowering. (*b*) Same, 24 hrs. later. (*c*) Diagram of the various layers of cellulose laid down in a mature cotton fiber: (1) outer primary cell wall, (2) concentric inner layers, revealing the different orientation of the cellulose in the secondary thickenings, (3) last inner layer of the secondary wall. (*d*) Same, in cross section, with (4) representing the remains of cell contents. (Redrawn and modified from H. B. Brown and J. O. Ware, *Cotton*, 3rd ed. New York: McGraw-Hill Book Company, 1958.)

times as long as it is wide. The primary wall then ceases to elongate, and the secondary wall forms as sugars in the cytoplasm are converted into cellulose fibrils (presumably the conversion takes place outside of the plasma membrane), and deposited on the inside of the primary wall. Deposition of cellulose continues until the fruit is mature. The cell then dies, collapses, and flattens to give the fiber used in the manufacture of cotton threads and cloths.

Cellulose is clearly formed by the activities of the cytoplasm of the cell, but how and where the fibrils are assembled and arranged in the secondary wall is not understood. In cross section, the cotton fiber has an area of about 300 μm^2 and is made up of approximately 1 billion cellulose chains. These are grouped into fibrils of several orders of size, each one running the length of the entire fiber in a parallel, or sometimes helical (spiral), fashion. The spaces between the fibrils give the fiber its flexibility and allow for the complete penetration of dyes, whereas the parallel orientation of the fibrils accounts for its great tensile strength (nearly that of steel). It has been estimated that each cotton fiber contains about 10 trillion cellulose molecules, which are built up from approximately 60 quadrillion glucose molecules. Also, a single fiber is but one of many thousands growing on the surface of each cotton seed. From these rough calculations, we can gain some appreciation of the activity of plant cells in transforming carbon dioxide and water through photosynthesis (see Chapter 4) into organic molecules; joining many of these molecules together in specific ways, the cell then builds them into an elaborate structure, the cell wall. The cell carries out the process of polymerizing repeating molecules to form its structural elements—proteins, fats, nucleic acids, and polysaccharides—in much the same way we form plastics and synthetic fibers, although the chemical reactions are different.

It appears that the arrangement of cellulose fibers initially laid down may be related to the kind of expansion a plant cell will undergo. When the fibers are randomly arranged, the cell expands uniformly in all directions. If the fibers are arranged in parallel fashion, expansion is at right angles to them, to produce an elongated cell (Figure 3.19). The cotton fiber is elongated and its fibers parallel, or nearly so, with its long axis, because they are laid down after elongation.

The organization of the cell wall also illustrates two sound construction principles. The strength of the cotton fiber, composed of pure cellulose, is gained by grouping the molecules into ever larger fibrils; this arrangement of parts is the principle used in the construction of cables and ropes. Other types of cell walls, however, are impregnated with different substances. One of these is lignin,

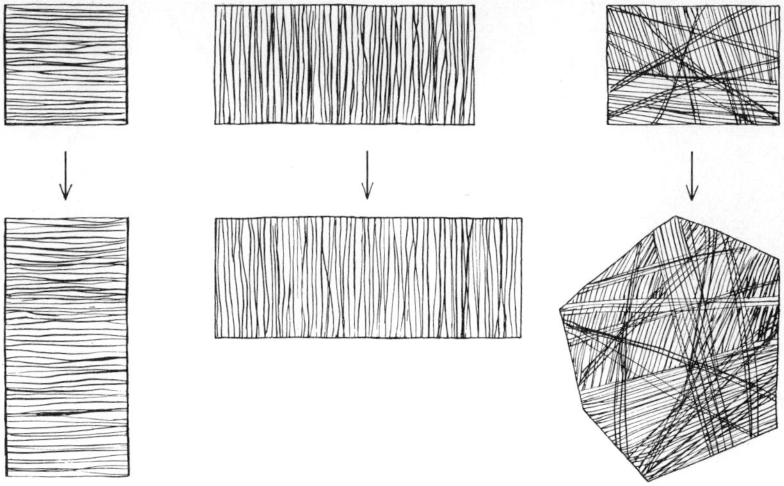

Figure 3.19
Schematic representation showing how the orientation of cellulose fibers in a young cell (top row) determines the axis of elongation of older cells. Elongation is essentially at right angles to the direction of the fibers. When the fibers exhibit no orientation (top right), the cell enlarges in all directions.

a complex, nonfibrous molecule, unrelated to the sugars, which forms in the spaces between the cellulose fibrils. This arrangement is also the principle of the reinforced concrete that is utilized in many modern buildings; the cellulose provides rods of high tensile strength and the lignin is a hard substance that is resistant to pressures and acts as a glue or cement. When lignin is absent, as in balsa wood, the material is soft and brittle. The cellulose does not have to be as well oriented as in the cotton fiber, and other substances, such as cutin and waxes, both derivatives of fatty acids, may replace lignin. In such instances, the strength of the wall is less, but the cell surface is water-repellent. Chitin, a polysaccharide found in the exoskeletons of insects, is also present in the cell walls of many fungi, while the cell walls of grasses and particularly the horsetails contain substantial amounts of silica, a major component of glass and sand.

Since most plant cells conduct water and dissolved substances as well as provide support, even after they have died (as in wood), the heavy wall must be interrupted at intervals to allow for passage from one cell to another. Interruption occurs in a variety of ways, and as Figure 3.20 indicates, the secondary wall may exist as rings, spiral bands, or sequences of thick and thin areas. Such gaps give flexibility as well as support and ease of conduction. In other cells, particular areas may be perforated by pits, or pit fields, or the end wall of a cell may be perforated or even missing to provide a connected column simulating a channel made up of short pieces of pipe (Figure 3.21).

Extremely elaborately sculptured walls form around pollen

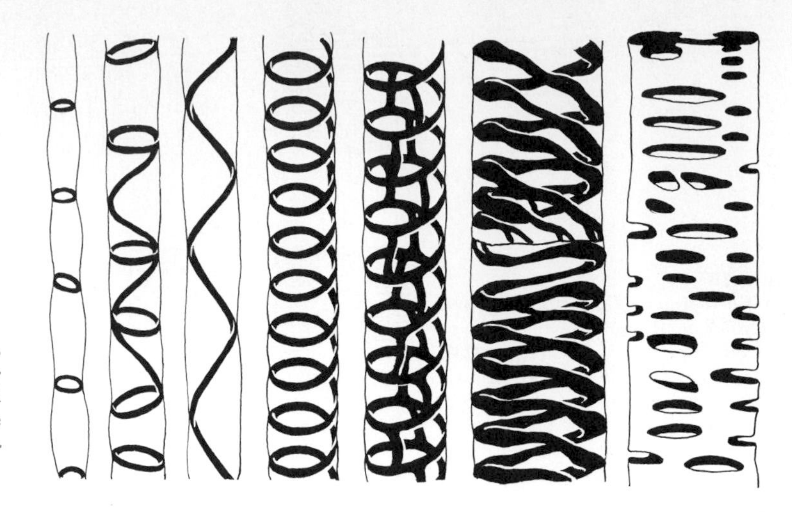

Figure 3.20
Cells in the wood of higher plants, exhibiting various patterns of secondary-wall formation. The interrupted areas are thin enough to permit the passage of water and dissolved materials.

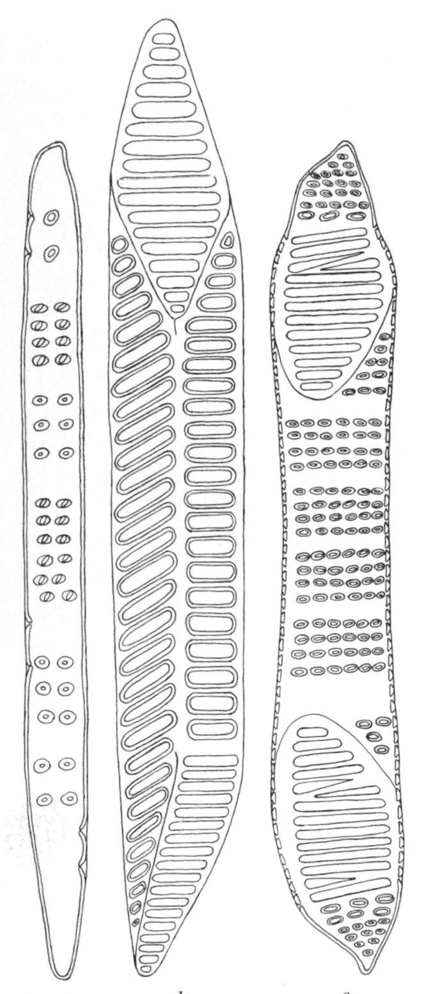

Figure 3.21
Water-conducting cells in the xylem of higher plants, showing different arrangements of pits on their side and end walls: (*a*) from Sequoia, (*b*) from bracken fern, (*c*) from alder.

grains of many plants (see Figure 1.21, Chapter 1). These structures consist mainly of sporopollenin, which is lipoidal in nature, and are extremely tough, affording much protection to the pollen grain. The walls can also contain antigenlike substances, which may play a role in determining whether the pollen grain is compatible or not with the stigma on which it lands. The antigens are responsible for the pollen allergies that can cause so much personal discomfort in humans, especially in spring and fall when pollen is shed in great quantities from the troublesome species.

Extracellular substances of animal cells The intercellular "glue" of animal cells is principally of two kinds: hyaluronic acid and chondroitin sulfate. Collectively, these are known as ground substances. Combined with protein and lacking sulfur, hyaluronic acid is a jellylike, amorphous, viscous polysaccharide of high molecular weight. It is able to retain water tenaciously. Functionally, hyaluronic acid serves a number of purposes. As a "glue" it binds cells together at the same time that it permits flexibility; in the fluids of joints it acts as a lubricant and, possibly, as a shock absorber; in the fluids of the eye it acts to retain water and keep the shape of the eye fixed.

The viscosity of hyaluronic acid is determined in part by the amount of calcium present; for example, the cells of young embryos of sea urchins tend to fall away from each other if kept in a calcium-free medium, and developmental processes cannot proceed normally. Another point about hyaluronic acid is that it can be dissolved by an enzyme, hyaluronidase, present in sperm cells and readily formed by some bacteria. The action of the enzyme in sperm cells permits sperm to penetrate the jelly coat that surrounds an egg; without it the sperm could not bring about fertilization. The ability of some bacteria to manufacture the enzyme means that they can effectively invade tissues and so spread an infection from the initial point of entry.

Chondroitin sulfate is a firmer gel than hyaluronic acid, and, like the latter, it is a polysaccharide combined with proteins. It is particularly evident in cartilage, where it is associated with fibrous elements such as collagen, and where it provides a matrix in which the cartilage-forming cells are nestled (Figure 3.22). The arrangement gives good support and adhesiveness while preserving a measure of flexibility (ears, nose, ends of ribs, new bones, respiratory tract, joints, and so on). Bone, in fact, appears first as a skeleton of cartilage, after which calcification of the intercellular substance takes place as one of the major processes of hardening.

The *basal lamina* is a somewhat more definitely organized and special ground substance. Found in a number of organs where it both binds cells together and helps to shape the particular organ, the basal lamina is prominent in vertebrate skin. There it lies be-

Figure 3.22
Electron micrograph of an osteoblast, or bone-forming cell, and the surrounding extracellular material. Note the characteristic axial periodicity of the collagen fibers every 64 nm. (Courtesy of Dr. Melvin Glimcher.)

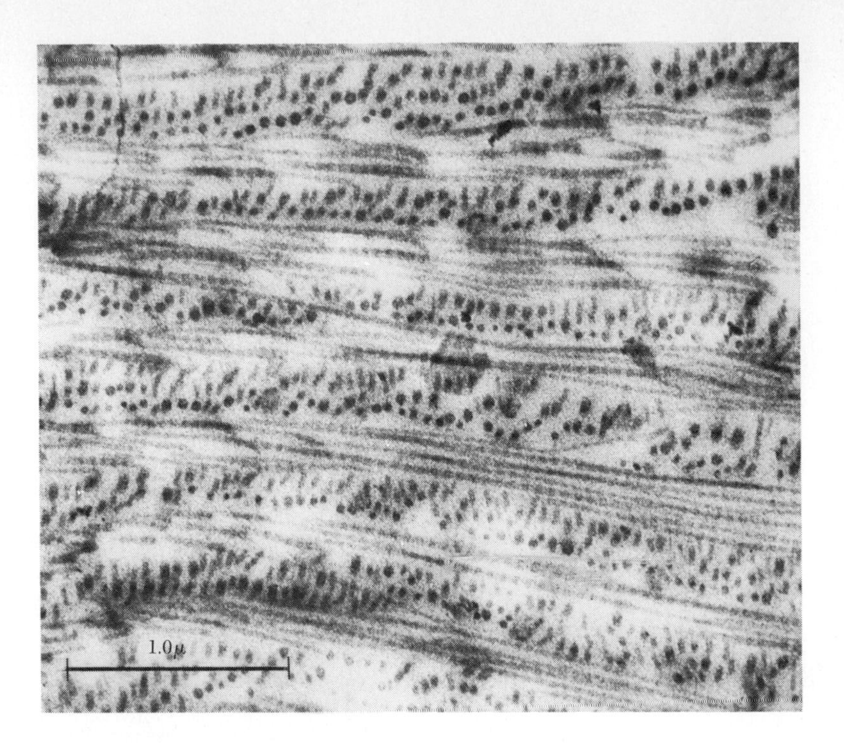

Figure 3.23
The basement membrane in amphibian skin, showing its laminated structure. The alternate layers are formed by collagen fibers running at right angles to one another. The periodicity of the collagen fibers can be seen clearly. (Courtesy of Dr. P. Weiss.)

tween the epidermis and the dermis as a condensation of intercellular substances. Unlike ground substances in general, it may be highly laminated, as in amphibians (Figure 3.23). There may be 20 or more laminae, with fibers in each lamina at right angles to the layers above and below it. Such an arrangement of fibers gives flexibility as well as strength.

The fibrous elements embedded in the ground substance are collagen, elastin, and reticulin. All are basically proteins of high molecular weight. It has been estimated that about one-third of the protein of a mammal is collagen (Figures 1.18 and 3.23), and it is located in those areas where a degree of firmness or rigidity is needed (muscles, bone, skin, tendons, and so forth). Collagen can be readily dissolved in dilute acid. Dissolved collagen will reaggregate spontaneously in solution if conditions are suitable; presumably this is the manner of its aggregation outside of the cell in the ground substance.

Elastin and reticulin, unlike collagen, show no periodic structure. Elastin is a stringy protein that has the capacity, as its name suggests, to stretch and snap back into its original state, much as an elastic band would. Consequently, it is prevalent where elasticity is required, as in skin and the tissues surrounding the major blood vessels.

Reticulin consists of much finer fibers than do collagen and elastin, but is probably closely related to collagen in all ways except aggregation. The fibers are finely branched, are found generally in the ground substance, and are particularly abundant in the basal lamina.

From what has been said, it should be clear that ground substances and their associated fibers serve a number of purposes. Cells aggregate into organs of definite shape and size, and organ systems are tied together to form the intact and functioning organism. Adhesiveness, lubrication, rigidity, and elasticity are but some of the required features of a functioning organism that are governed by the quality and quantity of ground substance. If the character of the intercellular products changes, the organism itself must similarly change.

A specific type of glycocalyx, the *vitelline coat,* is found external to the true plasma membrane of most animal eggs. The thick transparent surface coat of the mammalian egg is known as the *zona pellucida,* and is separated from the plasma membrane by a fluid-filled region, the *perivitelline space.* Before actual fusion of the sperm with the egg membrane proper, the sperm must attach to the exterior of, and then penetrate, this surface coat. The zona pellucida, which like the vitelline coat of other animal eggs contains carbohydrates, proteins, and glycoproteins, is thought to carry specific receptors to sperm cells on its surface as well as functioning to protect the egg from injury. Passage of the sperm through the vitelline coat is achieved by the release of enzymes by the sperm, and these dissolve away the protective barrier.

Bibliography

Albersheim, P. 1975. The walls of growing plant cells. *Scientific American* **V. 232**, No. 4, 80–95. The molecular architecture of cell walls and how it is related to cell wall properties.

Branton, D., and R. B. Park, eds. 1962. Papers on Biological Membrane Structure. Little, Brown, Boston. A collection of some of the classic papers dealing with membrane structure, with an excellent introductory chapter.

Bretscher, M. S. 1973. Membrane structure: some general principles. *Science* **V. 181**, 622–629. A discussion of erythrocyte membrane proteins and their relationship to membrane structural organization.

Capaldi, R. A. 1974. A dynamic model of cell membranes. *Scientific American* **V. 230**, No. 3, 26–33 (See Fox).

Fox, C. F. 1972. The structure of cell membranes. *Scientific American* **V. 226**, No. 2, 30–38. Both of the above are easy-to-read and well-illustrated accounts of membrane organization.

Kreger, D. R. 1969. Cell walls, in *Handbook of Molecular Cytology,* ed. Lima-de-Faria. A. North-Holland Publishing Co., Amsterdam and London. A detailed account of the chemistry, structure, and formation of cell walls.

Lehninger, A. L. 1975. Biochemistry (2nd ed.). Worth, New York. Chapter 28 of this comprehensive text provides an excellent discussion of mechanisms of transport across membranes.

Singer, S. J., and G. L. Nicolson. 1972. The fluid mosaic model of the structure of cell membranes. *Science* **V. 175**, 720–731. A consideration of properties of membrane components, and experimental evidence for the fluid mosaic model.

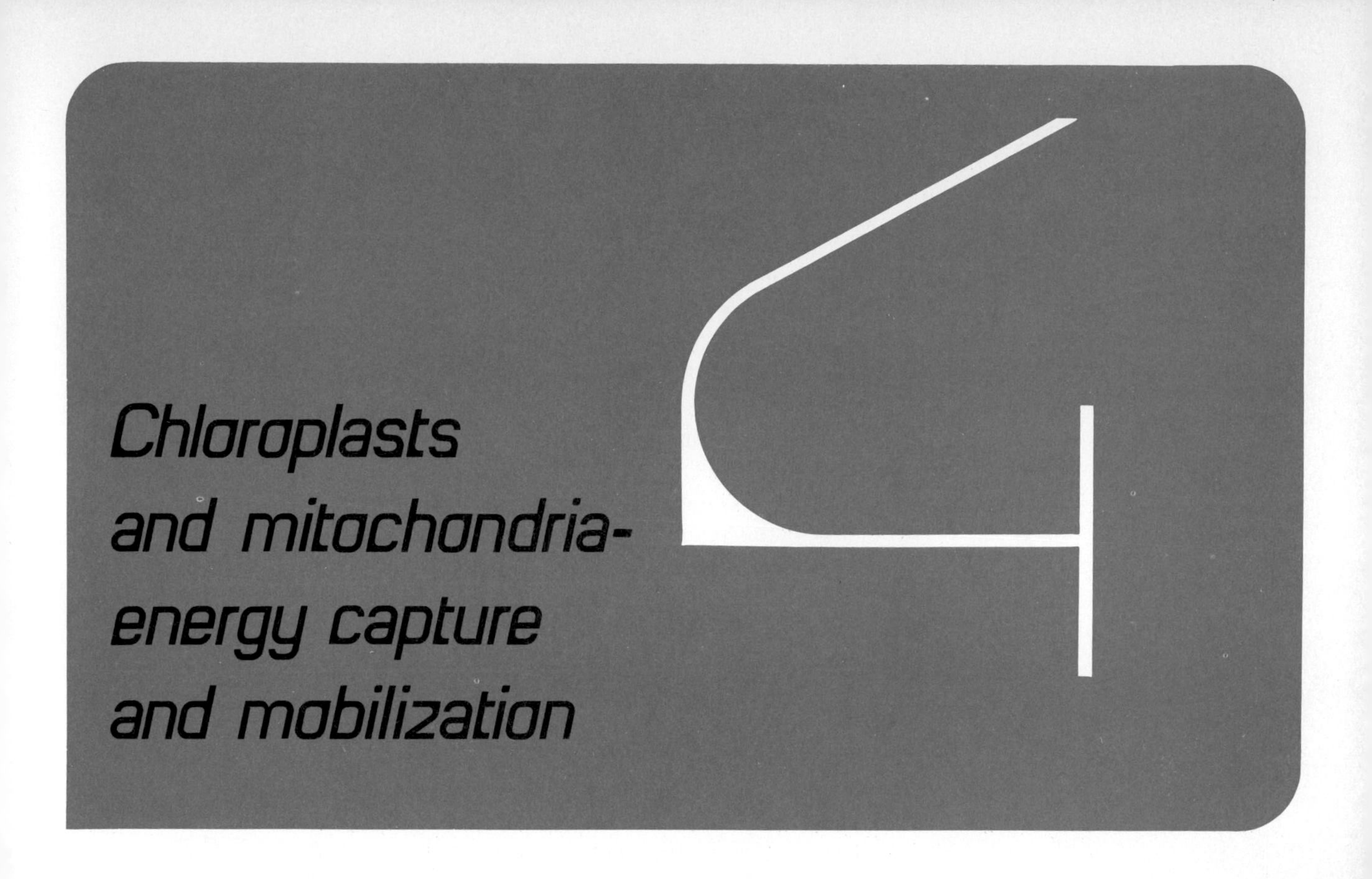

As we have seen in Chapter 1, life can be viewed as a process that organizes matter and energy into ordered states; as such it runs counter to the overall tendency toward an increase in entropy. To maintain and to increase the levels of order and complexity in the living world, therefore, energy must be acquired from the surroundings and funneled into biological processes. The ultimate source of energy for all life today is that part of solar radiation we call sunlight; this energy is captured by plants in the process of photosynthesis and converted into chemical energy that can be conserved in the storage molecules produced during the process. Animals, of course, cannot carry on photosynthesis, and depend for their energy supply on the storage products synthesized by plants; this dependence may be direct, as for plant-eating herbivores, or indirect, as for carnivores, which eat other animals, including the herbivores. The energy stored in these molecules can then be made available to the plant or animal cell in forms that can be utilized to carry out the work of the cell; for example, in the synthesis of other molecules needed by the cell, or for the mechanical, electrical, or osmotic work that must be done. The efficient release of the stored energy occurs in the process of *aerobic respiration,* in which the molecules containing the stored energy are broken down in a controlled series of reactions.

The complex series of reactions involved in the energy conversions of photosynthesis and respiration take place in highly specialized compartments of the cell, the chloroplasts and mitochondria respectively; these organelles can be thought of as the energy transducers of the cell. Before discussing the structures and activities of chloroplasts and mitochondria, however, we must briefly examine how energy is manipulated in chemical reactions within the cell.

All compounds are characterized by certain energy contents; some of this energy is free energy, that is, energy that can be used to perform work. (The rest of the energy is entropy, and it is no longer available to do work.) Since it is the useful energy that is important, we are primarily concerned with free energy changes that occur during chemical reactions. The amount of free energy available from a molecule depends on which atoms it contains and how they are arranged. When chemical bonds are broken and reformed during chemical reactions, energy is redistributed within the molecules. In some reactions (exergonic reactions), free energy is released and available to do work. Similarly, there are reactions that require an input of free energy to proceed (endergonic reactions). Such endergonic reactions must take free energy from other sources in converting compounds of a low energy content to products with a higher energy content. In the cell, the endergonic reactions, which constitute work, must be coupled to exergonic reactions in such a way that the free energy from the latter is made available to them.

The chemical reactions most frequently used by the cell to make energy available are oxidation-reduction reactions, which involve transfer of electrons from one compound (which, therefore, becomes oxidized) to another (which becomes reduced). These electron transfers, of course, result in changes in the chemical bonds within the molecules involved, and hence in the energy levels of the molecules. The capacity of one compound to give up its electrons (to act as a reducing compound) is a measure of its chemical potential. When a compound with a higher chemical potential reduces one with a lower chemical potential, free energy is released. The difference between the initial potential energy state and the final potential energy state of the electron(s) represents the amount of free energy "liberated" by the reaction. In the cell, of course, the energy is not liberated; rather some of it is conserved in a form that can subsequently be used by the cell.

Most of the work done by a cell, whether chemical, mechanical, electrical, or osmotic, depends on free energy supplied in a usable form by these oxidation-reduction reactions carried out in the cell. The transfer of energy from these energy-yielding chemical reactions to the energy-requiring processes, which constitute the work of the cell, is mediated by a very important energy-carrying molecule, adenosine triphosphate (ATP) (Figure 4.1). ATP is generated in the cell by the addition of a phosphate group to the compound adenosine diphosphate (ADP) (Figure 4.1). However, since the energy state of ATP is much higher than that of ADP, free energy must be supplied for the reaction to occur. It is the energy that is made available by the oxidation-reduction reactions of the cell that is used in the formation of, and hence conserved in, ATP. This is accomplished by coupling the phosphorylation reaction, involving transfer of a phosphate group, to oxidation-reduction reactions, involving transfer of electrons.

The energy now in the ATP molecule can be released, when required to do work in the cell, by the hydrolysis of ATP to ADP and inorganic phosphate (Figure 4.1). So that this free energy will not be completely dissipated as heat, the conversion of ATP to ADP is carried out enzymatically by the cell in such a manner that part of the energy can be used to do work. The energy-requiring reactions that constitute the work of the cell must, therefore, be coupled in some way to the energy-yielding hydrolysis of ATP.

We can think of ATP as an energy shuttle, acting as the intermediate carrier in energy transfers; it is the molecular currency of the energy economy of the cell. As we shall see, the oxidation-re-

Figure 4.1
ATP $\rightleftharpoons$ ADP + P_i. Adenosine triphosphate (ATP) is a nucleotide carrying three phosphate groups; adenosine diphosphate (ADP) has only two. The difference in free energy between the two compounds is approximately 7,000 kcal/mole. This free energy is used to do the work of the cell when ATP is hydrolyzed to ADP and inorganic phosphate.

Energy

+ Inorganic phosphate

Energy

ATP

ADP

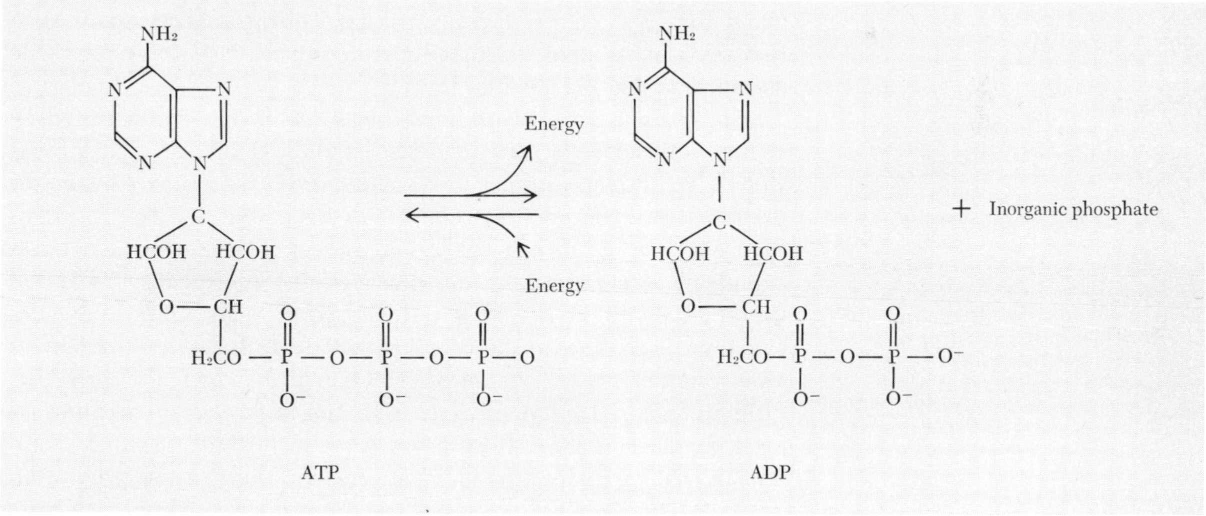

duction reactions and the coupled phosphorylations occur during respiration, when the stored energy of the cell is made available to form ATP, and during photosynthesis, when the energy trapped by the chloroplast is funneled through ATP into the final storage products.

Photosynthesis

The overall processes of photosynthesis involve the absorption and retention of light energy, its conversion into chemical energy, and the storage of this chemical energy in the final products of photosynthesis. The first process involves the absorption of light quanta by the pigment chlorophyll, the compound responsible for the green color of plants. When light strikes the chlorophyll molecule, one of the electrons in it is raised to a higher energy level and is said to be in an "excited state." If a solution of chlorophyll extracted from a plant is illuminated, it will fluoresce—it will give off light. This light represents the re-emission of the absorbed energy as the electron returns to its "ground state." In the chloroplast, however, the chlorophyll molecules are arranged in such a way that the excited electrons do not immediately return to their ground state, but are transferred to other molecules resulting in an increase in the free energy, or chemical potential, of the acceptor molecule. Thus, the light energy is converted into chemical energy. The "electron hole" left in the chlorophyll molecule is replaced by an electron donated by water; in this process the water is oxidized and molecular oxygen given off. Chlorophyll, therefore, is an intermediate in the pathway of electrons from a low energy level in water to a high energy level in the final electron acceptor. It is in the chlorophyll molecule that light energy is used to "pump" the electrons to that high energy level.

The current concept of how this is achieved is illustrated in Figure 4.2, which shows that two photosystems, or light-driven steps, are involved, connected by an electron transport pathway. In photosystem I, the excited chlorophyll molecules are able to transfer their energy-rich electrons, via intermediates, to a compound called nicotinamide adenine dinucleotide phosphate (NADP), thereby converting it to the reduced form. This reduced NADP can then itself donate the electrons required for the reduction of atmospheric CO_2 to glucose, the molecule in which the energy is ultimately stored. In photosystem II, the high energy electrons of excited chlorophyll are transferred to an electron transport system, and they are replaced by the low energy electrons derived from

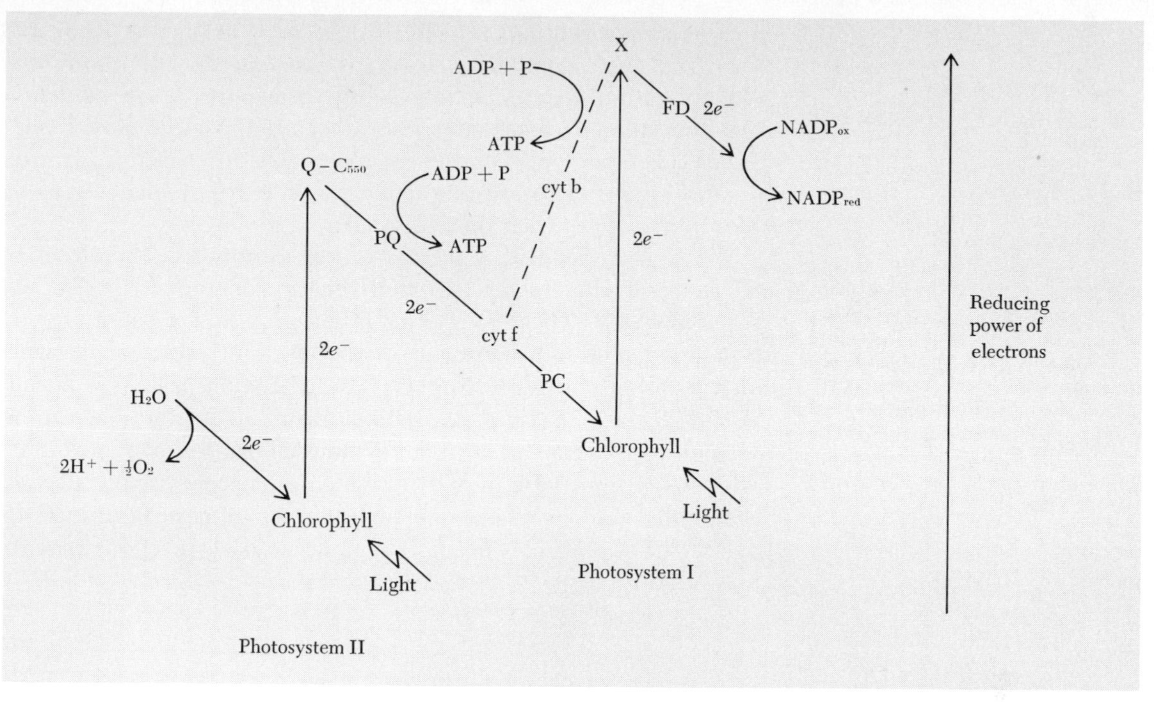

Figure 4.2
Diagrammatic representation of pathway of electron flow in the light reactions of photosynthesis. In photosystem I, light energy absorbed by chlorophyll raises the energy level of the electrons to the point where they can reduce an unidentified electron acceptor, X. The electrons can then be passed on, via ferredoxin (FD) to NADP, thereby reducing it. In photosystem II, electrons in chlorophyll are raised, as a consequence of the absorption of light, to an energy level sufficient to allow them to be transferred to an electron acceptor, Q-C550. These electrons can then be passed in a series of oxidation-reduction reactions involving plastoquinone (PQ), cytochrome f (cytf), and plastocyanin (PC) to the chlorophyll of photosystem I. The electrons lost by the chlorophyll of photosystem II are replaced by electrons donated by water, and molecular oxygen is liberated. Cyclic electron flow also occurs in photosystem I, with the electrons captured by X being returned, via cytochrome b (cytb), to the chlorophyll. The free energy made available in the "downhill" reactions of the electron transfer systems is conserved in the formation of ATP.

water, which is oxidized to molecular oxygen. The high energy electrons from photosystem II are funneled through a series of compounds commonly called the electron transport chain and ultimately replace the electrons lost by the chlorophyll of photosystem I. This electron transport chain involves a series of oxidation-reduction reactions, in each of which the electrons are passed to a slightly lower energy state (Figure 4.2). Some of the energy given up during this series of electron transfer steps is utilized to form ATP from ADP and inorganic phosphate; we can say, therefore, that ATP formation is *coupled* to electron transport. A second site of ATP formation is believed to be associated with photosystem I. This "cyclic" photophosphorylation depends on electrons captured by "X" (Figure 4.2) returning, via electron carrier molecules, to the chlorophyll from which they originally came.

The light-dependent reactions of photosynthesis, therefore, result in the formation of two high energy compounds, ATP and reduced NADP. These compounds then provide the energy for the subsequent, non-light-requiring reactions, the reduction of atmospheric carbon dioxide to sugars. The reduction of CO_2 is accomplished by the fixation of molecules of CO_2 into "acceptor" molecules, and the subsequent cycling of the carbon through a series of reactions (Figure 4.3) in which the acceptor molecules are regenerated and in which one molecule of glucose is formed for each six CO_2 molecules fixed. It is in these dark reactions that the

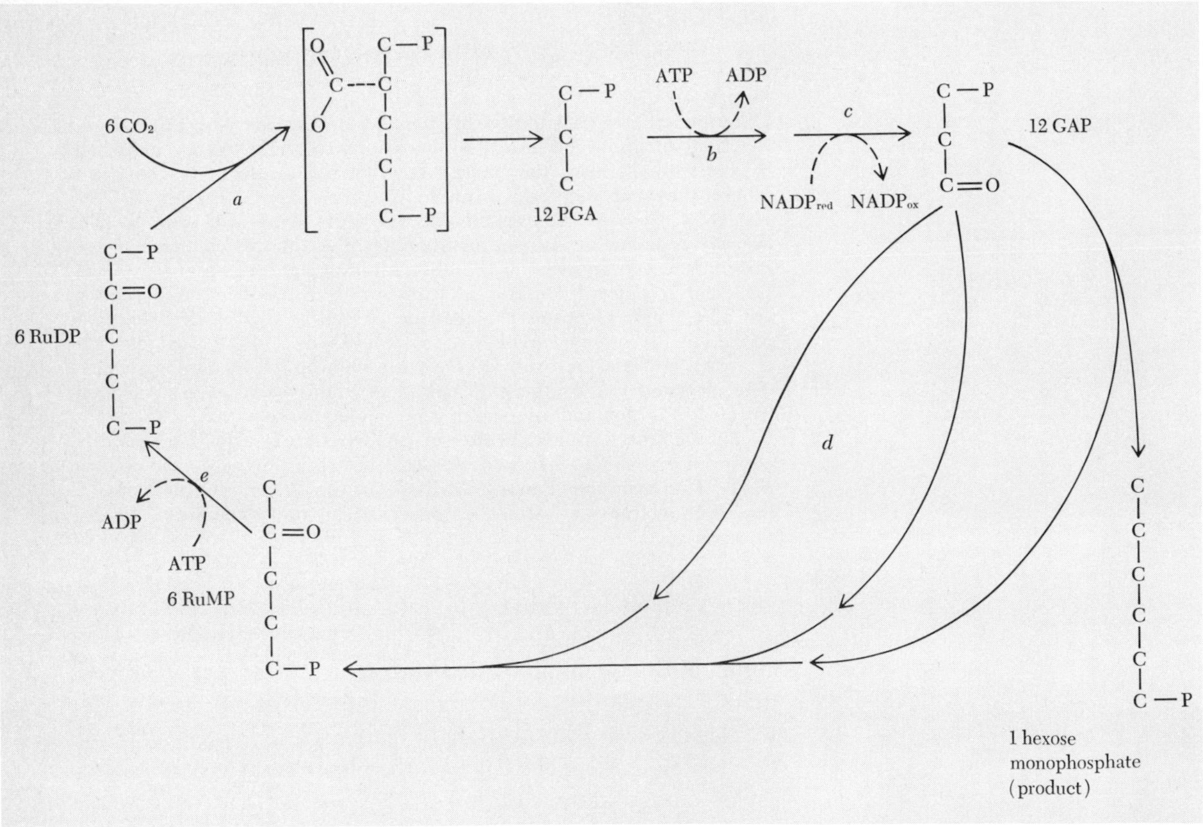

Figure 4.3
Major steps in the dark reactions of photosynthesis. Atmospheric CO_2 is attached to a 5-carbon sugar, ribulose-diphosphate (RuDP), in a reaction catalyzed by the enzyme RuDP carboxylase (*a*). The unstable 6-carbon intermediate breaks down to form phosphoglyceric acid (PGA), which is in turn converted to glyceraldehyde phosphate (GAP). In a complex cyclic series of reactions (*d*), the 5-carbon acceptor sugars are regenerated from GAP, and a final product, glucose, is formed. The ATP and reduced NADP produced by the light reactions provide the energy and the electrons necessary for the dark reactions (*b*, *c*, and *e*). (The -H and -OH groups present on the carbon atoms have been omitted for simplification.)

chemical energy generated by the light reactions is stabilized in the glucose molecule. Finally, the glucose molecules are converted into starch, in which both the energy and the carbon can be stored.

The importance of photosynthesis to life in general and to mankind in particular cannot be overstated. Not only does photosynthesis capture and transform energy, and thereby provide the fuel on which all forms of life depend, it has also been responsible for providing, in the form of fossil fuels, most of the energy reserves currently available to mankind and our increasingly industrialized society. Approximately 840 trillion kilowatt hours a year are funneled through photosynthesis, more than 10 times our current world energy consumption. Furthermore, the atmospheric oxygen on which higher forms of life depend has been produced as a by-product of photosynthesis, liberated by the oxidation of water.

Chloroplasts

Both the light and dark reactions of photosynthesis take place in discrete, membrane-bound organelles within the cell, the chloroplasts (Figure 4.4). Each chloroplast is enclosed by a double-membraned envelope, inside of which is a complex internal membrane system. These inner membranes are organized into flattened sacs, or *thylakoids*, embedded in the aqueous matrix, or *stroma*, of the chloroplast. In higher plants the thylakoids are arranged in stacks, rather like a pile of coins, as shown in Figure

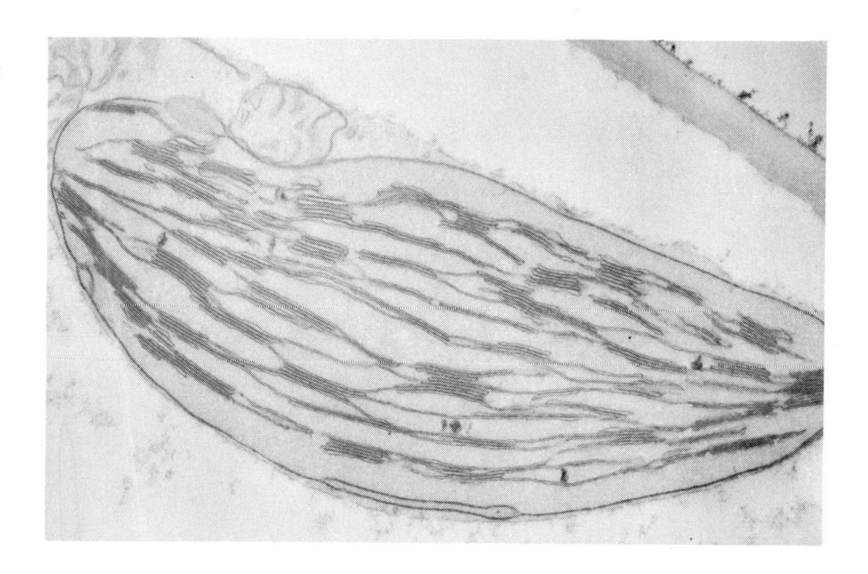

Figure 4.4
Electron micrograph of a chloroplast from the duckweed, *Lemna minor*. An outer membrane surrounds the structure, and the grana do not consist of many layers. Compare with Figure 4.5. (Courtesy of Dr. H. T. Arnott.)

4.4. These stacked thylakoids constitute the *grana,* which are interconnected by channels enclosed by membranes, the stroma thylakoids. Thus, within the chloroplast itself we see a further example of compartmentalization, the thylakoid space being separated by its enclosing membranes from the stroma. Starch grains and lipid bodies, representing storage products of photosynthesis, also may be present (Figure 4.5).

It is relatively easy to fractionate chloroplasts by differential centrifugation; in this way it can be shown that while the enzymes that catalyze the dark reactions are localized in the stroma, the membrane fraction contains the chlorophyll and is capable of carrying out the light reactions of photosynthesis. The molecular composition and organization of the thylakoid membrane must, therefore, be such that photosystems I and II, the electron transport chain, and the enzymes necessary for the formation of ATP and reduced NADP, are capable of functioning in the membrane in an integrated manner.

Electron microscopic examination of freeze-fractured and freeze-etched thylakoid membranes clearly demonstrates their structural

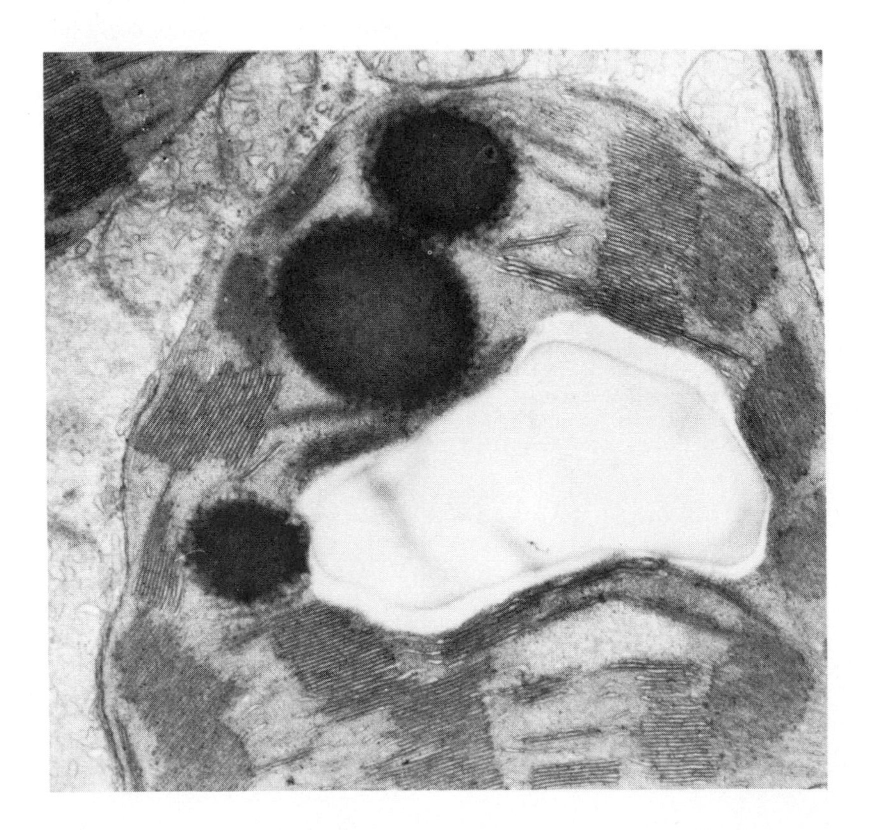

Figure 4.5
Portion of a chloroplast of a leaf of the rubber plant, *Ficus elastica*, showing grana, starch grain, and lipid bodies. (Courtesy of Dr. J. G. Duckett.)

complexity. As we have already mentioned, this technique provides not only surface views of membranes, but also representations of internal organization, since the membranes can be "cleaved" along the plane of the relatively weak bonds that hold the two lipid layers of the membrane together.

Two major size classes of particles, or subunits, are embedded within the grana membranes; the larger of these is approximately 180 Å in diameter and is oriented toward that surface of the membrane facing the interior of the thylakoid, while the smaller subunit, approximately 115 Å in diameter, is a component of that "half" of the membrane whose surface faces the stroma. This structural asymmetry of the internal chloroplast membrane is related to a corresponding functional asymmetry. There is evidence that at least part of photosystem I occurs at the outer, or stroma, surface of the membrane, and that photosystem I activity is associated with the small particles, while the larger particles are involved in photosystem II toward the innerfacing surface. Although the molecular organization of the membrane subunits is not understood, we can think of photosynthetic membranes as including functional aggregates of the pigments, enzymes, and other molecules involved in energy trapping and conversion processes. The functional integrity of the membrane must depend on specific and highly ordered spatial and structural relationships that exist within and between these molecular aggregates.

The pattern of development of chloroplasts illustrates the importance of thylakoid organization in photosynthesis, and the importance of both light and chlorophyll to that organization. Chloroplasts (as well as other types of plastids) develop from *proplastids*, small, double-membrane-bound structures seen in meristematic cells. In the presence of light, proplastids develop into normal chloroplasts, invaginations of the inner membrane forming thylakoids, which subsequently become organized into functional grana. At the same time, synthesis of chlorophyll and the rest of the photosynthetic machinery takes place. In the dark, however, no chlorophyll is synthesized and very few thylakoids form; instead an aggregate of membraneous tubules, the *prolamellar body*, appears (Figure 4.6). When such dark grown cells are exposed to light, chlorophyll is formed, and the prolamellar body is converted into functional thylakoids, which themselves form grana.

Chloroplasts may assume many forms and vary widely in number per cell in different plants. In some algae, such as the filamentous *Spirogyra*, only a single spiral chloroplast is present in each cell; when the cell divides, it does so at the same time. In contrast, a cell in the spongy part of a grass leaf may have 30 to 50 chloro-

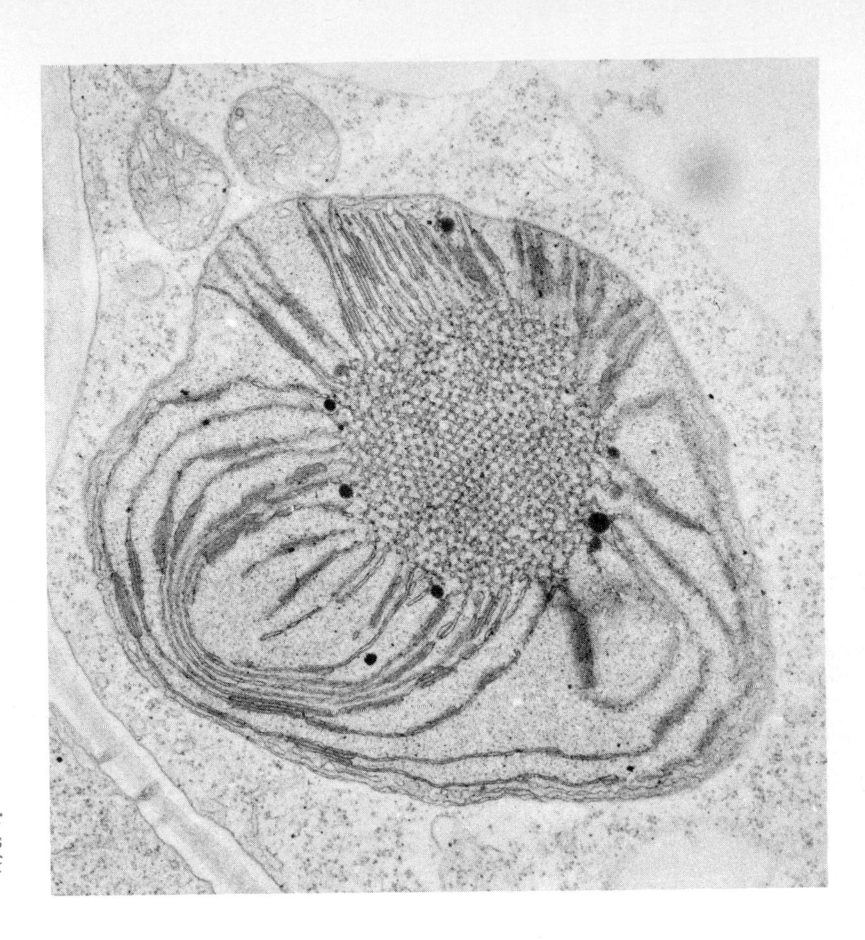

Figure 4.6
Electron micrograph of prolamellar body. The photosynthetic membranes are beginning to form. (Courtesy of Dr. C. L. F. Woodcock.)

plasts; their division, which occurs in the immature, or proplastid, state, is not correlated with cell division in any exact way. The stacked grana are missing in some chloroplasts, as in some brown algae, to be replaced by long membranes running the length of the chloroplast, but these presumably function in the same manner as the grana. The blue-green algae, on the other hand, lack definite chloroplasts; instead they possess loosely arranged membranes in the cytoplasm on which the photosynthetic pigments are layered (Figure 4.7). Only in some bacterial cells do we find a photosynthetic capacity unassociated with an obvious membranous structure. However, the vacuolelike chromatophores (Figure 4.8), which are the photosynthetic units, are bounded by membranes; unfortunately, we know little about the molecular arrangement of the light-absorbing pigments. However, bacteria kept in the dark lose their chromatophores and are no longer photosynthetic; the chromatophore thus behaves as the chloroplast of *Euglena,* and is its func-

Figure 4.7
Electron micrograph of a blue-green algal cell showing photosynthetic membranes. Particles in the membranes can be seen. (Courtesy of Dr. S. C. Holt.)

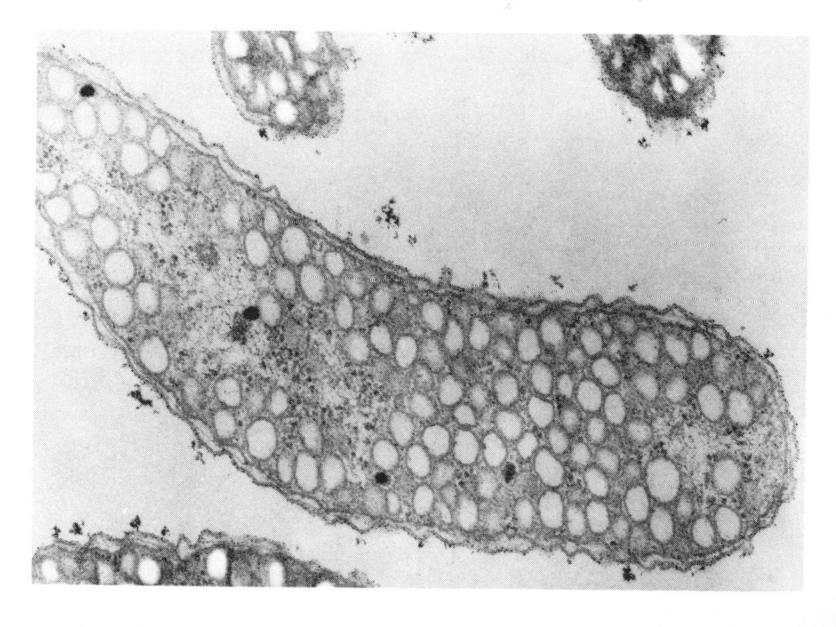

Figure 4.8
Electron micrograph of photosynthetic bacterium showing membrane-enclosed chromatophores. (Courtesy of Dr. S. C. Holt.)

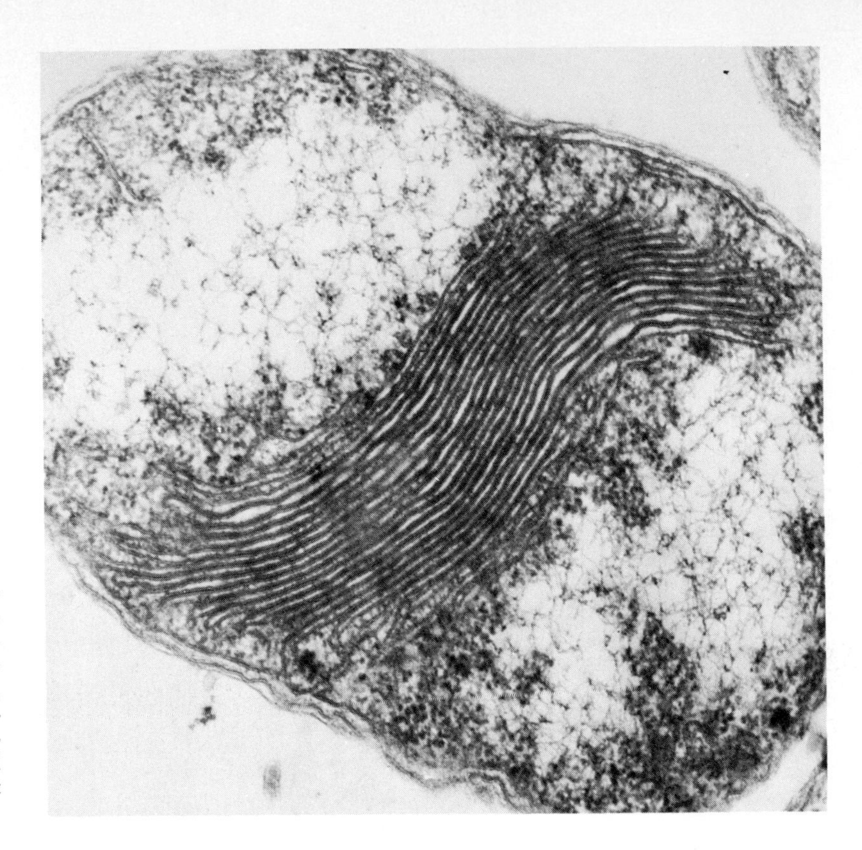

Figure 4.9
Electron micrograph of the marine bacterium, *Nitrosocystis oceanus.* The elaborate membrane system is comparable to the photosynthetic membrane of plastids, but the former, however, is engaged in chemosynthesis, a process in which energy is obtained from the alteration of chemical compounds rather than from light sources. (Courtesy of Dr. S. Watson.)

tional but not structural equivalent. Other bacteria, on the other hand, may have beautifully organized layers of membranes comparable to grana, but in that depicted in Figure 4.9, the membranes are chemosynthetic rather than photosynthetic, the energy for synthesis being derived from the breakdown of sulfur compounds.

Not all plastids contain chlorophyll and function photosynthetically. Some, as in the potato tuber, are for starch storage (Figure 4.10); others may contain oil or protein. These lack the lamellar construction of the chloroplast. However, they are derived from a common type of plastid in the cell and what each becomes is related to the kind of cell in which it is found at maturity. For example, starch-containing plastids found in root cap cells (Figure 4.11) are thought to be involved in the perception of gravity, to which most roots respond. Several such structural and functional modifications are possible, but it is in the chlorophyll-containing, thylakoid type of organization of the chloroplast that occur the reactions allowing life to reverse temporarily the trend toward thermodynamic equilibrium.

Figure 4.10
A starch-storage plastid from the tuber of the sweet potato. The large light masses are stored starch; no grana are present. (Courtesy of Dr. H. J. Arnott.)

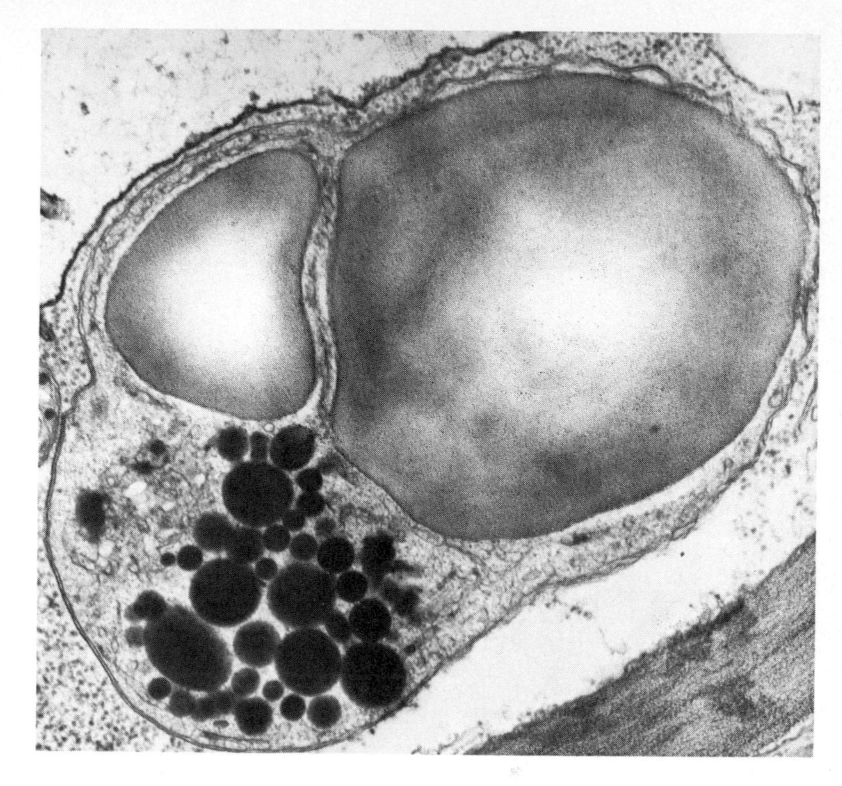

Figure 4.11
Starch grains in root cap cells of pea. Note the orientation toward the bottom of the cells. (Courtesy of W. McDaniel.)

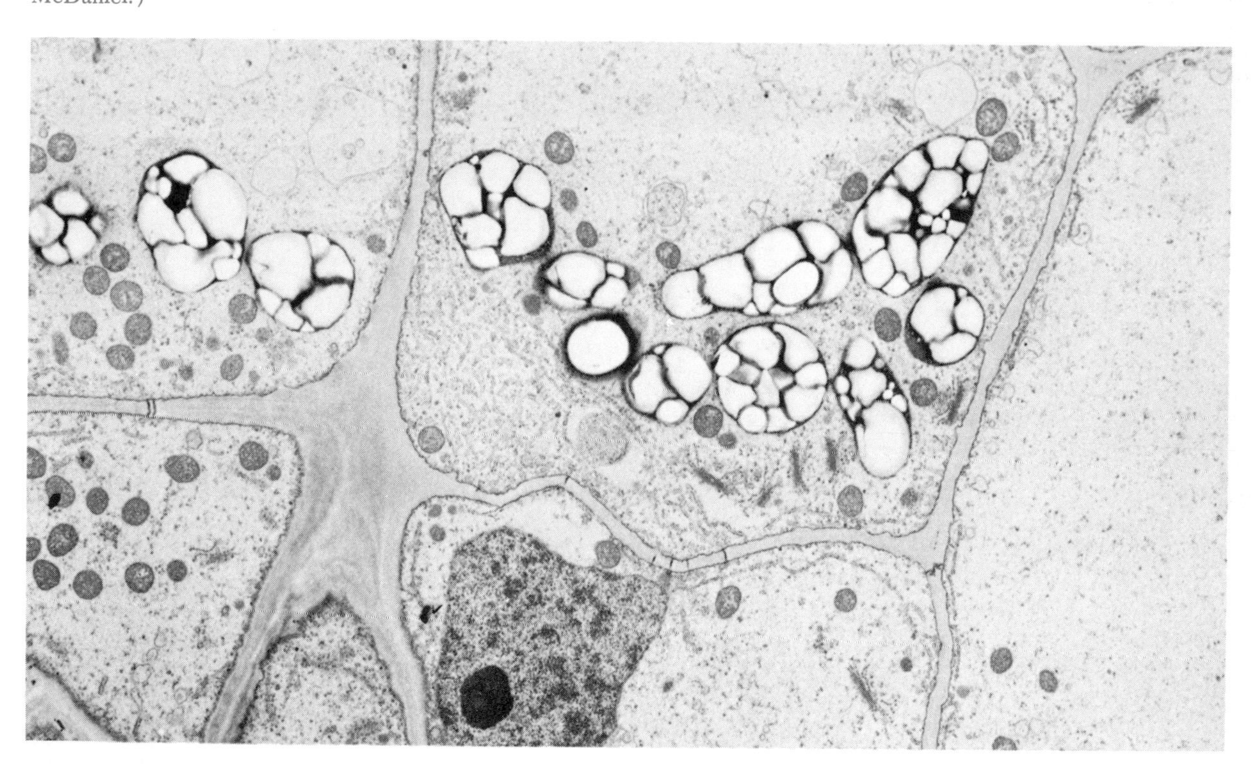

Respiration

We have seen how the energy of sunlight is trapped by photosynthetic plants and converted into chemical energy stored mainly in carbohydrates. This stored energy can then be used by cells of both the plants and the animals, which derive their fuel by eating the plants. Respiration is the process by which this energy is released and made available to the cell in the form of ATP, and involves the controlled breakdown, or oxidation, of the molecules in which the energy is stored. The overall process of respiration requires oxygen as an electron acceptor, but the first steps in the breakdown of sugars do not. In these first steps the 6-carbon sugar molecules are only partially broken down to two 3-carbon molecules of pyruvic acid; in the absence of oxygen, or in cells that do not require oxygen, the pyruvic acid can be converted to either lactic acid or alcohol (Figure 4.12). This anaerobic process, fer-

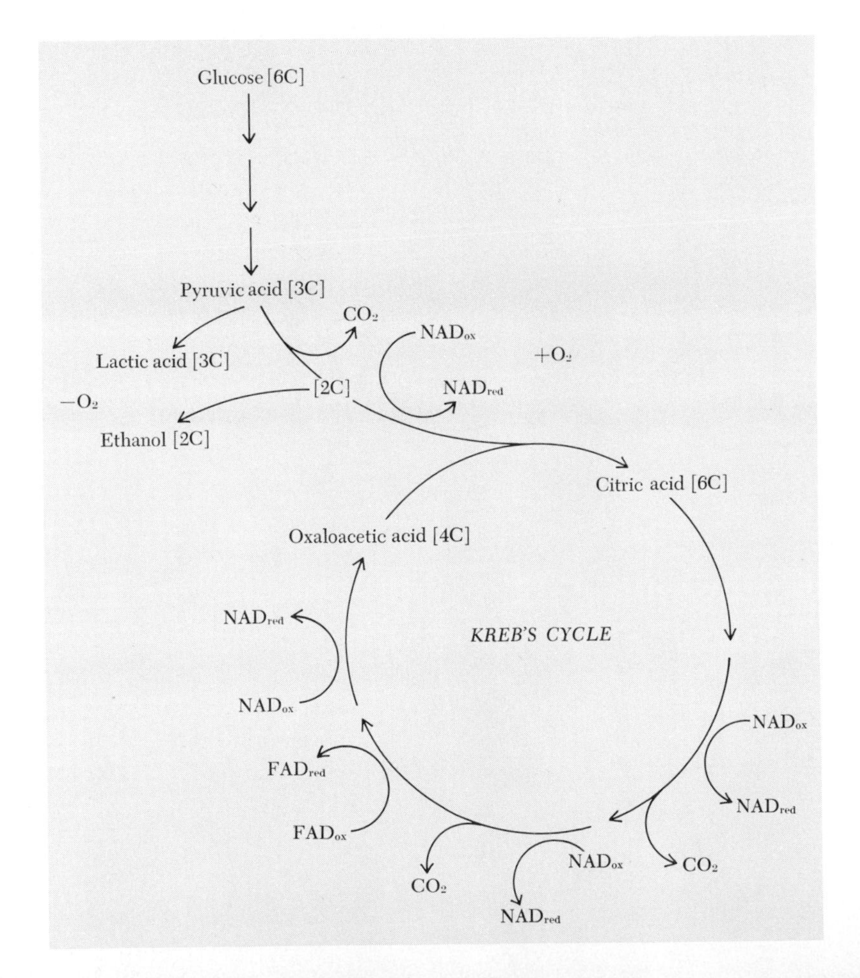

Figure 4.12
Pathway of oxidation of glucose in respiration. Glucose is first oxidized to pyruvic acid, which in the absence of oxygen is converted to either ethanol or lactic acid. In the presence of oxygen the pyruvic acid is decarboxylated and fed into a series of reactions, the Kreb's cycle, being added initially to a 4-carbon acceptor molecule, oxaloacetic acid, to form the 6-carbon molecule, citric acid. Further oxidations regenerate oxaloacetic acid, and in these reactions electrons from the Kreb's cycle substrates are used to reduce NAD and the related compound FAD. Some of the energy in the original glucose molecule is thus conserved in the formation of reduced NAD and reduced FAD, and can be used to form ATP (see text and Figure 4.13).

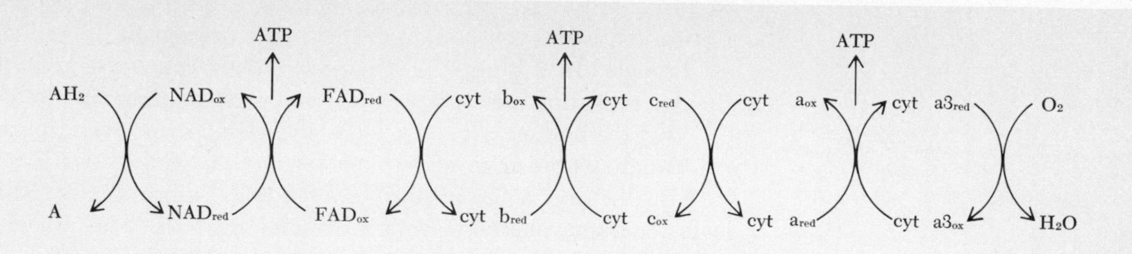

Figure 4.13
Electron transport system in oxidative respiration. Oxidation of Kreb's cycle substrate (A) is accompanied by reduction of NAD and FAD. In a further series of oxidation-reduction reactions involving cytochromes b, c, a, and a3, the electrons are passed down to oxygen, the terminal electron acceptor. At three of these steps the free energy change is sufficient to form ATP from ADP and inorganic phosphate.

mentation, is involved in the production of wine and beer; fungi utilize the sugars of the grapes or barley as substrates for anaerobic respiration, and oxidize them to alcohol. Anaerobic respiration is very inefficient in terms of energy yield, since much of the energy is still locked up in the alcohol molecules. In the presence of oxygen, however, the pyruvic acid is further oxidized to CO_2 in a cyclic series of reactions known as the Kreb's cycle (Figure 4.12); it is in the oxidation-reduction reactions of the Kreb's cycle that electrons are given up and used to reduce NAD (a compound similar to NADP, the electron acceptor in photosynthesis). This high energy intermediate then funnels its electrons down an electron transfer chain (Figure 4.13), in which they drop to successively lower energy levels and are ultimately accepted by oxygen, which is reduced to water. (The role of the oxygen taken in by plants and animals, including that oxygen we inhale in breathing, is to accept these electrons derived from respiration.) As the electrons pass down this electron transfer chain, much of their free energy is conserved in the formation of ATP from ADP and inorganic phosphate. Thus, the supply of ATP necessary for the work of the cell is provided. Once again we can see that the generation of ATP is coupled to an electron transport system, as in the light reactions of photosynthesis.

Mitochondria

While the reactions involved in the breakdown of the substrates of respiration to pyruvic acid occur outside the mitochondria, it is in these organelles that the major energy conversions of the Kreb's cycle and the electron transfer chain occur. Thus, the mitchondria can be considered as the "power plants" of the cell, responsible for providing energy from the metabolic fuel.

Mitochondria are found in every type of cell except those of bacteria and blue-green algae. They are also absent from the mature red blood cell, although they were present in the immature erythrocyte. In Figure 1.23, Chapter 1, they can be seen as long, slender rods, but their shape and number are characteristic of the cell in which they are found. In the living cell, seen in tissue culture, they appear to be in constant motion.

Mitochondria range in size from 0.2 to 7.0 μm in diameter, and, in form, from spheres to rods to branching rods. The unicellular green alga, *Microsterias*, has a single mitochondrion per cell, whereas the giant amoeba, *Chaos chaos*, may have up to 500,000. A mammalian liver cell, 25 μm in diameter, may possess as many as 1,000, a kidney cell about 300, and a sperm cell as few as 25. Plant cells appear to have relatively few when compared to an animal cell of similar size, but they tend to cluster where cellular activity is high. Such aggregation is not accidental, for, as we have seen, these organelles provide ATP, the currency of the cellular economy.

Mitochondria as seen in the electron microscope consist of a smooth outer membrane separated by a space from an inner membrane; the inner membrane is folded to form projections, called *cristae,* which extend into the *matrix* of the mitochondrion and separate the matrix from the intracristal space (Figures 4.14, 4.15). The cristae may be few in number, leaving most of the volume of the mitochondrion occupied by the matrix, or there may be many cristae packed very closely together. Figure 4.15 shows a schematic drawing of a liver mitochondrion, which is representative of the arrangement of cristae seen in mammalian cells. Cristae may also assume tubular form, as in *Paramecium* and in some plant cells, or reticulate form, as in flight muscle cells of some insects. These various ways of increasing the available surface area of the inner membrane system appear to be related to the degree of metabolic activity of the mitochondrion. Dramatic conformational changes of the inner membrane are also known to occur as mitochondria change from an inactive to an active state.

It is possible to separate the inner and outer membranes from each other and from the internal matrix, and to analyze these components for activities of various enzymes. None of the major respiratory enzymes are located in the outer membrane, although other enzymes are present. Most of the enzymes that catalyze the Kreb's cycle reactions are within the matrix, while the electron-carrier molecules of the electron transport chain, along with the enzymes for the electron-transport-dependent formation of ATP, are located in the inner membrane. Indeed small fragments of mitochondrial inner membranes are still capable of carrying out electron transport and ATP formation, suggesting that the inner

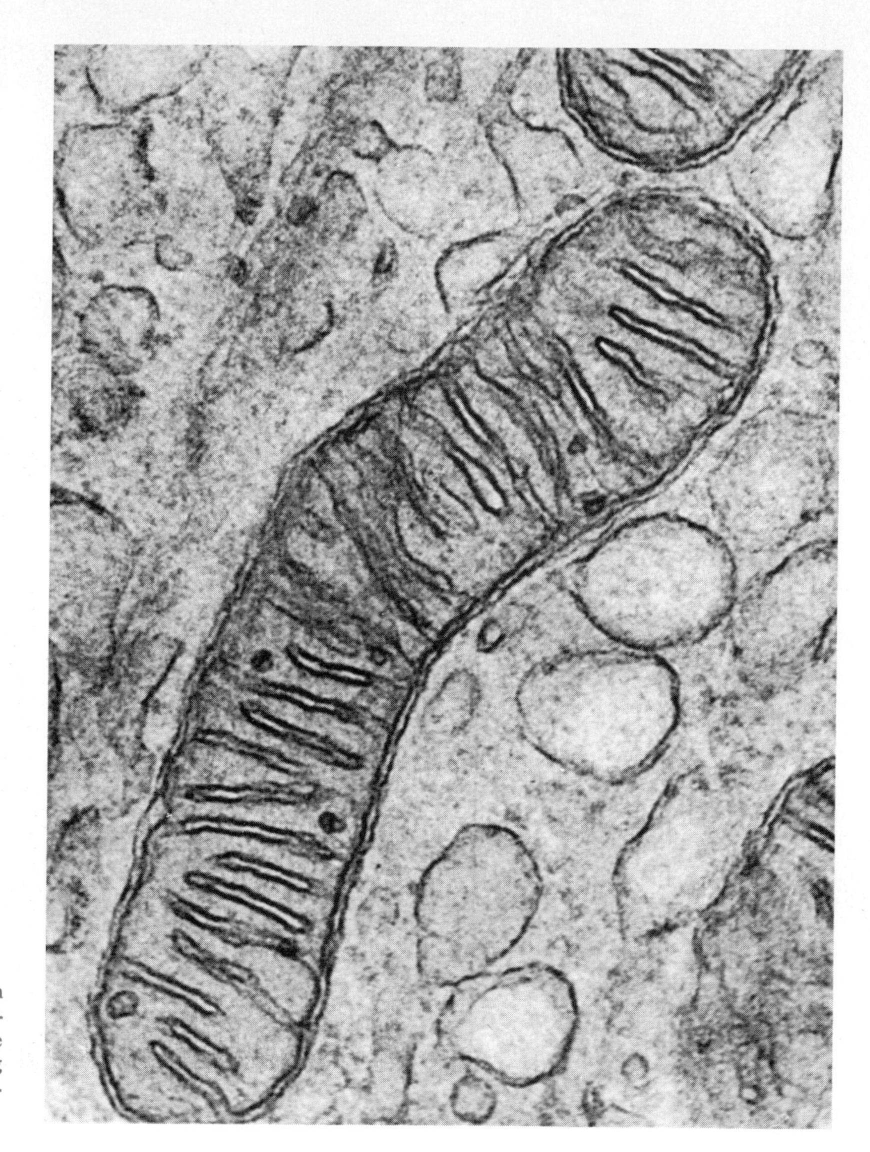

Figure 4.14
Highly magnified electron micrograph of a mitochondrion. The outer boundary of the mitochondrion is a double structure, with the inner layer being continuous with the inner cross membranes (cristae).

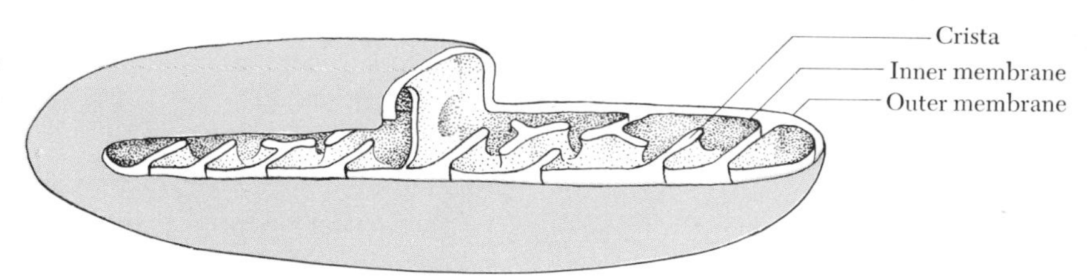

Figure 4.15
Schematic drawing of a typical mitochondrion. (Courtesy of Dr. A. H. Lehninger.)

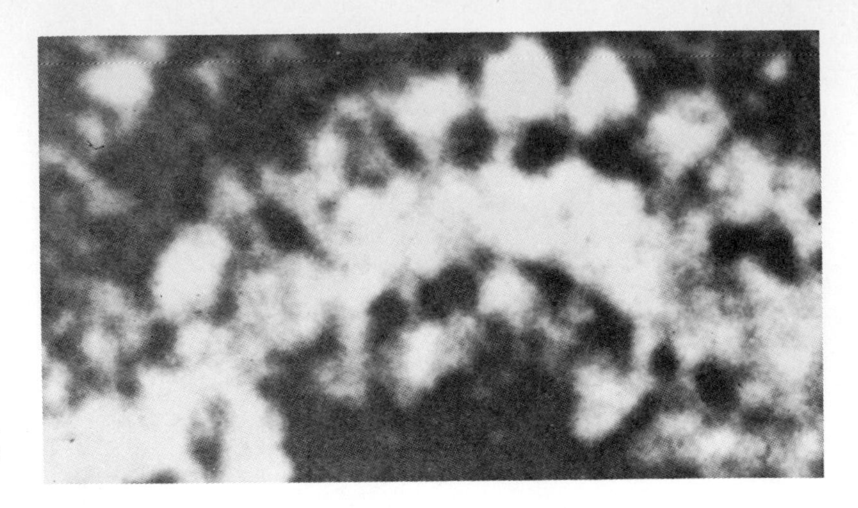

Figure 4.16
Portion of inner mitochondrial membrane showing stalked particles that contain ATP-ase activity.

membrane must consist of many repeating subunits of the orderly array of enzymes necessary for these reactions. Although structural, particulate subunits can be seen by various electron microscopy techniques, their relationship to any functional molecular organization is not clear. One exception, however, is the stalked particle attached to the matrix face of the inner membrane (Figure 4.16). When these are present on membrane preparations, both electron transport and ATP formation occur; upon removal of these particles electron transport continues but no ATP is formed. Thus, these particles, of about 100 Å diameter, represent the ATP-forming enzyme associated with electron transport.

The inner membrane of the mitochondrion is also known to be extremely impermeable to most of the metabolites found within the matrix. For example, the ATP formed within the mitochondrion must be transported to the surrounding cytoplasm by a specific carrier, which is also thought to be responsible for movement of ADP into the matrix. Other carrier systems are present, and as in the case of the plasma membrane, these presumably constitute part of the membrane structure. Thus, the mitochondria not only function to provide ATP for use by the rest of the cell, but also selectively control the transport of material into and out of themselves.

Autonomy of mitochondria and chloroplasts

These energy-converting organelles are integral components of eukaryotic cells, but certain features indicate that they display a considerable amount of structural and functional independence.

For example, both their internal membrane systems, which are organized in specific and orderly ways to carry out specific and vital reactions of making energy available to the cell, and their aqueous matrices are separated from the rest of the cytoplasm by other membranes. Furthermore, neither mitochondria nor chloroplasts arise *de novo,* but from growth and division of other mitochondria and plastids. We also know that mitochondria and chloroplasts contain DNA which can replicate; RNA which can support protein synthesis; and ribosomes and the rest of the machinery required for protein synthesis (see Chapter 2). However, both the DNA and the protein-synthesizing apparatus are different in several significant ways from their nuclear and cytoplasmic counterparts, being more similar to those found in prokaryotic cells. For example, some mitochondrial DNA molecules exist as circular forms, and both the organelle ribosomes and their constituent RNA components are smaller than those found in the cytoplasm outside the organelles. It has been conclusively demonstrated that organelle DNA does code for at least one of the organelle proteins, and certain mutations that affect mitochondrial or chloroplast structure and activity are known to show patterns of cytoplasmic rather than nuclear inheritance. There is also good evidence that at least the ribosomal and transfer RNA molecules of the organelles are transcribed from organelle DNA, and that synthesis of some organelle proteins occurs on organelle ribosomes. Whatever the source of the messenger RNA involved, be it transcribed from nuclear or organelle DNA, the chloroplasts and mitochondria are clearly delegated some control over the synthesis of their own components.

Both the degree of autonomy exhibited by chloroplasts and mitochondria, and the structural and biochemical features they share with blue-green algae and bacteria, have led to the suggestion of a symbiotic origin of these organelles; that is, that they are actually invaders that have become not only established in but also essential to the host cell. During the course of evolution, according to this hypothesis, their role in the symbiotic association has become refined to the point where their function is now crucial (see Chapter 11).

Bibliography

BASSHAM, J. A. 1962. The path of carbon in photosynthesis. *Scientific American* **V. 206**, No. 6, 88–100. A description of the dark reactions of photosynthesis and of the elegant experimental procedures used to follow the fate of photosynthetically fixed carbon.

GOODENOUGH, U. W., and R. P. LEVINE. 1970. The genetic activity of mitochondria and chloroplasts. *Scientific American* **V. 223**, No. 5, 22–29. An account of the genetic systems of mitochondria and chloroplasts, their genetic autonomy, and their resemblance to free-living prokaryotic cells.

GOVINDJEEK and R. GOVINDJEEK. 1974. The absorption of light in photosynthesis. *Scientific American* **V. 231**, No. 6, 68–82. A detailed but clearly written account of how light energy is trapped and transformed in the primary events of photosynthesis.

LEHNINGER, A. L. 1965. *Bioenergetics*. N.A. Benjamin, New York. A first-class elementary treatment of how energy is managed by living systems.

———. 1975. Biochemistry (2nd ed.). Worth, New York. Chapters 14–22 provide detailed and extensive discussions of respiration and photosynthesis.

LEVINE, R. P. 1969. The mechanism of photosynthesis. *Scientific American* **V. 221**, No. 6, 58–70. A clearly explained account of the major stages in the light reactions of photosynthesis.

MCELROY, W. D. 1971. *Cell Physiology and Biochemistry* (3rd. ed.). Prentice-Hall, Englewood Cliffs, N. J. An introductory-level text that contains easy-to-follow accounts of the biochemical reactions of photosynthesis and respiration.

RACKER, E. 1968. The membrane of the mitochondrion. *Scientific American* **V. 218**, No. 2, 32–39. How the inner mitochondrial membrane can be taken apart and put back together, and the functions of its components analyzed.

Organization of the cytoplasm

As we saw earlier, the information necessary for the manipulation of energy flows through the enzymes of the cell. Each cell has many thousands of enzymes, catalyzing the various reactions involved in the activities in which the available energy is utilized. These reactions, which are carried out in a controlled and precise manner, do not occur in a homogeneous milieu, but in one that is highly structured and compartmentalized. In discussions of the nucleus, the plasma membrane, and the chloroplasts and mitochondria, those components whose activities confer on the cell its characteristic autonomy as the unit of life, we saw how the structural basis for these activities represents a high degree of molecular organization and of compartmentalization. Such structural complexity, which certainly requires information and energy for its maintenance, is itself necessary for the manipulation of information and energy.

We must now examine the structural basis for the other cellular activities, including synthesis and degradation, transport of materials within and out of the cell, and both cellular and intracellular movement. As we shall see, membraneous structures again play a major role in cellular architecture. The significance of this ubiquitous type of macromolecular organization to the cell presumably lies in the capacity of membranes to restrict diffusion, and therefore to compartmentalize the cell, and to provide surfaces in which specific orientation of multienzyme systems can be maintained and on which specific types of reactions can be localized.

Endoplasmic reticulum

All nucleated cells contain a network of cytoplasmic membranes that delimit a series of often-interconnected regions of the cell. This membrane-enclosed, vacuolar system, the endoplasmic reticulum (ER), is of great importance in the synthesis and transport of materials within the cell. The ER, which can vary considerably in amount from cell to cell, can also assume various configurations within the cell, forming flattened sacs (cisternae), tubules, and vesicles (Figure 5.1). Both the morphology of the ER and the total and relative amounts of its components appear to reflect the state of metabolic activity of the cell. As we mentioned previously, the ER is often continuous with the outer membrane of the nuclear envelope (Figure 5.2).

The ER is also classified according to whether ribosomes are attached to the outer surfaces of the membranes (rough ER) or not

Figure 5.1
The endoplasmic reticulum (ER) in parotid (salivary-gland) acinous cells of the mouse. (*a*) The ER above the mitochondrion is of the rough, or granular, variety, containing ribosomes, and is much more highly organized than in the area immediately below; many of the membranes end blindly in the cytoplasm. (*b*) The mass of rough ER membranes is a continuously branching and interconnected system, which is also connected with the outer portion of the nuclear membrane (arrow). (Courtesy of Dr. H. F. Parks.)

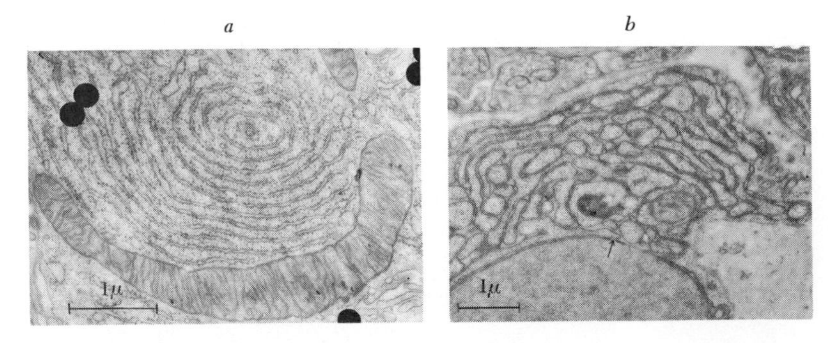

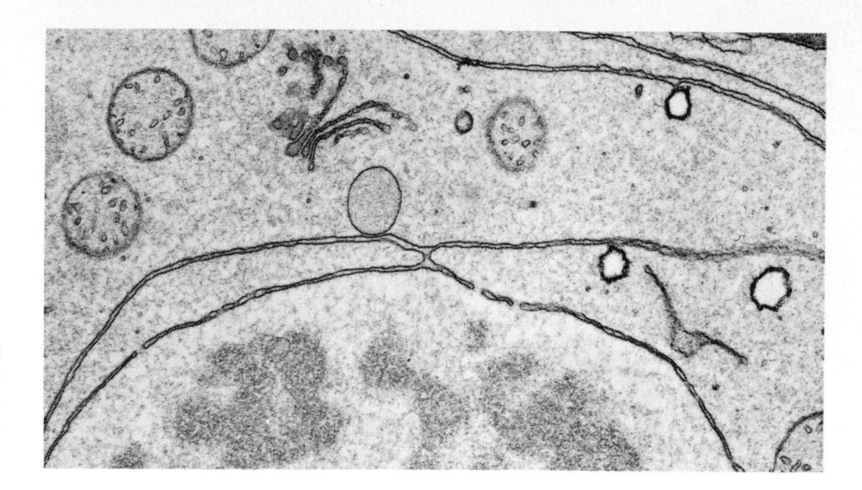

Figure 5.2
Electron micrograph of part of plant cell showing continuity between the outer membrane of the nuclear envelope and the ER. (Courtesy of Dr. G. Whaley.)

(smooth ER). Rough ER (Figure 5.3) is particularly abundant in cells that are actively synthesizing proteins, which might be expected in view of what is known of the role of the ribosomes in protein synthesis (see Chapter 2); however, ribosomes are also present in the cytoplasm free of any association with membranes. The bound ribosomes, which are often arranged to form spiral polysomes, are attached to the ER membrane by their large subunits. At least part of the significance of this attachment of ribosomes to membranes of the ER appears to lie in the subsequent fate of the proteins synthesized on these ribosomes; many such proteins, including collagen fiber proteins and several digestive enzymes, are ultimately secreted by the cells in which they are formed. These proteins are believed to pass directly from the ribosomes on which they are synthesized across the membrane and into the cavity of the ER, from where they can be subsequently transported to the outside of the cell. Thus, the ER appears to provide a system of communication in the cell that permits segregation of the products formed and channels for the transportation of these products either to other parts of the cell or to the outside. However, rough ER is also present in nonsecretory cells, and furthermore not all of the protein synthesized on attached ribosomes passes into the cavity of the ER; it is likely, therefore, that attachment of ribosomes to the ER membranes can serve some function other than to facilitate secretion.

Smooth ER (Figure 5.4), whose membranes usually form tubular elements free of attached ribosomes, also has been implicated in synthesis and transport. Such smooth ER is prevalent in cells that synthesize and secrete steroid hormones, such as the cells of the testes and the adrenal gland. Furthermore, enzymes of

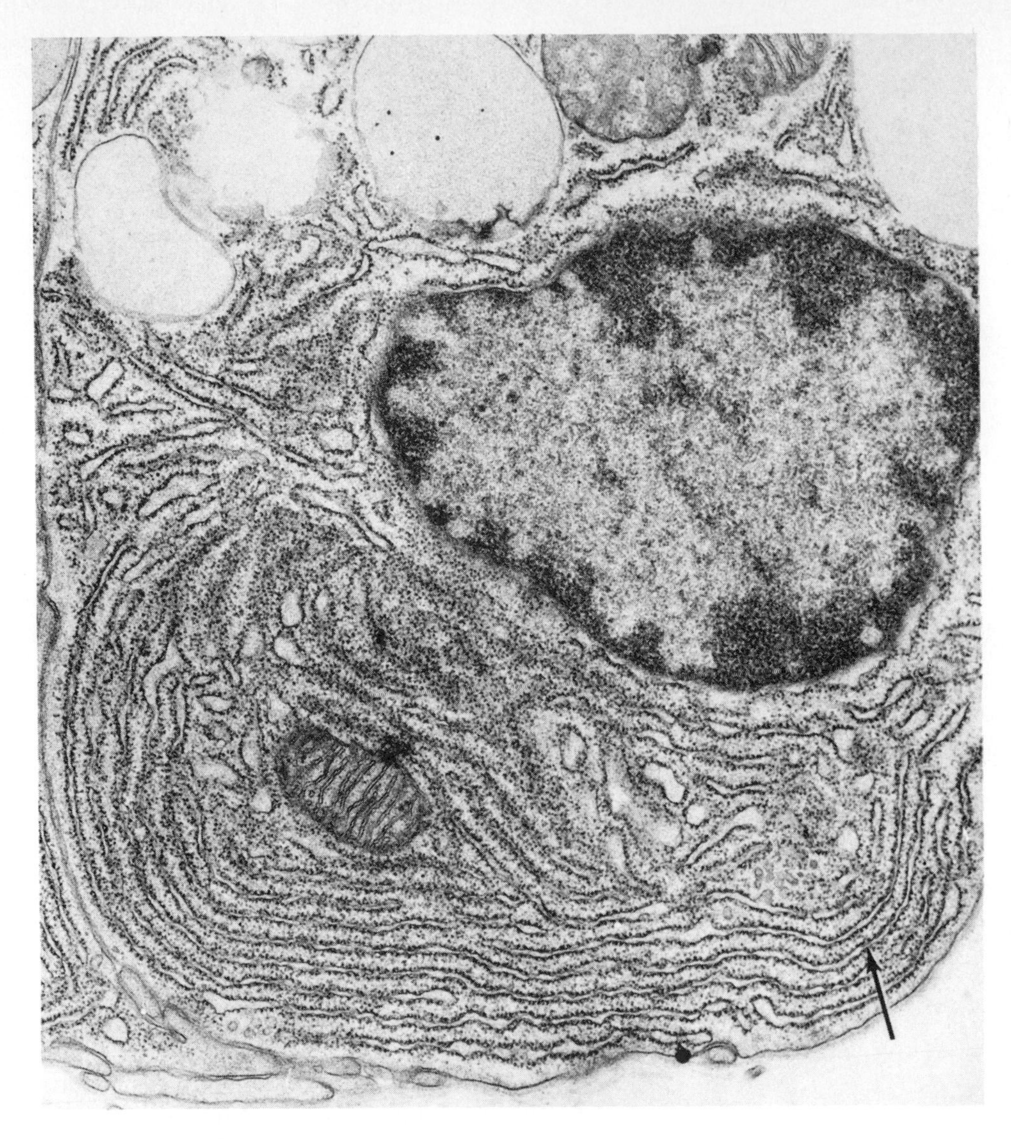

Figure 5.3
Electron micrograph of portions of "chief" cells of the bat, showing the extensive rough ER. The ER forms flattened saclike arrangements, with the ribosomes (arrow) attached to the outer surfaces of the membranes enclosing the sacs, or cisternae. These cells are active in synthesis and secretion of the digestive enzyme, pepsin. See also Figure 6.5. (Courtesy of Dr. K. Porter.)

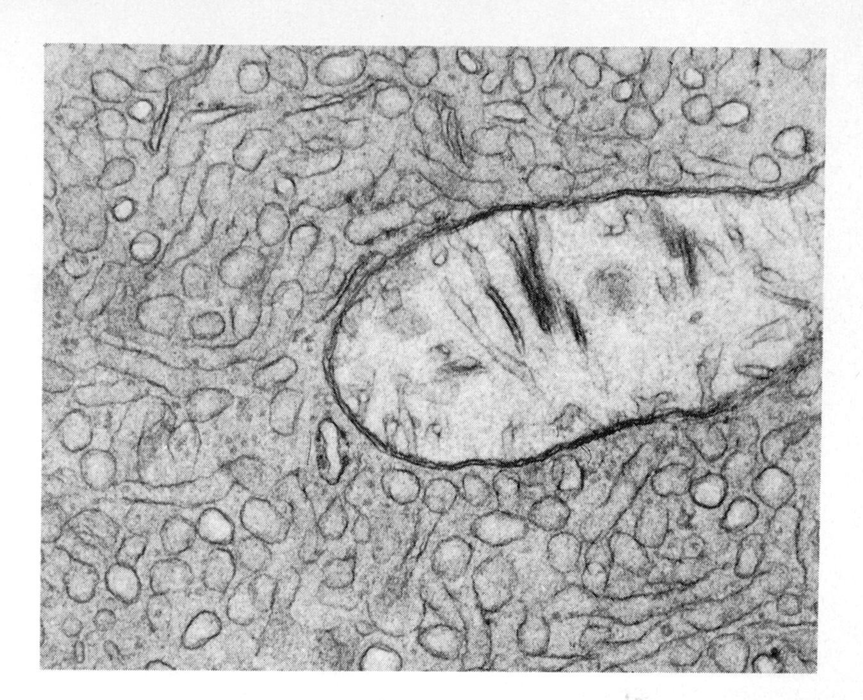

Figure 5.4
Electron micrograph of a portion of a cell from the testis of an opossum, showing the smooth, or agranular, form of ER. (Courtesy of Dr. D. Fawcett.)

glycogen metabolism, lipid synthesis, and even of electron-carrier chains, are associated with certain types of smooth ER. Thus, a range of functions can be associated with the smooth membranes of the ER; even in one cell the homogeneous appearance in electron micrographs of the smooth ER may mask a considerable degree of functional heterogeneity.

The spatial continuity and morphological similarities between membranes of rough and smooth ER systems suggest that these may be parts of the same basic membrane system. There is also evidence that smooth ER may be derived from rough ER, the constituent proteins of the former being synthesized on ribosomes that become detached once the protein has been inserted into the membrane. Thus, the ER can be thought of as a relatively labile membrane system that can display intracellular differentiation, becoming modified in different ways to facilitate the various activities associated with it.

The Golgi complex

Like the endoplasmic reticulum, the *Golgi complex* (named after its discoverer) is a system of membranes found in almost all eukaryotic cells. The Golgi system of a cell frequently consists of

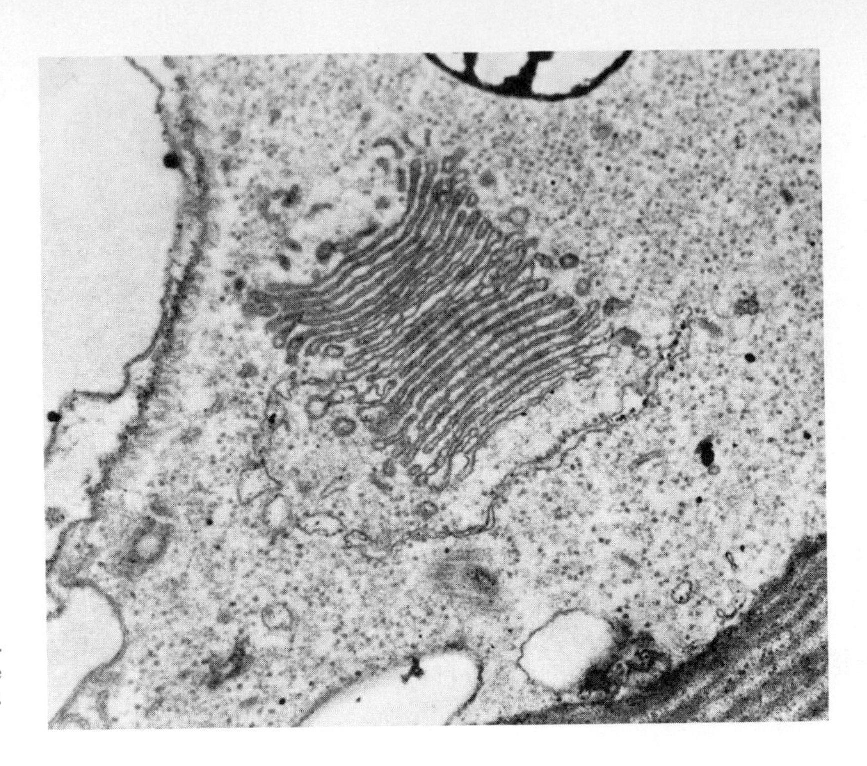

Figure 5.5
Electron micrograph of a Golgi complex in *Euglena*. Note the large number of cisternae. (Courtesy of Dr. S. C. Holt.)

several apparently distinct Golgi complexes, or *dictyosomes*, each of which is made up of a parallel array of membrane-enclosed, flattened cisternae, ranging in number from 3 to 20 depending on the cell type (Figure 5.5). The cisternae, which are usually equally spaced in the stack, are frequently curved, giving the Golgi complex a distinct polarity with both a convex (outer) face and a concave (inner) face. A network of tubular and vesicular membraneous components radiates out from the cisternae (Figure 5.6). These vesicles originate from the cisternae by a "pinching-off" process.

The most widely established role of the Golgi complex is in the packaging and transport of certain materials, notably proteins and polysaccharides, out of the cell; in some cases actual synthesis of material may occur within the Golgi complex. The formation of the cell wall of plant cells clearly depends on the activity of the Golgi complex. Following mitosis (see Chapter 7), vesicles derived from the Golgi apparatus move to the region where the new walls will form, and they fuse with each other (Figure 5.7). As a result of this fusion, two sheets of membrane form, with the former contents of the Golgi-derived vesicles between them. The fused membranes then become the plasma membranes of the newly

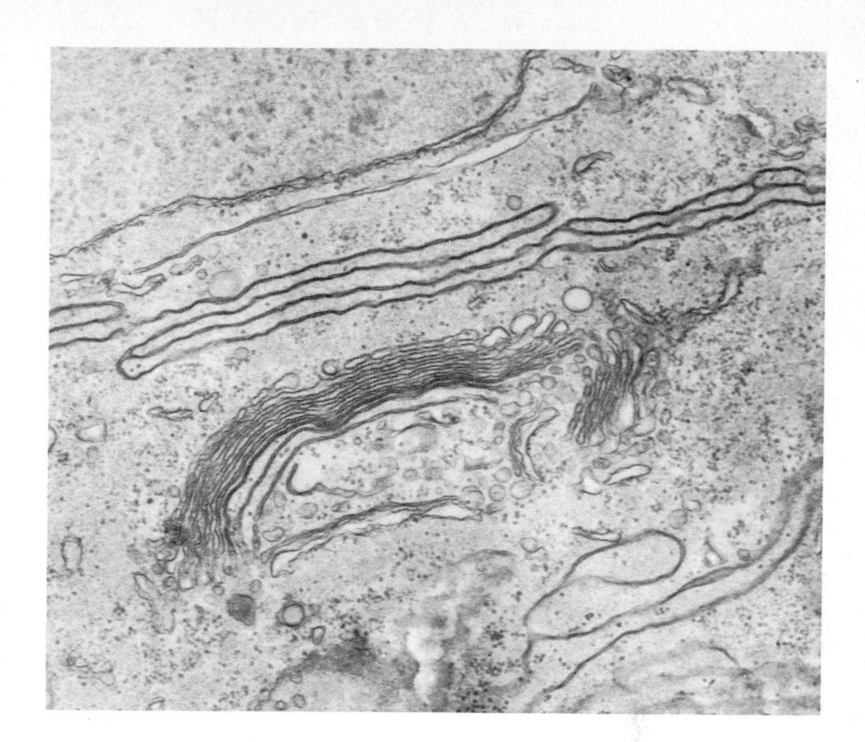

Figure 5.6
Electron micrograph of a Golgi complex in an animal cell. The parallel flattened sacs are usually curved slightly, and they tend to bud off vesicles at the ends of the sacs.

Figure 5.7
Formation of phragmoplast following mitosis in a root tip cell. Vesicles derived from Golgi complexes fuse with each other to form new membranes and wall. (Courtesy of W. McDaniel.)

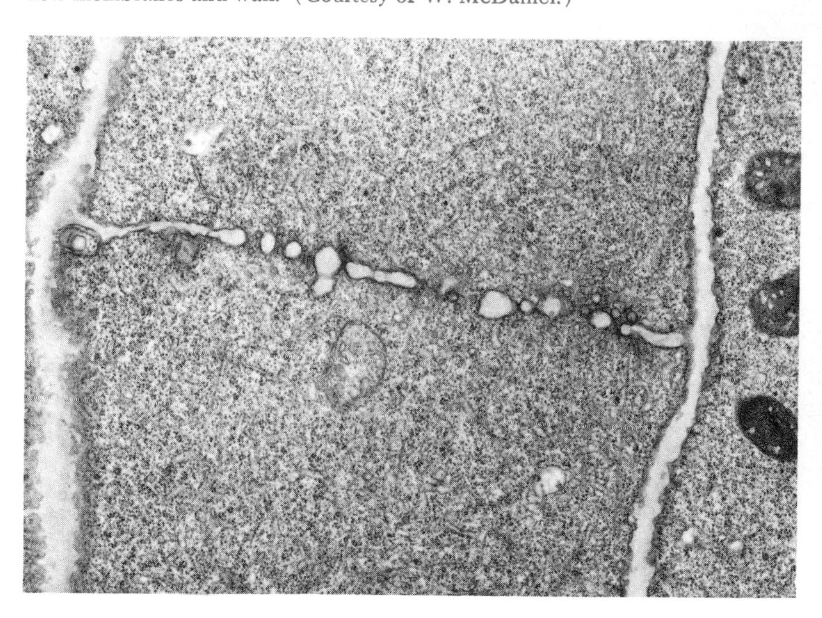

formed cell, and the material between them forms the matrix of the new cell walls and the middle lamella. Subsequent wall growth involves the fusion of vesicles containing polysaccharide material with the plasma membrane, and the resulting deposition of this material, consisting of pectin and hemicellulose, into the wall matrix. Some of the enzymes necessary for the synthesis of these cell wall constituents are believed to be closely associated with the Golgi complex. We can see, therefore, that the Golgi complex is involved not only in synthesis and transport of cell wall material, but also in the formation of the plasma membrane itself. The formation of the cleavage furrow during division of animal cells also represents an increase in total plasma membrane surface area, and Golgi complexes appear to contribute material to the expanding membrane.

The Golgi complex is also involved in secretion of proteins from the cell; a clear relationship between the Golgi system and membranes of the ER in this role is evident. As discussed earlier, some of the digestive enzymes to be secreted pass from their site of synthesis on the rough ER into the cavity of the ER system. From there they are transported by a process thought to involve fusion of ER membranes with the Golgi complex into the Golgi complex itself. These proteins are then concentrated into the vesicles, which pinch off from the cisternae and ultimately fuse with the plasma membrane, discharging their contents to the outside of the cell. Glycoproteins also can be secreted in this way, but with the assembly and attachment to the protein of the carbohydrate component taking place within the Golgi complex.

It is clear that a close relationship can exist between the Golgi complex and the ER. As we shall see, some of the other membraneous components of the cell also may be related to these membrane systems.

Lysosomes

Many of the degradative hydrolytic enzymes of animal cells are localized within lysosomes, small spherical structures bounded by a single membrane. The contents of lysosomes provide a clue to their function, which is in the breakdown and degradation of material in the cell. Lysosomes are thought to be derived from vesicles formed either by the Golgi complex or the smooth ER, the hydrolytic enzymes first being synthesized on rough ER. The *primary* lysosomes can then participate in digestion either of material brought into the cell from outside or of other cellular components.

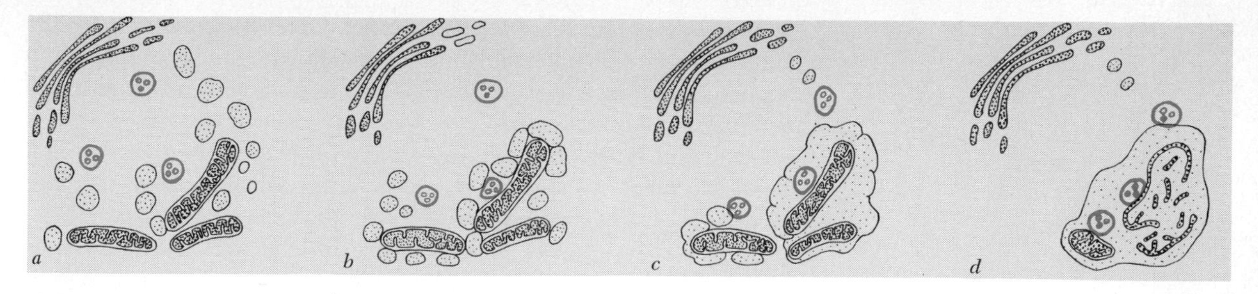

Figure 5.8
Suggested origin for lysosomes from vesicles derived from the Golgi complex. The vesicles (*a*) surround cellular materials, (*b*, *c*) coalesce, and then (*d*) digest the enclosed materials by means of hydrolytic enzymes. (Modified from Dr. D. Brandes.)

In the former case, the lysosomes fuse with pinocytotic or phagocytotic vesicles (see Chapter 3) to form *secondary* lysosomes, in which digestion of the incorporated material by the lysosomal enzymes takes place. In autophagic processes, that is, self-digestion, other cellular components, such as mitochondria, may enter the lysosome and themselves be digested (Figure 5.8). The significance of such autodigestion is not clear, although presumably some cellular processes require controlled breakdown of certain regions of cytoplasm. Even cell death can be part of normal developmental processes, and indeed dying cells often appear to contain many lysosomes; release of the lysosomal hydrolytic enzymes would certainly accelerate the dissolution of a cell.

Lysosomes can be thought of as disposal units of the cell, being involved in digestion of material brought into the cell from outside and in removing foreign bodies or elements of cellular architecture no longer needed. The necessity of keeping the powerful lysosomal enzymes packaged in a membrane—and hence isolated from the rest of the cell, which they could digest—is obvious. How the cell controls the time and place of activity of these enzymes, however, is not known.

Microbodies

Another class of membrane-enclosed organelles found in both plant and animal cells is that comprised of the *microbodies*. This term is used to describe structures that, like lysosomes, are bounded by a single membrane and characterized by the presence within them of specific enzymes. The function of microbodies in general is to compartmentalize specific enzymes and, therefore, specific biochemical reactions in various regions of the cell. These microbodies frequently contain proteinaceous crystallinelike bodies, which may represent an orderly array of some of the enzymes present.

Microbodies are believed to originate from evaginations of the ER, as indicated by visual observation and by the fact that many biochemical similarities between ER membranes and microbody membranes can be detected.

Two main types of microbodies have been described, depending on which reactions are catalyzed by the enzymes they contain; some enzymes, however, may be common to both types of microbody. Peroxisomes were first described in liver and kidney cells of the rat, where they carry out various oxidative reactions involving the formation and subsequent destruction of hydrogen peroxide. Plant peroxisomes contain similar enzymes, and one major role of the peroxisomes in leaf cells is in the oxidation of certain intermediates formed in the dark reactions of photosynthesis; close associations between leaf peroxisomes and chloroplasts are commonly encountered. A second class of microbody, known as *glyoxysomes*, is also present in plants, particularly in cells of some germinating seedlings. In addition to peroxisome-type enzymes, the glyoxysomes contain enzymes involved in the conversion of the stored fats of the seeds to carbohydrates; following germination of the seedling and the utilization of the storage fats, the glyoxysomes are no longer present in the cells.

Vacuoles

Vacuoles are characteristic of mature plant cells and consist of an aqueous solution surrounded by a single membrane, the *tonoplast*. As much as 90 percent of the internal volume of the cell is likely to be occupied by the vacuole, with the cytoplasmic components and the nucleus being appressed against the plasma membrane and cell wall (Figures 5.9 and 5.10). The vacuole functions in the maintenance of turgor of plant cells, in their growth, and in providing an aqueous environment for the accumulation of water-soluble compounds.

The tonoplast membrane, like the plasma membrane, is differentially permeable and hence can maintain concentrations of materials very different from those found in the cytoplasm. High concentrations of salts, sugars, and acids are found in most vacuoles, and many water-soluble pigments, such as the red pigment of beets and many flower color pigments, also may be present.

The tendency of water to move into the hypertonic vacuole exerts pressure on the surrounding cytoplasm and hence on the cell wall. This turgor pressure not only allows plant cells to remain relatively rigid, but is also responsible for enlargement of the cell

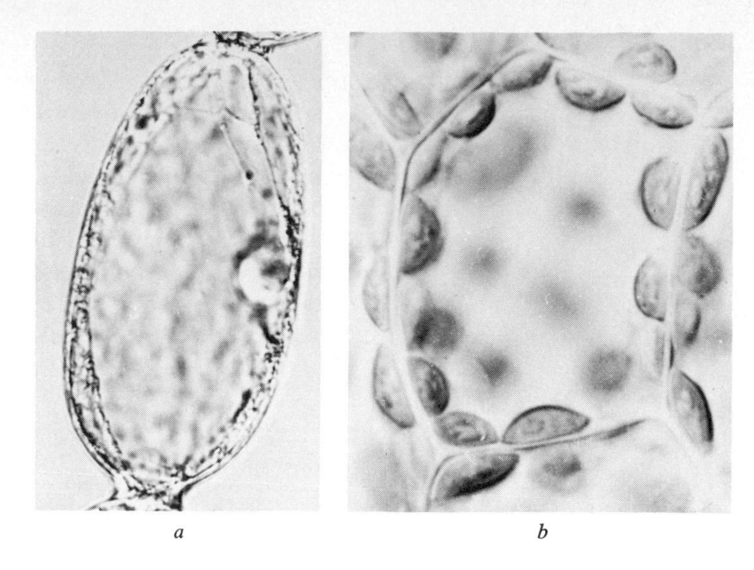

Figure 5.9
Two examples of plant cells, illustrating the manner by which the formation of a vacuole pushes the cytoplasm to the outside, thus increasing the exchange of materials between the cytoplasm and the exterior of the cell. (a) Cell from the stamen hair of the spiderwort, *Tradescantia*, showing nucleus appressed to cell wall, and strands of cytoplasm traversing the vacuole. (b) Cell from "leaf" of a moss, showing the cytoplasm and chloroplasts forced to the outside.

Figure 5.10
Root tip cells of pea showing large vacuole occupying most of the volume of the cell. (Courtesy of W. McDaniel.)

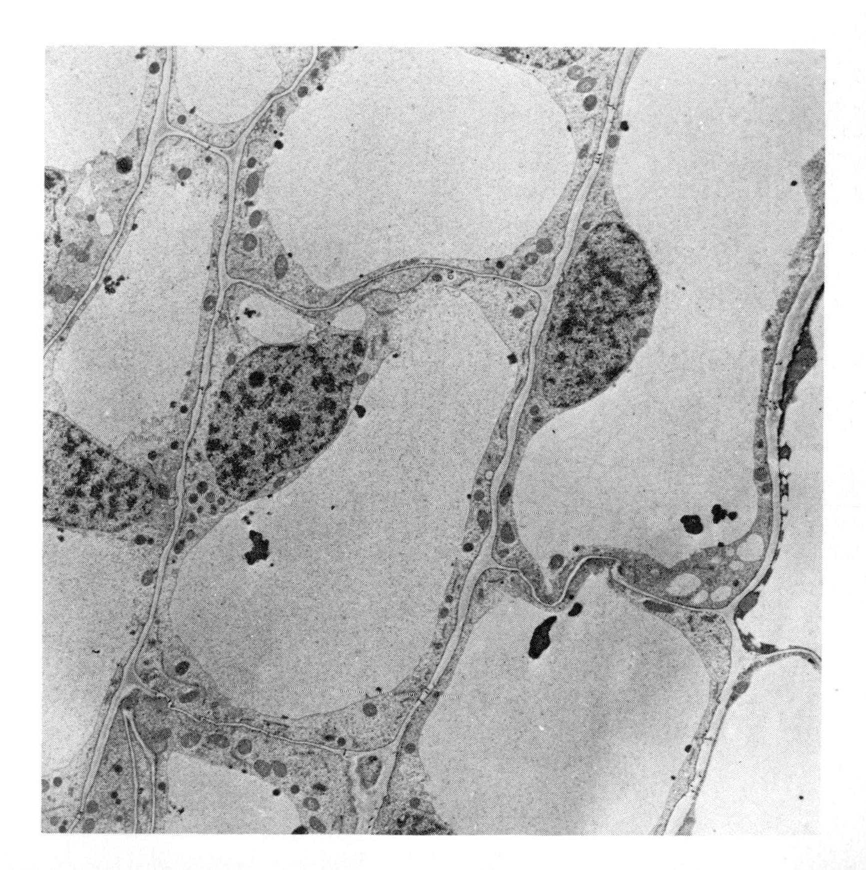

before the outer wall becomes too restricting. The vacuole also may serve as a waste deposit box into which unwanted materials can be shunted. There is also evidence that some degradative enzymes are associated with the tonoplast, suggesting a possible lysosomal role for the plant vacuole.

The vacuole forms from the enlargement and fusion of smaller vacuoles present in meristematic cells; these smaller vacuoles are themselves believed to originate from the endoplasmic reticulum.

Relationships between membrane systems

From our discussion of ER, Golgi, lysosomes, microbodies, and vacuoles, it is apparent that all of these represent membrane-enclosed compartments of the cell, specialized in different ways related to their various functions. Although direct physical continuity between these different compartments is only rarely observed in the electron microscope, it is likely that there is a

Figure 5.11
Diagram of possible interrelationships between membrane systems of the cell. NE: nuclear envelope; RER: rough endoplasmic reticulum; SER: smooth endoplasmic reticulum; L: primary lysosomes; G: Golgi complex; V: Golgi-derived vesicles; PM: plasma membrane.

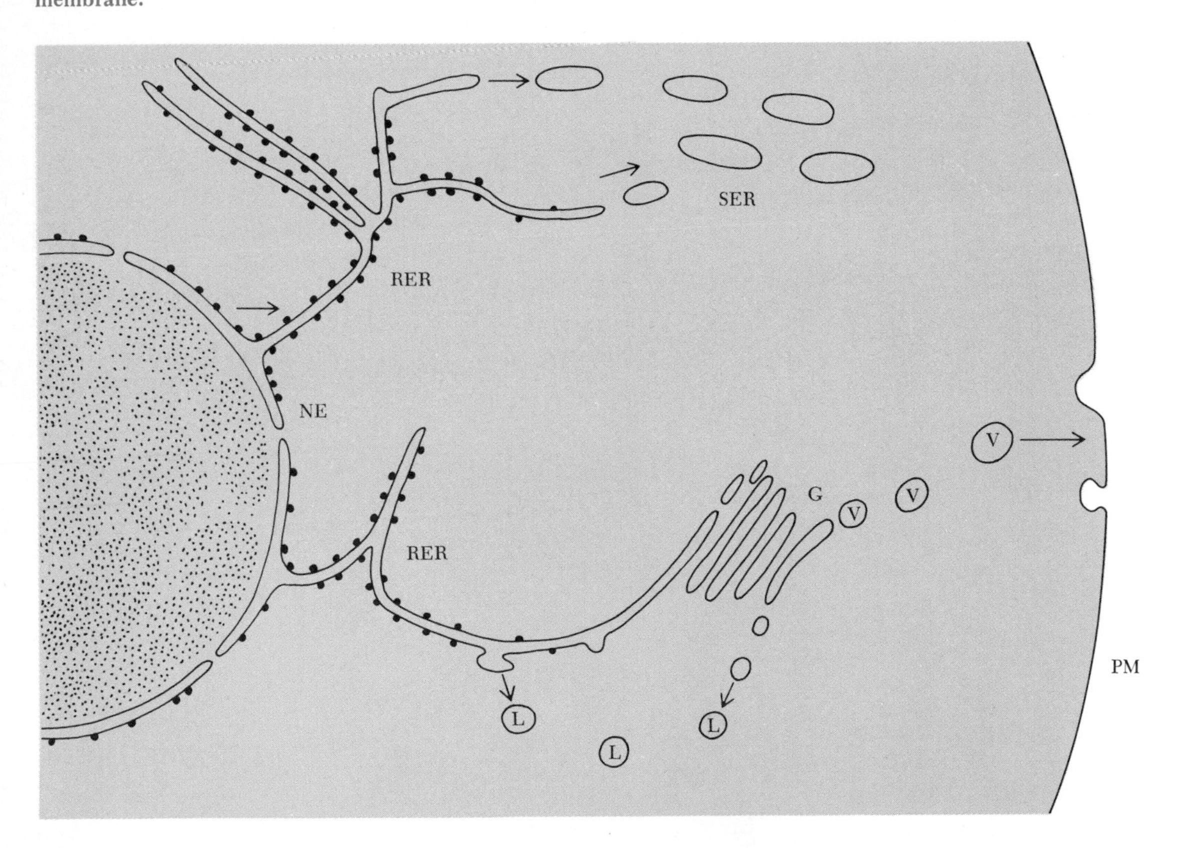

temporal continuity between the ER membranes and those of the other components, the latter being derived directly or indirectly from the former. Even the cell membrane itself is at least in part derived from the vesicles of the Golgi complex. Thus, we can think of all of these membrane systems as being in a state of dynamic equilibrium in the cell, with a basic membrane organization being modified at different times and in different regions of the cell to fulfill various roles. In this way the distribution of materials and labor in the cell would be accomplished by a flow of membranes through the different cellular compartments (Figure 5.11).

Microtubules and microfilaments

Two nonmembraneous structural elements—*microtubules* and *microfilaments*—are recognized as being characteristic of eukaryotic cells. These elements are very long, narrow structures involved in both cellular and intracellular movement and in the determination of cell shape.

Microtubules, as seen in the electron microscope, consist of a cylindrical wall approximately 50 Å in thickness surrounding a central space of about 100 to 150 Å in diameter (Figure 5.12). The

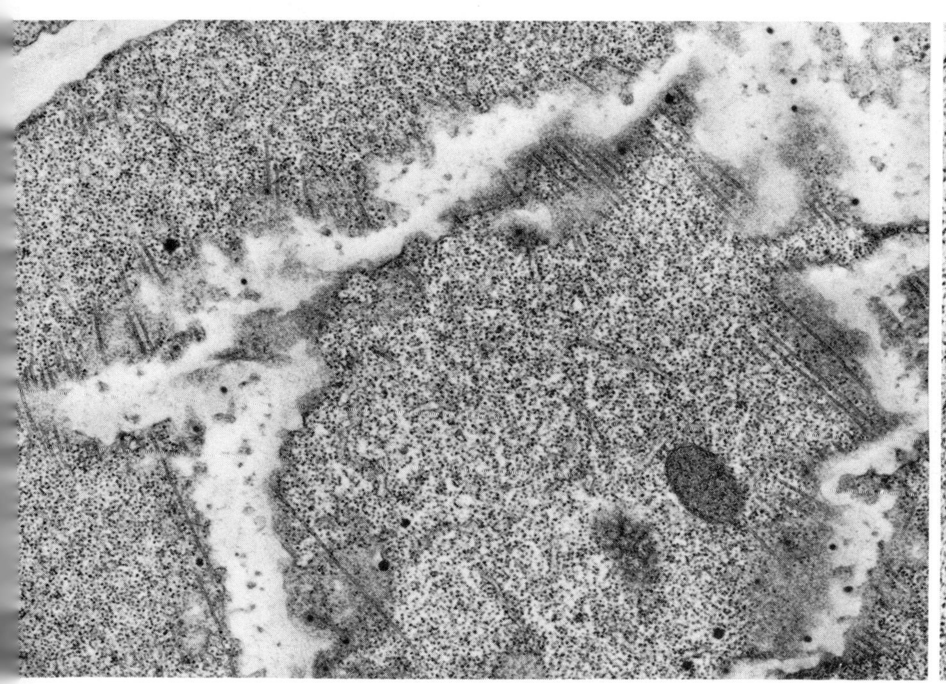

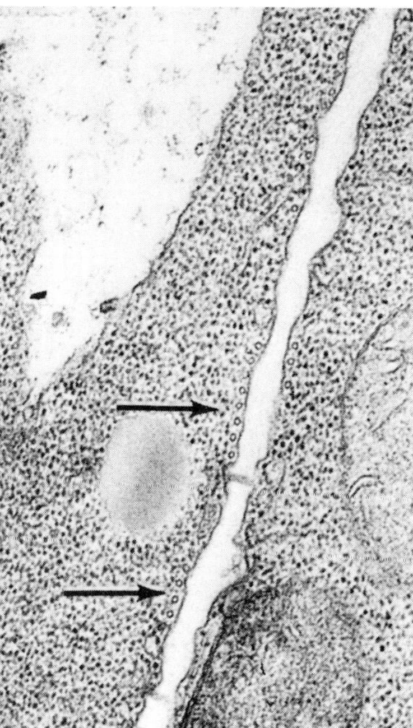

Figure 5.12
Electron micrographs showing microtubules in pea root cells. Left: Oblique section through cell wall showing microtubules in longitudinal section. Right: Microtubules are seen in cross-section along the periphery of the cytoplasm of each cell. (Courtesy of W. McDaniel.)

walls are composed of subunits, of which usually 13 are seen in cross sections. Occasional linkages, or cross bridges, between adjacent microtubules are also observed. Microtubules are remarkably uniform in chemical composition, being formed by the polymerization of molecules of tubulin, a dimer protein with a molecular weight of about 110,000.

Aggregates of microtubules form the *spindle fibers* of dividing cells, on which chromosome movement takes place (see Chapter 7). Spindle microtubules can run from pole to pole of the spindle, or from pole to chromosome, to which they are attached by a specialized structure, the *kinetochore*. The nature of chromosome movement on the spindle fibers is not understood, but it may involve a balance between growth by polymerization and breakdown by depolymerization at opposite ends of the microtubules (see also Chapter 7).

To varying degrees, microtubules also can be organized to function as "cytoskeletons," conferring rigidity and maintaining shape, such as in the pseudopods of certain protozoans or in the extremely elongated axons of nerve cells. In plant cells, the orientation of microtubules in the peripheral cytoplasm is frequently correlated with the orientation of the cellulose microfibrils in forming cell walls, and these microtubules may play a role in determining cell shape and cell wall patterns.

Microtubules are also found in the *cilia* and *flagella* of motile cells, where they are arranged in a specific and characteristic manner. Two central microtubules are surrounded by nine pairs of microtubules, as shown in Figures 5.13 and 5.15. This 9 + 2

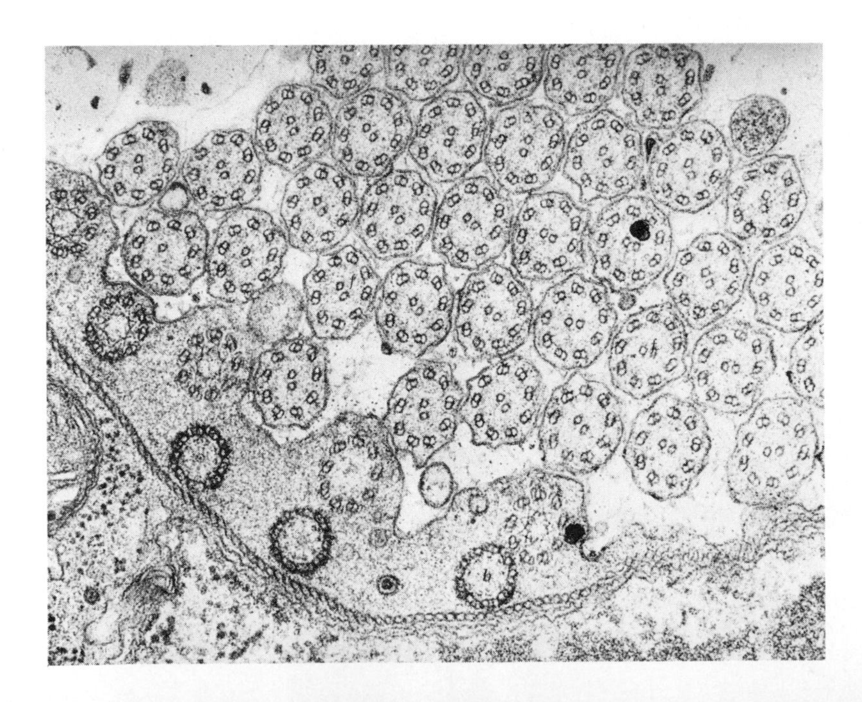

Figure 5.13
Electron micrograph of spermatid of *Equisetum*, showing cross sections of many flagella (f), with the typical 9 + 2 arrangements of microtubules. The basal bodies (b) are also shown. (Courtesy of Dr. J. G. Duckett.)

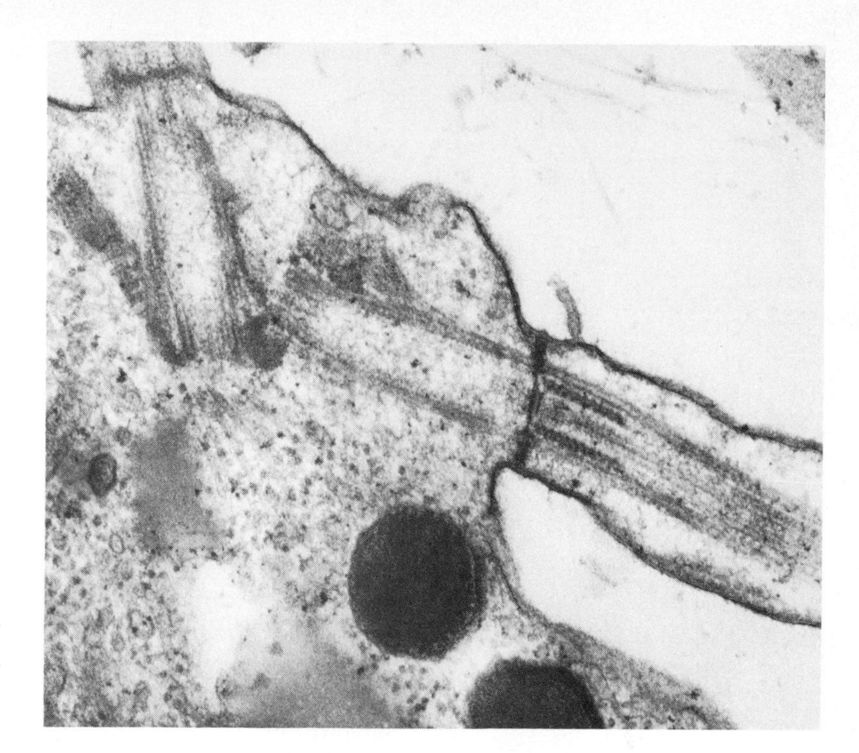

Figure 5.14
Longitudinal section through basal bodies and flagella of motile zoospore of *Phytophthora*, the potato blight fungus. (Courtesy of Dr. J. G. Duckett.)

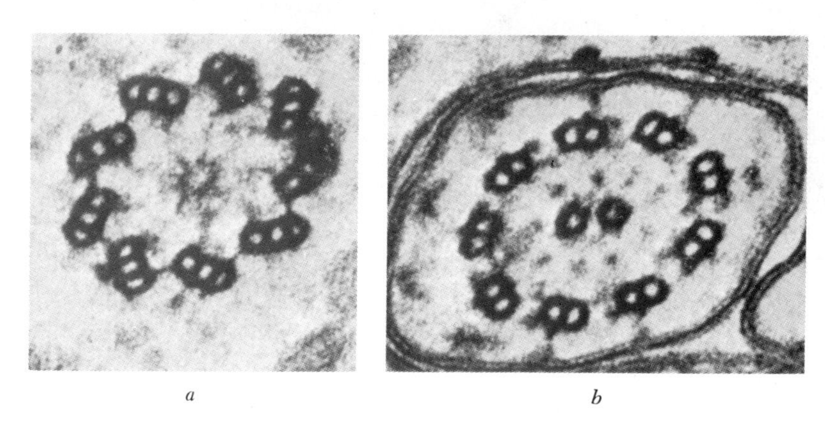

a *b*

Figure 5.15
The arrangement of tubules characteristic of centrioles (*a*), and cilia and flagella (*b*).

arrangement of microtubules is embedded in a matrix enclosed by an extension of the cell membrane. The beating of flagella and cilia provides the forces for propulsion of the cells and for movement of material past the cell surface.

At the bases of cilia and flagella are structures called basal bodies (Figure 5.14), the microtubules of which are arranged to form nine triplets (Figures 5.13 and 5.15). The basal bodies are involved in the formation of cilia and flagella, and they are thought to provide a template on which the 9 + 2 arrangement of microtubules is assembled.

Basal bodies are frequently derived from *centrioles,* remarkably similar organelles found in cells of organisms that have a motile stage in their life cycles. The centriole microtubules are also present in nine sets of three. Centrioles are found in close association with the nucleus and can replicate by the formation of a new centriole perpendicular to the old. Centriole duplication is frequently coordinated with cell duplication, and prior to cell division the centrioles move to the regions to be occupied by the poles of the spindle. Centrioles were formerly thought to be necessary for spindle formation; this is not the case, however, since in cells of higher plants, which do not have centrioles, and in animal cells from which the centriole has been removed, spindle formation and cell division occur normally. The behavior of centrioles during cell division may simply reflect a mechanism to ensure the distribution of centrioles to all of the daughter cells. Lower plants also have centrioles, although usually in cells that are motile, such as sperm cells of ferns and zoospores of algae. The absence of centrioles from other cells of the organism suggests that they may arise de novo during formation of the motile cells.

Microfilaments are typically smaller than microtubules, ranging for the most part from 40 to 80 Å in diameter. They are also made up of protein subunits, this protein being very similar to actin, the protein found in the smaller filaments of muscle cells. Contraction of muscle (see Chapter 6) results from interactions between the actin filaments and the larger myosin filaments, ATP being required to provide the necessary energy.

Microfilaments in nonmuscle cells are involved in changes in cell shape during development and motility, and in protoplasmic streaming in plant cells. They generally occur in bundles, often being located just beneath the cell membrane. In some cells myosinlike proteins are also present, and the movement generated by microfilaments is thought to be related to actin-myosin interactions similar to those involved in muscle contraction. However, the nature of the interactions that must take place between the cytoplasm and the microfilaments for movement to occur is not clear.

Actin molecules also have been shown to be present in close association with the spindle in dividing cells, and they may interact with the spindle microtubules during chromosome movement. Actin-containing microfilaments are also associated with the cleavage furrow following mitosis and appear to be responsible for the changes in cell shape occurring at this time.

Cellular and intracellular movements which are dependent on microtubules and microfilaments represent work, and as such involve the expenditure of energy. ATP is required for most such movement, and the enzyme ATP-ase has been shown to be associated with most microtubular and microfilamentous systems examined. For example, the beating of cilia is dependent on the activity of ATP-ase molecules which connect the outer doublets of microtubules of each cilium. The doublets can then slide past one another; since the doublets are also connected to the central microtubules, resistance to such sliding is encountered, and this shear resistance is believed to be responsible for the bending of the cilium.

Bibliography

DeDuve, C. 1963. The lysosomes. *Scientific American* **208** (5), 64–72. A description of the functions of the so-called "suicide bags" of the cell.

Fawcett, D. W. 1966. *An Atlas of Fine Structure.* W. B. Saunders Co., Philadelphia and London. This collection of micrographs includes descriptions and beautiful illustrations of cell organelles.

Hepler, P. K., and B. A. Palevitz. 1974. Microtubules and microfilaments. *Annual Review of Plant Physiology* **25,** 309–362. An extensive review of the characteristics of those structures of the cell involved in motility.

Morre, D. J. 1975. Membrane biogenesis. *Annual Review of Plant Physiology* **26,** 441–481. An extensive review of how membrane structures are formed within the cell, including descriptions of functions and inter-relationships of membrane compartments.

Northcote, D. H. 1973. The Golgi Complex, in *Cell Biology in Medicine*, ed. E. E. Bittar, John Wiley and Sons, New York. An excellent brief review of the structure and functions of the Golgi complex.

Palade, G. 1975. Intracellular aspects of the process of protein synthesis. *Science* **189,** 347–358. The Nobel lecture describing the relationship between intracellular membranes and protein synthesis, transport, and secretion.

Comparative cytology

As we read in Chapter 1, living systems—cells or organisms—do not violate the second law of thermodynamics in reversing the trend toward randomness of both energy and matter. In contrast to most systems in the universe, life is an ordering process, achieving thereby a high degree of structure and behavior. The ability to achieve and maintain a high degree of order, or a low entropy status, resides in the fact that living systems are open-ended; they have the capacity to undergo a continuous exchange of energy with their immediate surroundings and with the rest of the universe. In addition, the capacity of a living system to do work, represented by the amount of its free energy at a constant temperature, will decrease with time unless energy other than heat is added from an outside source. Life has this ability to maintain its level of free energy by making use of radiant energy from the sun or chemical energy from molecules taken in and manipulated.

Even a casual acquaintance with the world of cells leaves one with the impression of an almost bewildering array of cell types. This is not a haphazard array, however, for order exists at all levels, but the more complex the form the higher the degree of order and the greater the expenditure of energy necessary to achieve and maintain that form. The fact that complexity is energy-requiring is more obvious if we examine human societies, but it is equally true of cells. As Chapter 2 indicates, this order has its origin in the coded information of DNA, with proteins giving the information expression in either structural situations or as behavior.

As cells perform different functions, and as they vary in shape, size, and complexity, we should expect these features to be reflected in their intimate structure and the manner in which energy relates to these structures. Common features of cells should be reflected in their commonly shared attributes, while unique features are to be explained by some peculiarity of structure or chemical pathway. Here we shall concentrate primarily on the structural differences between cells, sometimes only their exterior appearances, and consider only briefly their varied biochemistry.

Cell shape and size

The attainment of a given form, whether of organism or a cell, presents an intriguing problem to which there are very few answers. Among unicellular species, form is the principal means of classification, although as in bacteria, it may be reinforced by the grouping of organisms according to their staining properties (gram negative or gram positive), their by-products of metabolism, or, if pathogenic, the symptoms they induce in the invaded host. The protozoa and unicellular algae are especially rich in their variety of shapes (Figure 6.1). We might ordinarily assume that a single cell, existing in a liquid environment, would possess a somewhat spherical form, because this would be the simplest shape for an equitable distribution of surface tensions, much as a soap bubble is rounded when free-floating. Many cells of this sort are spherical, for example, eggs of many organisms, some algae, and many bacteria. Others, however, are not, and many of these possess distinctive shapes of extraordinary intricacy of markings. In particular, the diatoms and dinoflagellates among the algae have hard, silicious coverings that are fixed in shape; the algal desmids lack the hard wall but possess shapes of great variety and much beauty; the amoeba changes its shape continuously; and many protozoa possess a thick pellicle, variously smooth, ciliated, or flagellated. It would be fascinating, for example, to know how the code of DNA determines the meticulous precision of a diatom such as that in Figure 6.2.

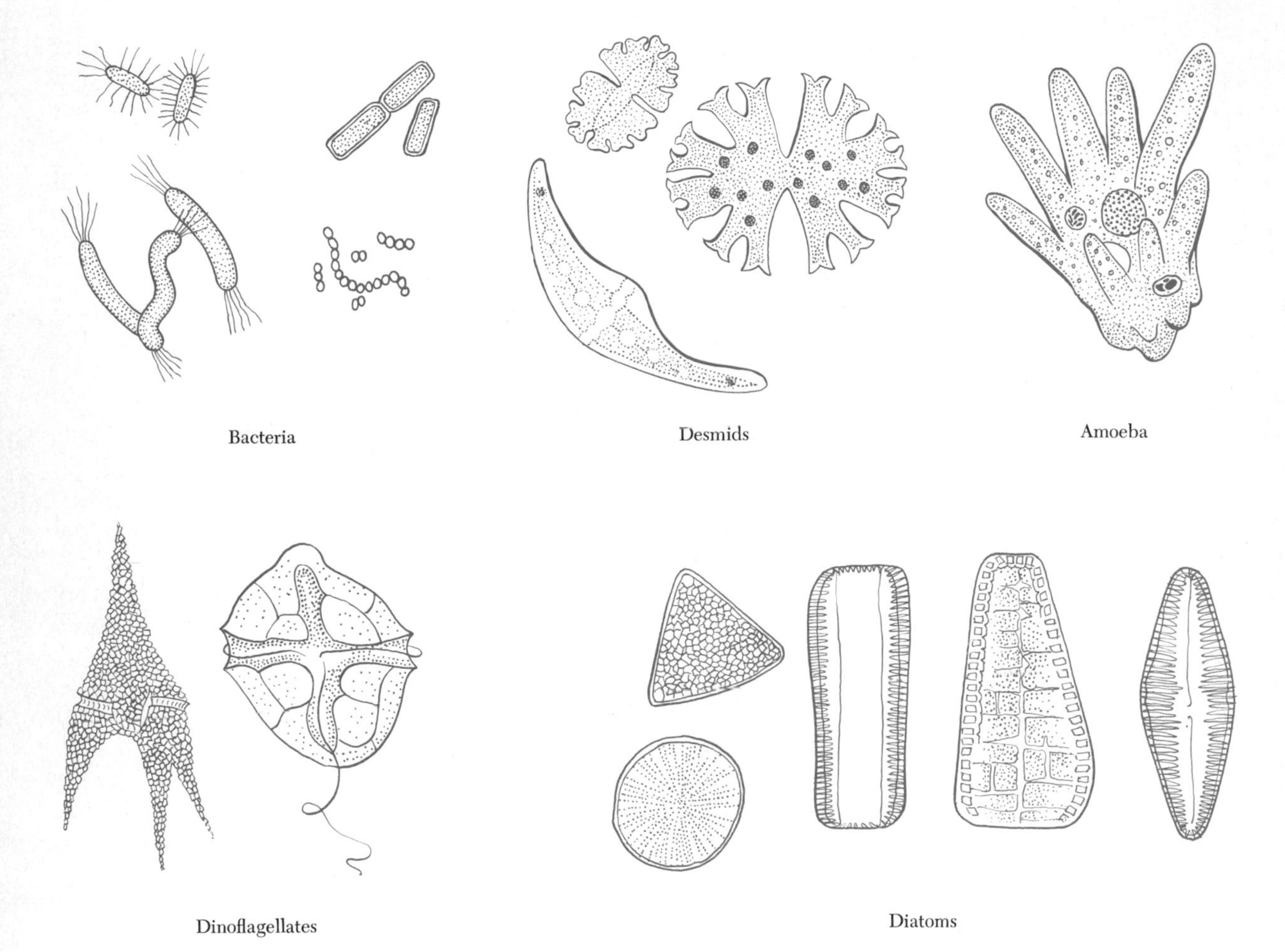

Figure 6.1
Examples of cell shape among unicellular organisms. The bacterial cells are prokaryotic, the others eukaryotic. The shape of each cell is determined by the genetic information contained within its DNA; those of the dinoflagellates and diatoms (see Figure 6.2) result from the shapes of their external shells, which are composed of hemicellulose impregnated with silica. Some of the dinoflagellates are toxic and are the cause of the "red tides" in oceanic waters when they appear in great numbers.

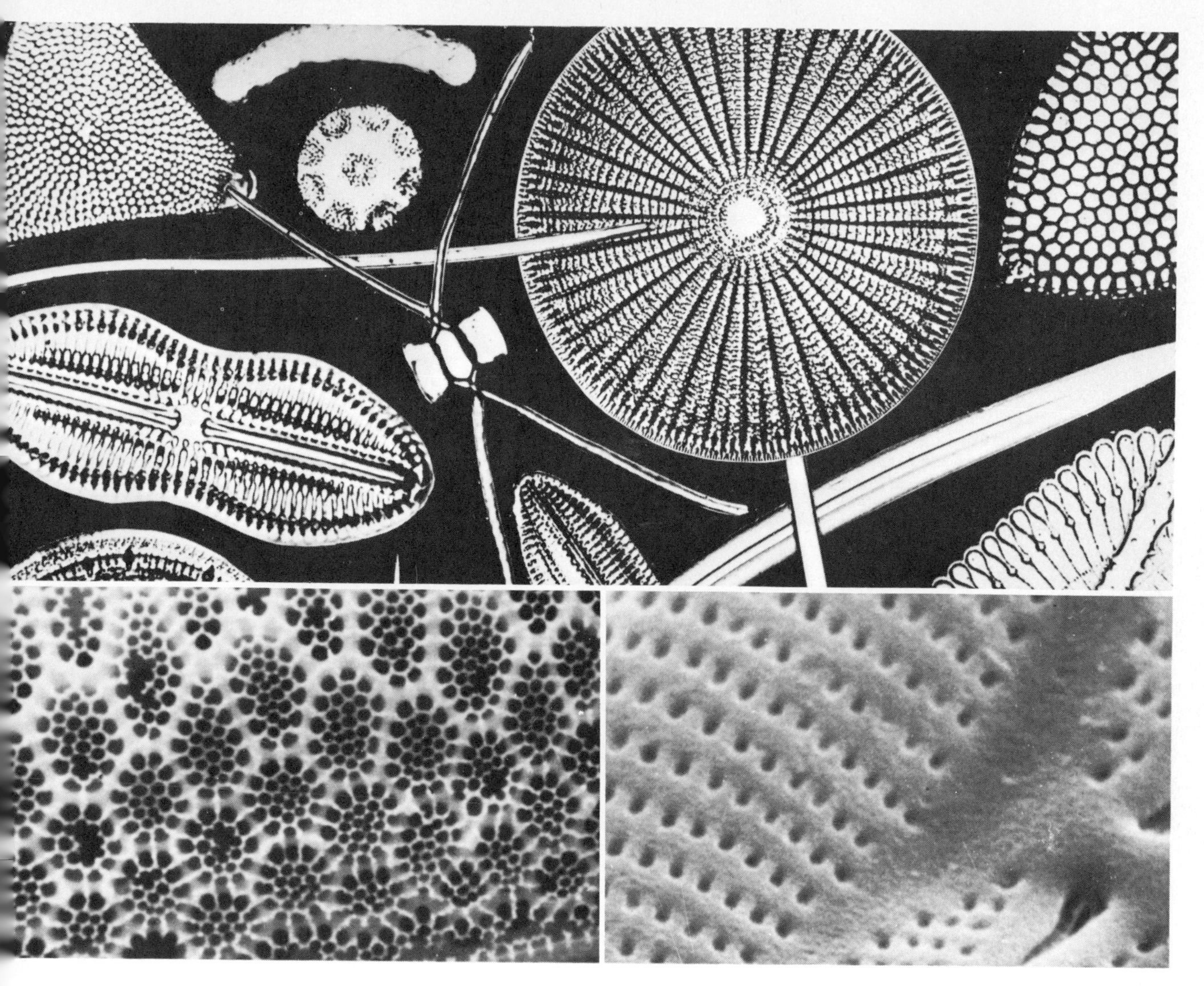

Figure 6.2
Above, a collection of diatoms from diatomaceous earth, as seen in the light microscope. Below, scanning electron micrographs, at high magnification, showing a portion of the walls of two species. The patterns of sculpturing are so precise and so symmetrical in their details that lens grinders use them to test their lense for imperfections. Diatomaceous earth consists of the shells of dead diatoms that come to rest on the bottom of the ocean, and over millions and millions of years of accumulation, vast deposits are formed; one of the largest is at Lompoc, California, and occupies an area of 12 square miles in extent, reaching in some places to a depth of 3,000 feet. Since the diameter of the largest of the diatoms is no more than a few micrometers, one can imagine how many must die to form such vast deposits. (Courtesy of the Johns-Manville Co.)

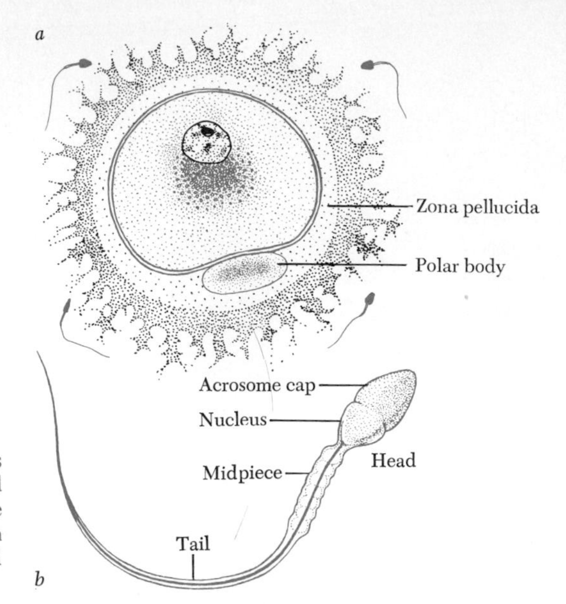

Figure 6.3
The human egg and sperm, with the egg many thousands of times larger. The egg has already gone through one meiotic division, and a single polar body has formed outside of the egg proper and in the zona pellucida. The acrosome of the sperm is formed largely from the Golgi complex; the midpiece develops from mitochondria and provides the necessary energy for the movement of the tail.

Among the cells of multicellular forms, a variety of shapes are present in the same organism. Here we must assume that shape is related to function and position within the body since all cells would have a comparable amount and kind of DNA. In the human, the egg and sperm provide the most sharply contrasting forms: the egg spherical, the sperm elongate, flagellated, and distinctively organized (Figure 6.3). The shape of each is appropriate to its function and activity.

Figure 6.4 illustrates a number of other cells in the human body. Those of the outer skin tend to be longer and wider than they are deep; they form several distinctive layers over the surface of the body, with the cells becoming more and more compacted and flattened as they approach the exterior where they will eventually die and be sloughed off. Cells lining the arteries and veins are somewhat similar in shape to epidermal cells in being flattened, and they are able to maintain the longitudinal shape of these organs at the same time that they can be stretched or contracted as the blood flow varies in volume or pressure. The cells lining the digestive tract differ among themselves; the absorptive cells of the intestine are columnar, the absorptive surface covered with tiny projections, or microvilli, but other cells of the stomach, for example, vary quite markedly from each other (Figure 6.5). Nerve and muscle cells are strikingly elongate in appearance, melanocytes can vary their shape and assume many forms, and cells of the blood are rounded or

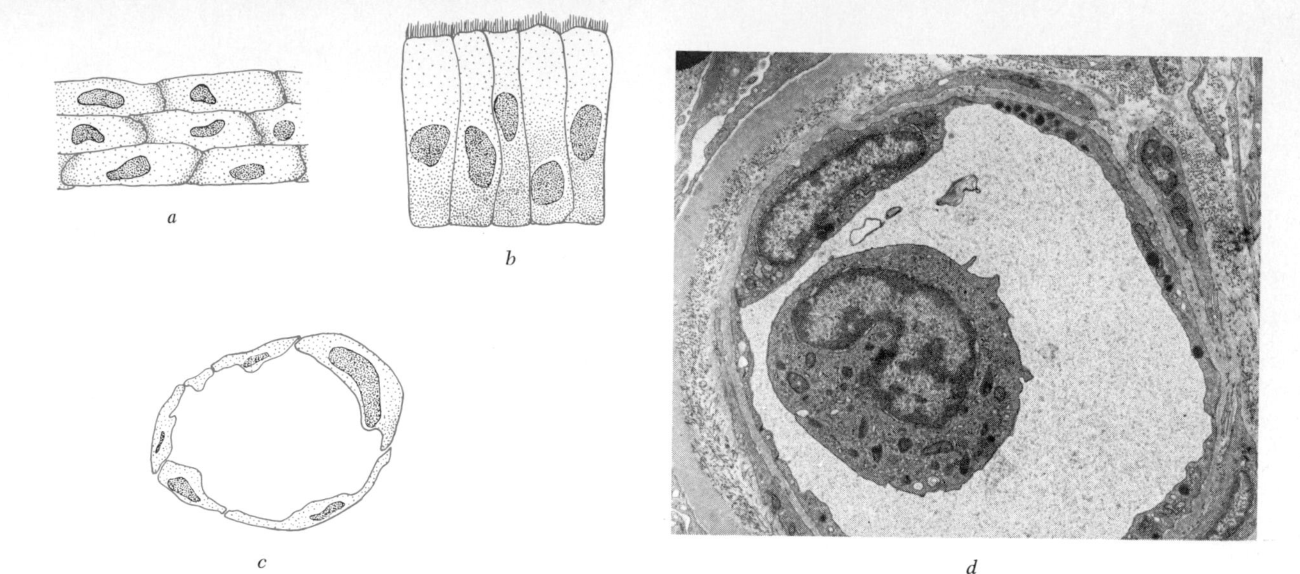

Figure 6.4
Cells of different shapes from the human body. (*a*) Cells of the skin are longer and wider than they are deep. (*b*) Cells of the digestive tract, which aid in the absorption of food from the intestine, are columnar and far deeper than they are wide; their inner surface is lined with microvilli for increased absorptive capacity (see also Figure 3.11). (*c, d*) Cells lining an artery are flat, thin, and capable of expansion and contraction as the heartbeats cause the pulse and the arteries to expand and contract. The large cell in the electron micrograph is one of the leucocytes of the bloodstream.

amoeboid (Figure 6.6). It is obvious that the transmission of nerve impulses or the contraction of muscles is more easily accomplished by elongated cells than either function would be by cells having the shape of a red blood cell. The red blood cell, however, needs the largest possible surface-to-volume ratio, and this is achieved maximally by its biconcave disc-like shape.

The cells of a higher plant, for example, those of a flowering species, also vary according to their position and function. Where the body of the plant is elongate, as in the stem or the vein of a leaf, so too are the cells, although conducting cells are more so than the protective cells of the outer portions (Figure 6.7). Those of a leaf possess other shapes, consistent with their function, while dividing and storage tissues, as in an apical growing region of a stem or in the pith of an elderberry stem, are somewhat isodiametric but faceted by the pressure of adjoining cells (Figure 6.8).

Just as cell shape varies both within and between organisms, so too does cell size. The human egg (diameter of 0.1 mm), for

Figure 6.5
A *chief cell* from the stomach lining of the bat. This cell differs appreciably in shape from the columnar cells of the digestive tract, and it also differs in function, being a source of pepsin, one of the digestive enzymes. The bottom of the cell is packed with membranes of the endoplasmic reticulum, while the spherical masses are packets of pepsinogen. These packets will be discharged through the top of the cell (not shown) and into the stomach, and the pepsinogen will be converted into pepsin. (Courtesy of Dr. Keith Porter.)

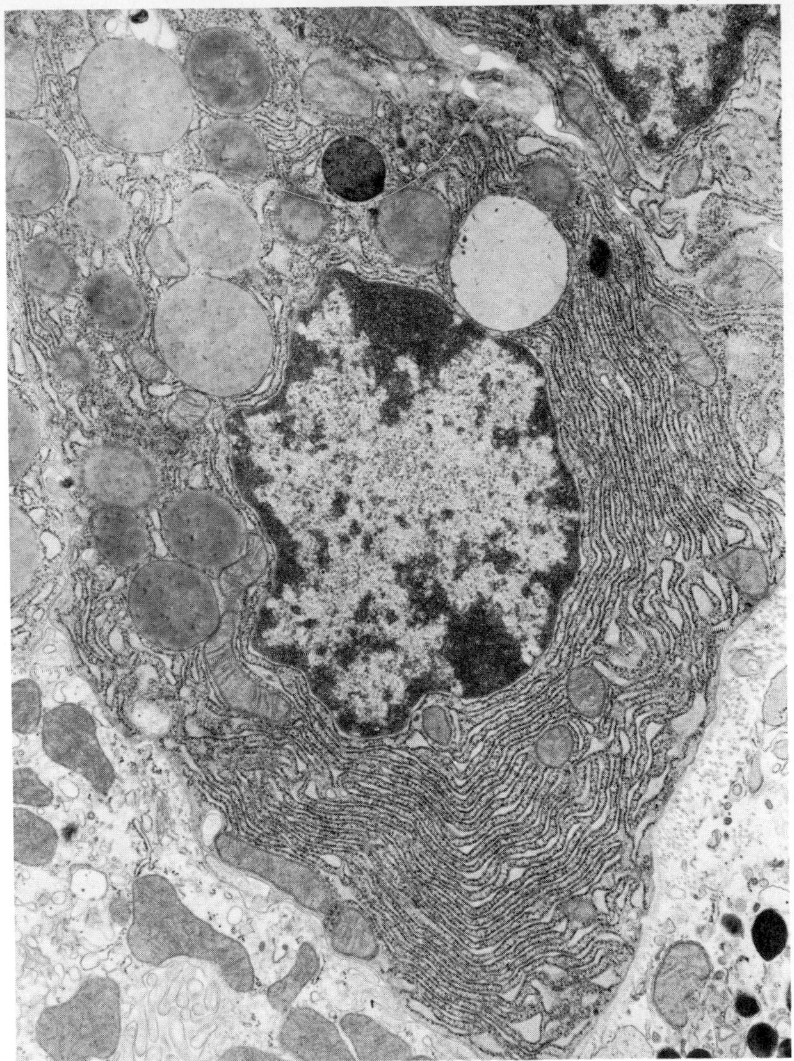

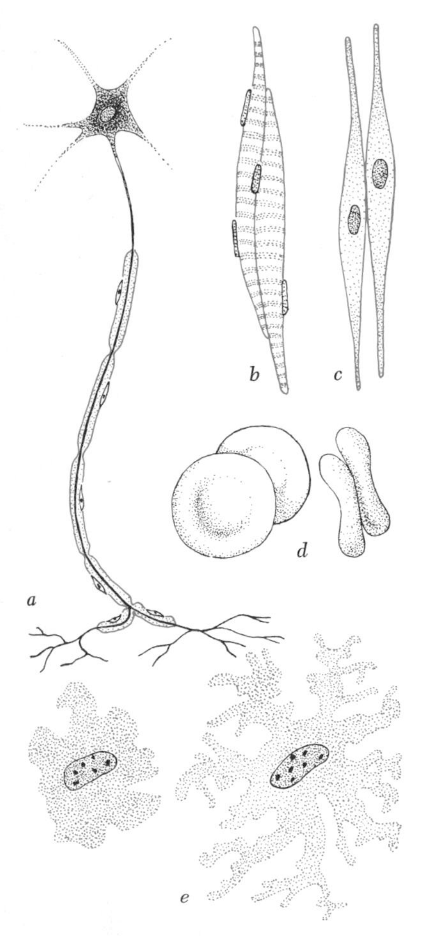

Figure 6.6
Animal cells of different shapes and functions. (*a*) Nerve cell consisting of three major portions: the cell body containing the nucleus, the axon (long branch) with a good portion of its length covered by Schwann cells (see Figures 3.14, 3.15), and the dendrites (the finer, uncovered terminal branches), which make contact with other dendrites or with the muscles they enervate. (*b*) Several cells from striated muscle. (*c*) Smooth muscle cells. (*d*) Human red blood cells, face-on and edge-on views. (*e*) Pigment-containing melanoxytes in contracted and expanded phases.

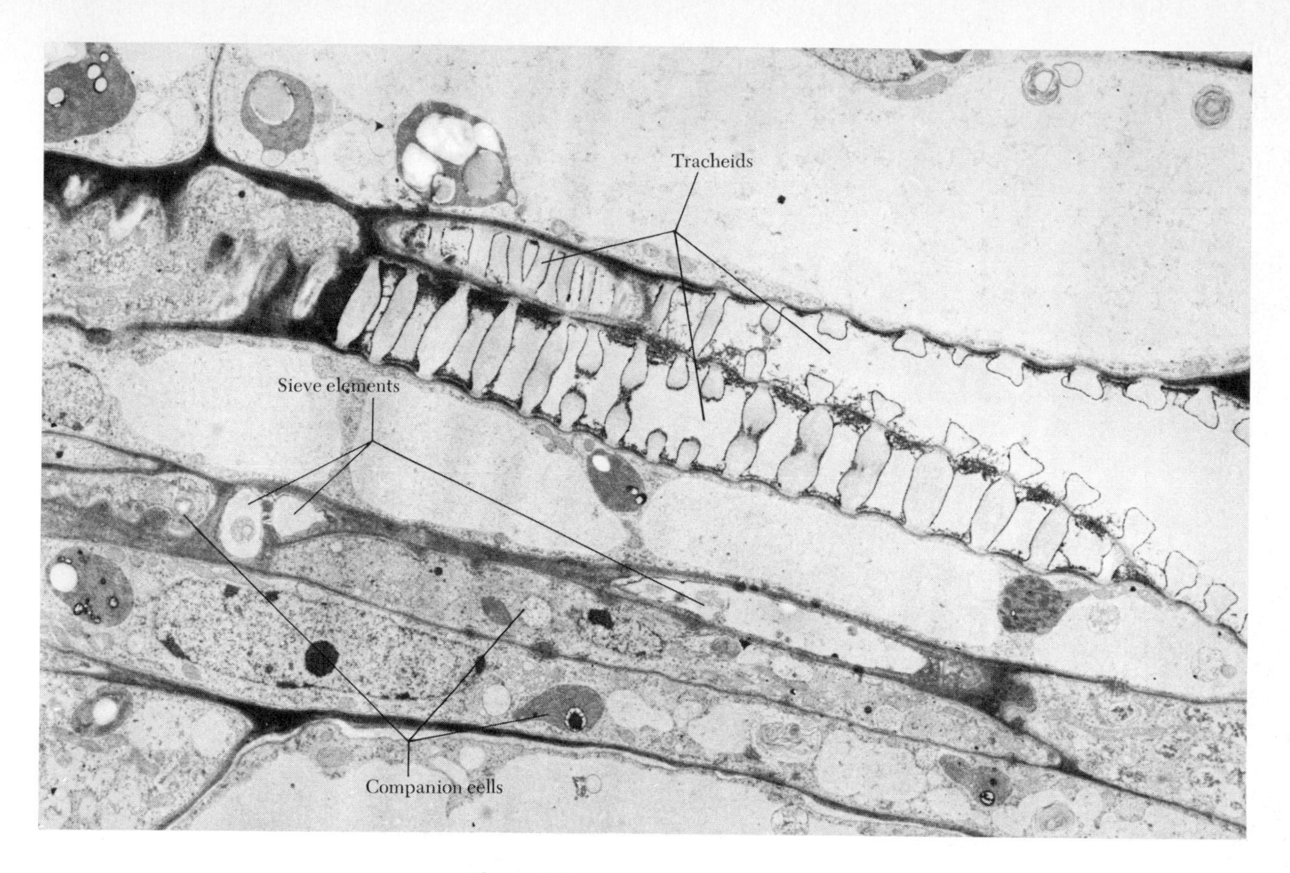

Figure 6.7
Electron micrograph of a longitudinal section through the vascular bundle of the leaf of *Ficus elastica*, a member of the fig family. The tracheids with their ladderlike wall thickenings, the sieve elements largely devoid of internal contents, and the companion cells are identifiable. The large, elongated central cell is of an uncertain nature: it may be a cambial element, an immature xylary element, or an immature cell of the phloem. The top and bottom cells are from the photosynthetic mesophyll. (Courtesy of Dr. J. D. Duckett.)

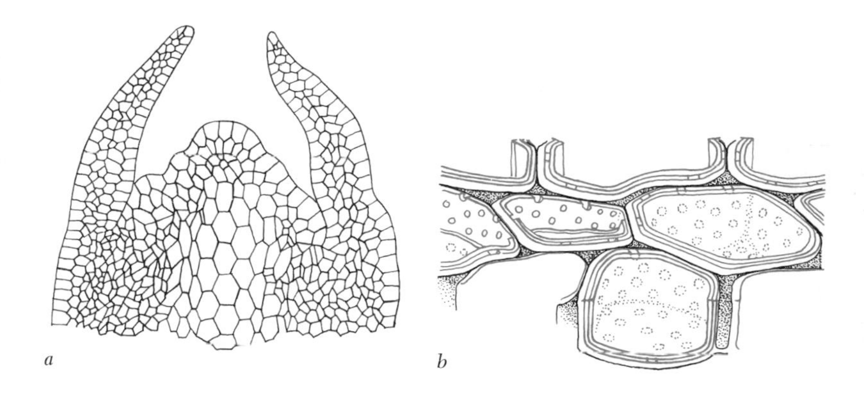

Figure 6.8
Isodiametric cells found in the growing tips (apical meristems) of plants (*a*) and in the pith (storage tissue) of an elderberry stem (*b*). The walls of the pith cells are dotted with small interconnecting pits. The angular aspect of these cells is caused by the pressure of adjoining cells, much as soap bubbles would have faceted figures if confined in an enclosed space.

example, has a volume over a million times that of a human sperm, and some nerve cells may be much greater in volume than the egg. An ostrich egg differs in volume from a pneumococcus bacterium by a factor of approximately 10^{16} or 10^{17} (Figure 6.9). Yet all these structures are intact and functional cells. This would suggest that there is no relation between the amount of DNA per cell and its volume, but the difficulty of determining DNA-volume ratios in cells that store food material (as the yolk of an egg) makes any such calculation of dubious value, although it is probably generally true that the more DNA there is in a cell the larger its volume.

In the human body, nerve cells are the largest, at least in longitudinal dimensions; their neuronal extensions may be a yard or more in length. Cells from striated muscle are also large, but those of the skin, liver, kidney, and intestine, for example, probably average 30 μm in diameter. The *small leucocyte* (white cell) of the blood is at the other end of the size scale, having a diameter of 3 to 4 μm.

The smallest cells visible under the light microscope are the bacteria; the dimensions of some are at the lower limits of visibility (between 0.2 and 0.3 μm). They are not the smallest cells, however. This smallest category includes a group of organisms called pleuro-pneumonialike organisms, belonging to the genus *Mycoplasma* (see Figure 1.7, Chapter 1). They are somewhat like viruses in size, but their organization and behavior are bacterial:

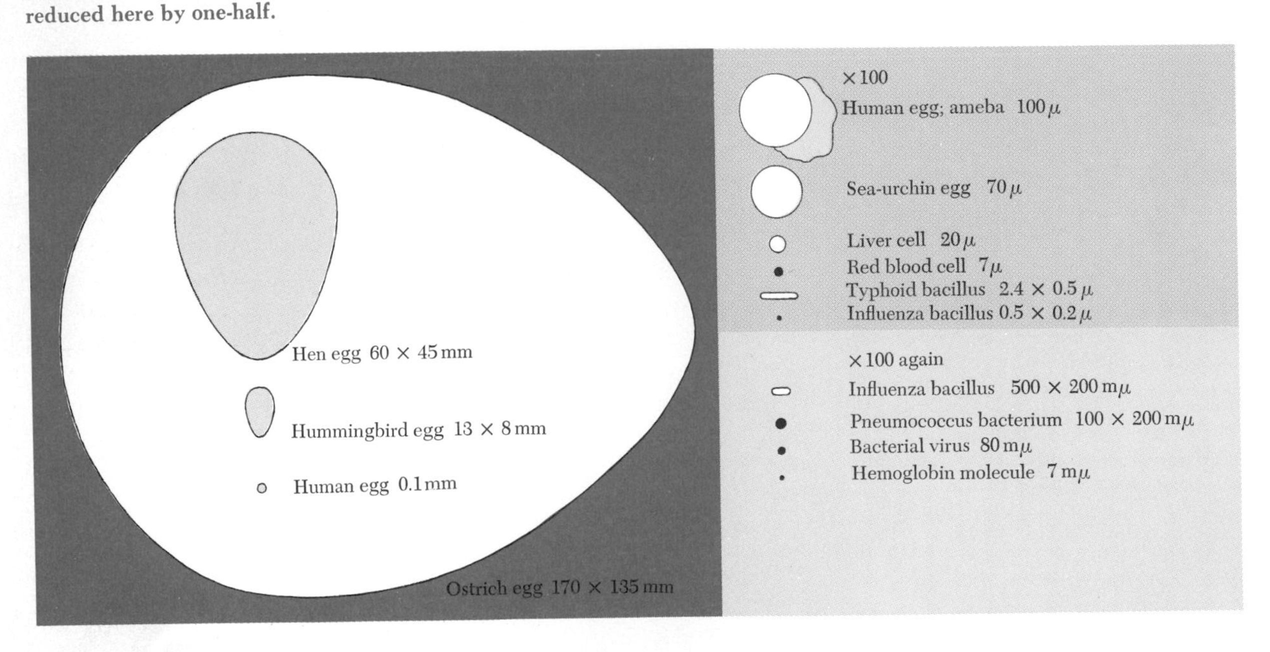

Figure 6.9
A scale of sizes of different kinds of cells, with the bacterial virus and the hemoglobin molecule included for comparative purposes. The ostrich egg and the avian eggs within in it are reduced here by one-half.

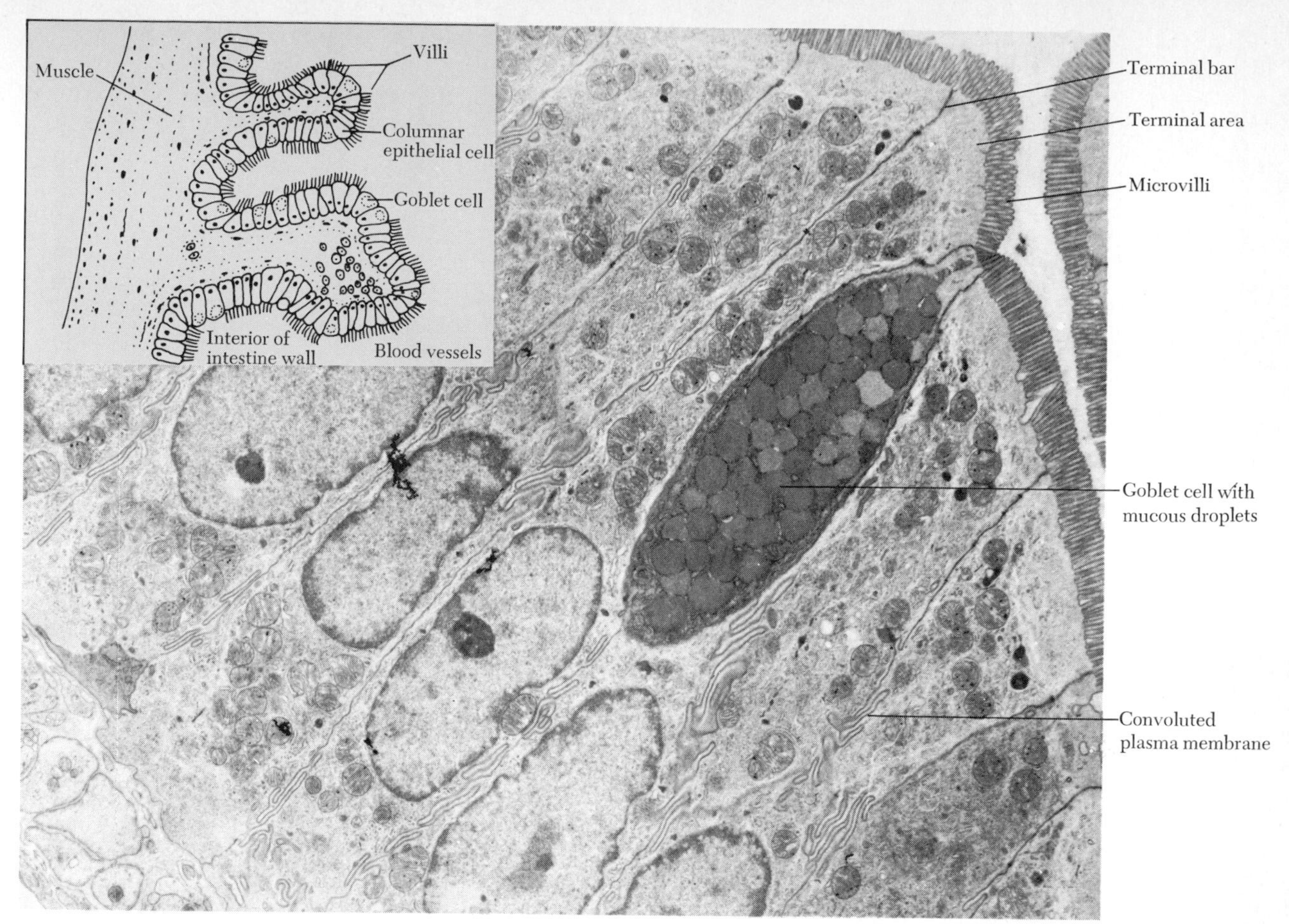

Figure 6.10
An electron micrograph of columnar epithelial cells from the intestine of the bat. The microvilli at the top right-hand side greatly increase the absorptive surface of each cell. Just below the microvilli of each cell is an area of dense microfibrils, the *terminal web*, which gives support to the cell; where the terminal webs of adjacent cells meet, a *terminal bar* is formed. The plasma membranes of the cells are variously convoluted along their sides. The cell packed with dark material is a *goblet cell* filled with mucous, which eventually will be emptied into the lumen of the intestine. The inset indicates the relation of these cells to the wall of the intestine and to the blood vessels. (Electron micrograph, courtesy of Dr. Keith Porter.)

outer cell membrane, ribosomes, hereditary material lying free in the nuclear area, and a complete metabolic system of enzymes (about 40 are known), all of which permits them to be grown in a test tube independent of other cells.

The size of a cell determines the amount of surface it can present to the surrounding environment. Metabolic reactions can occur throughout the cytoplasmic mass, and, all other things being equal, the amount of cell surface governs the amount of materials that passes into or out of the cell. If the cell is too large, the interior may be starved for nutrients or be otherwise inefficient, but limitations of this sort can be overcome in a variety of ways in addition to controlling the rate. The columnar epithelial cells of the gut (Figure 6.10, see also Figure 3.9, Chapter 3), for example, enormously increase their absorptive area through the plasma membrane by the formation of microvilli on the face of the cell toward the lumen of the intestine and convolutions where the cell abuts

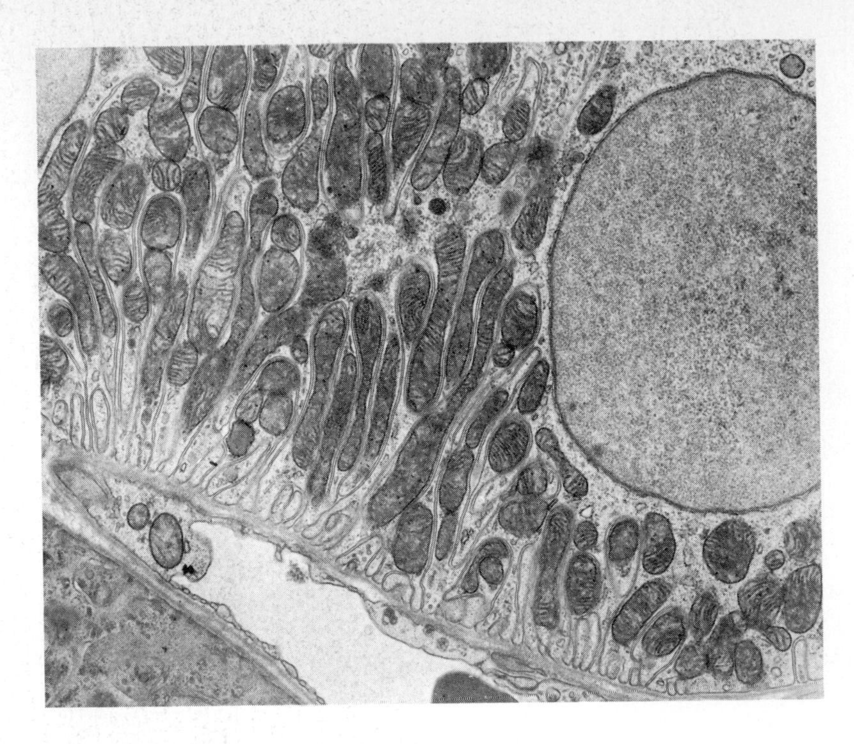

Figure 6.11
An electron micrograph of the basal side (that opposite the lumen of the kidney tubule) of a bat epithelial cell, which lines the tubule at its distal end. The function of this cell is to transport water, and it is believed that the many convolutions of the plasma membrane projecting into the cytoplasm, and the numerous, closely associated mitochondria assist in this process, the convolutions providing an increased surface area and the mitochondria providing the energy needed for active transport. The nucleus with its double membrane is at the right, and the light space at the bottom is the lumen of a blood vessel. The flow of transported water would be from the blood vessel, through the cell, and out the other end and into the lumen of the tubule. (Courtesy of Dr. Keith Porter.)

other cells. Kidney tubule cells, governing the passage of liquids or solutes in or out of the secreted urine, have the basal cell membrane folded in a highly complex manner (Figure 6.11). Plant cells solve this problem in still another way: through the formation of large vacuoles the cytoplasm is forced to the exterior where a ready exchange of gases, liquids, or nutrients can take place (Figure 6.12).

Cell size also can be assessed by metabolic control and rate, features that are governed, at least in part, by the flow of materials across the cell and nuclear membranes. As we know, the nucleus is the controlling center of the cell. Like the cell membrane, the nuclear membrane is also a barrier to the free passage of materials; although it can be folded or lobed to increase the surface without alteration of nuclear volume, the nucleus itself can probably exercise control over only a limited amount of cytoplasm. A high rate of metabolism, for example, requires not only a constant supply of energy-yielding foodstuffs from outside of the cell, but also a continuing supply of ribosomes, tRNA, mRNA, and enzymes that are nucleus-derived or nucleus-determined. The more rapid the rate of metabolism, therefore, the smaller the cell is likely to be, although the correlation is no more exact than is that of overall cell size and amount of DNA. In the animal kingdom, hummingbirds, shrews, flies, and bees metabolize at far higher rates than such organisms as humans, elephants, amphibia, and grasshoppers. They also have smaller cells, which are necessary so that a proper nucleocytoplasmic ratio is maintained. In fact, if a human metabolized at the same rate as a hummingbird, the heat resulting from the chemical reactions would not be able either to be dispersed quickly enough or to escape the cells or the body, and the individual would literally roast him- or herself to death.

Optimum cell size is, therefore, determined by a variety of factors, any one of which can set upper limits to size. Lower limits

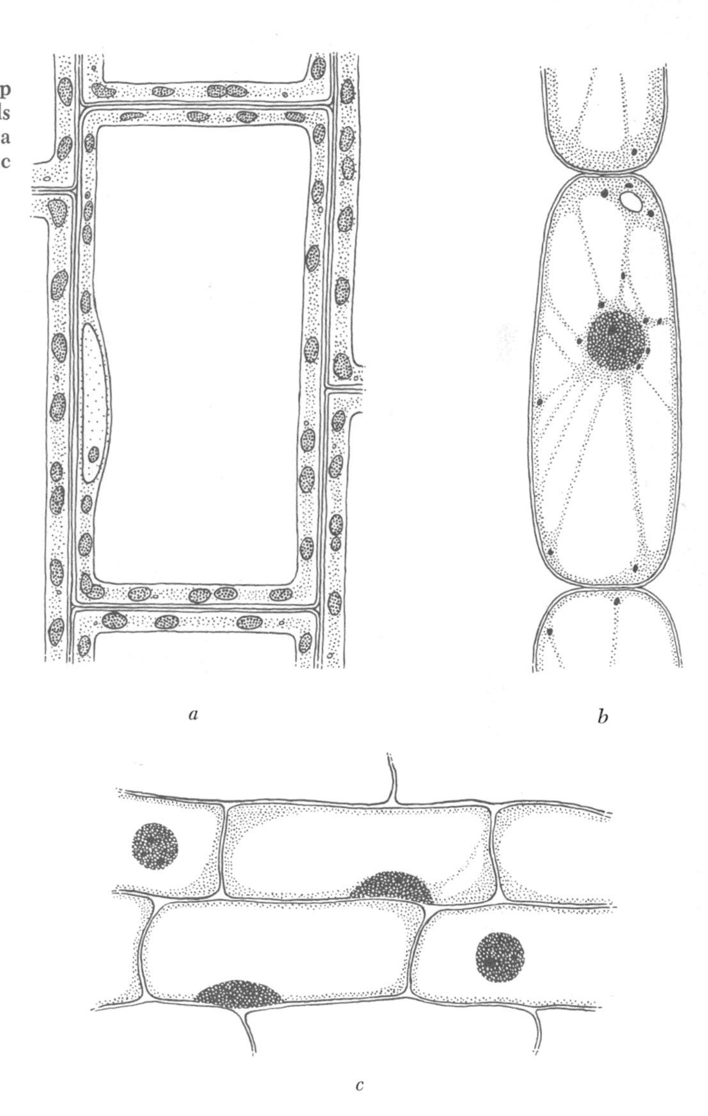

Figure 6.12
Plant cells that reveal the existence of large sap vacuoles. (*a*) onion skin cells; (*b*) stamen hair cells of the spiderwort, *Tradescantia virginiana;* (*c*) a diagrammatic representation of a photosynthetic cell in a leaf.

must exceed a 200-Å diameter, otherwise there would be no space between the cell membranes for the cytoplasm. But the cell is an amazingly flexible unit, and the problems of size in relation to function and space have been met in a great variety of ways.

Comparative cytology

If we assume, as we have been, that the various structural features of the cell are related to the functions these features perform in the general economy of the cell, we would expect each kind of cell to reflect its performance in its particular structure. In considering a variety of cells, we have been dealing with comparative cytology in an oblique manner; here we shall confront the subject directly by comparing selected cells within a given organism—a vertebrate—to show how the special functions relate to the presence and behavior of one or more structural features.

Before dealing with comparisons, however, we need to remember the central fact that *all cells are instruments of energy transformation* (see Figure 1.5, Chapter 1), and that structural differences are closely correlated with the ways energy is obtained and the use to which energy is put.

Let us visualize the cell as a factory, a not unreasonable comparison. We know that this factory may have many sizes and shapes, and that it may perform either general or specialized functions; that is, it may carry out all of the necessary tasks of being alive, as would be true for a unicellular bacterium or alga, or it may exhibit a division of labor and, as part of a multicellular organism, carry out only a very particular task. Its internal structure must, however, be consistent with the tasks demanded of it. The energy problems are basically similar in most cells.

Figure 6.13 is a schematic representation of simple factory operations in terms of energy input and output. The fuel is generally coal, gas, or oil, its energy locked up in the chemical bonds of these fossil fuels. During combustion, O_2 is utilized and the energy is released as heat. The heat, in turn, makes steam to drive a generator, thereby converting heat energy into mechanical energy, which in a generator is converted into electrical energy. The waste products are CO_2, H_2O, and unused heat, which is difficult to recover and put to work. Electricity, in the form of electrons flowing through wires, performs work of three basic kinds: *mechanical work*, or lifting; *transport*, or moving materials from one place to another; and *assembly*, or the manufacture of a product. It also may be converted into radiant energy.

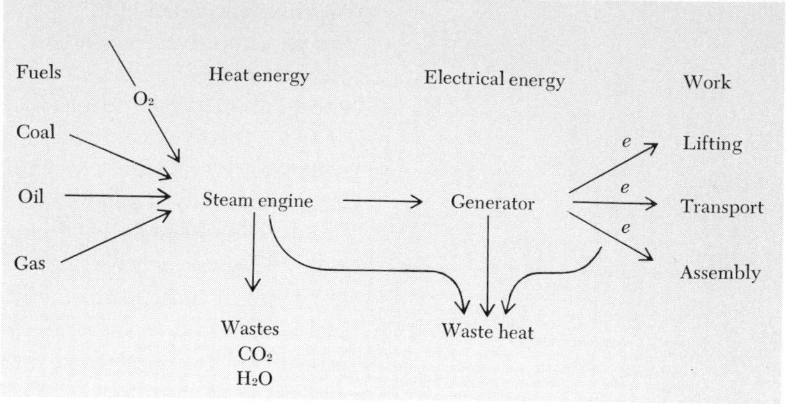

Fuels
O_2
Heat energy
Electrical energy
Work
Coal
Oil
Gas
Steam engine
Generator
e
e
e
Lifting
Transport
Assembly
Wastes
CO_2
H_2O
Waste heat

Figure 6.13
Schematic representation of the flow of energy in a simple factory, from its bound form as fuel to its final utilization to do work. The *e* represents electrical energy in the form of electrons flowing along a wire. Compare with Figure 6.14.

The cell is basically similar in operation, with one major exception: it cannot use heat to do work since it has no mechanism for doing so. The primary source of energy is the sun; chloroplasts capture light energy and transform it into chemical energy in the form of carbohydrates. All other energy resides in the chemical bonds of molecules, and cells conduct their energy transactions by changing molecules through the medium of enzymes and at a relatively constant temperature.

Figure 6.14 depicts the basic steps in the energy cycle of the cell. The fuels are mainly carbohydrates, but fats and proteins also may be used when carbohydrates are not available. These are broken down into smaller units by the degradative enzymes in lysosomes and, in the presence of O_2, by the respiratory enzymes

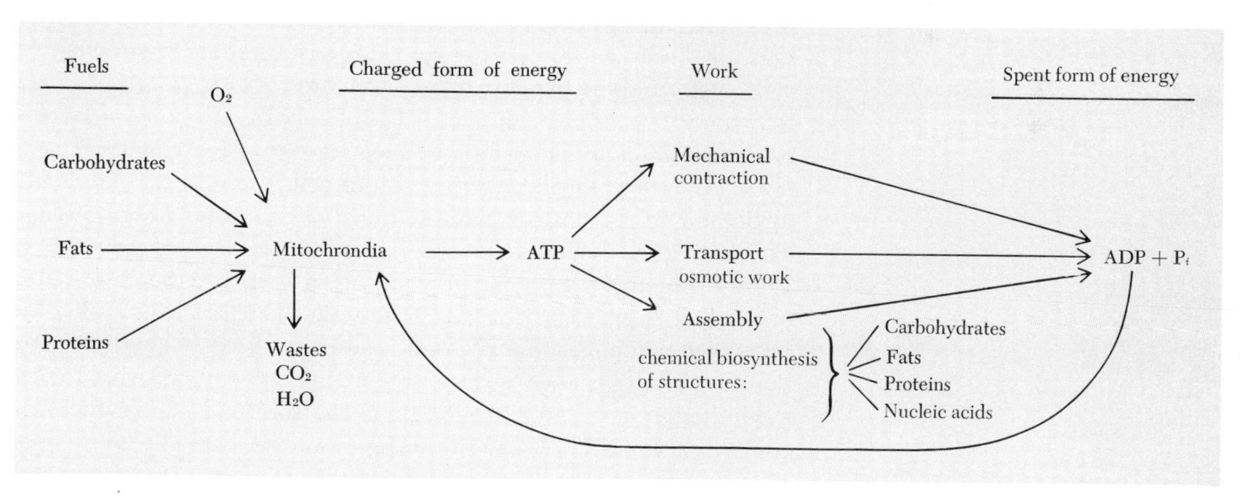

Figure 6.14
Schematic representation of the flow of energy in a eukaryotic cell, from the combustible fuels to its ultimate utilization for specific kinds of work. ATP and ADP + P_i are, respectively, the charged and spent forms of energy.

in the membranes and matrix of mitochondria. As in our factory, the principal wastes are CO_2 and H_2O, but in the cell the chemical energy of the fuels is largely repackaged as chemical energy with good (about 50 percent) efficiency. Part of the energy is dissipated as heat; this is waste energy in a sense, but it is also what keeps a warm-blooded animal warm. As a *charged form of energy* in the form of readily available, easily tappable, high energy phosphate bonds, the molecule ATP can yield its energy to other parts of the cell, and work can be performed. The tasks again are comparable in nature to those performed by our hypothetical factory: mechanical work, as when muscles contract; transport, as when substances are moved from one area to another, often across membranes and against concentration gradients; and assembly, or biosynthesis, of other molecules and cellular structures. Even illumination (bioluminescence) may result, as in the case of fireflies, when a substance, luciferin, in the presence of O_2, is acted upon by an enzyme, luciferase. Energy is used in all of these cellular endeavors, and in the process ATP is degraded into ADP (adenosine diphosphate) and inorganic phosphate (P_i). ADP and P_i are *spent forms of energy*; to be recharged, they pass back into the mitochondria where they emerge once again as ATP. The cycle must go on constantly in the living cell, and it has been estimated that a molecule of ATP in an active cell is recycled every 50 seconds.

Cells for movement The cells we will consider now are those capable of performing specifically the several kinds of work that we have depicted in Figure 6.14. Mechanical work, or contraction, can be either sudden and violent or more gentle and continuous: the throwing of a baseball as opposed to rhythmic motion of breathing; the wink of an eyelid as opposed to the slow contractions of the stomach and intestines during digestion. Two kinds of cells perform these two kinds of contraction.

Striated, or *striped,* muscle cells are capable of sudden coordinated contraction; like a muscle itself, the cells are elongated and tapered, are multinucleate as a result of cell fusion, and are difficult to delimit in an electron microscopical preparation (Figure 6.6). The individual cells may be from 1 to 40 mm in length and from 10 to 40 μm in width; their several nuclei, found at the outer edge of a cell, are apparently needed to govern this large mass of highly organized cytoplasm in a synthetic sense and keep it supplied with those molecules needed by a muscle cell. Each cell is covered with a thin, tough membrane, the *sarcolemma.* Internally, the cytoplasm consists primarily of longitudinal filaments, called myofibrils, arranged systematically; they appear under an electron microscope

Figure 6.15
Electron micrographs of portions of striated muscle. Top: a sarcomere is a region between two Z bands (the dark vertical lines), and lines of sarcomeres are separated from each other by closely packed masses of mitochondria. Bottom: a higher magnification of a region between two Z lines, with the larger, darker lines of myosin and the thinner, lighter lines of actin.

as alternating light and dark areas (Figure 6.15). Different bands are clearly distinguishable, and one block of myofibrils between two Z bands is known as a *sarcomere*. As Figure 6.15 shows, the appearance of the fibrils depends upon the axis of the preparation; in cross section, the regular arrangement of elements is evident (Figure 6.16).

During contraction the sarcomeres become shorter and broader, thereby shortening the cell. The united action of many cells,

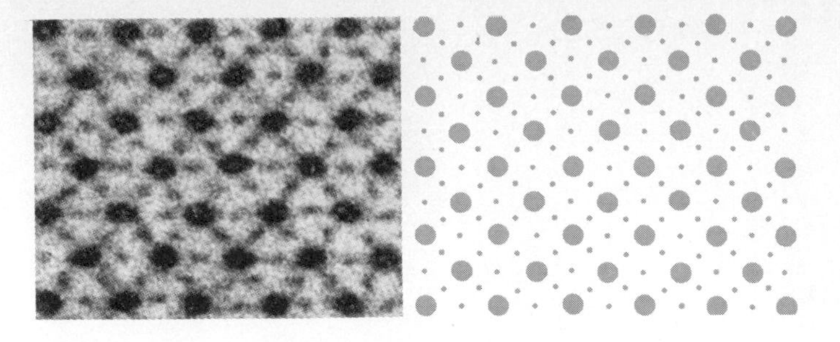

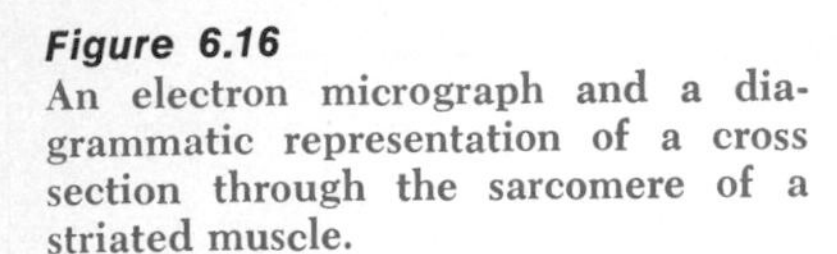

Figure 6.16
An electron micrograph and a diagrammatic representation of a cross section through the sarcomere of a striated muscle.

coordinated by nerve impulses, causes the whole muscle to contract, and thus perform work. The myofibrils are the agents of contraction. When viewed in both longitudinal and cross section, the fibrils are alternately thick and thin. Both types are protein, the former being myosin (about 100 Å in diameter), the latter, actin (about 30 Å in diameter). The fibrils are believed to oscillate back and forth rather constantly along the long axis of the cell, but that movement can be altered by the formation of chemical cross links between the two kinds of fibrils. When cross links form between myosin and actin, they slide toward each other and thus shorten both the cells and the muscle as a whole (Figure 6.17). Only the dual protein molecules, called actinomyosin, is capable of this sliding action, and ATP is also needed to convert actin from a globular to a filamentous form before it can interact with myosin. The number of cross links at any given time is related to the amount of additional ATP present. Thus, when the amount of ATP is high,

Figure 6.17
Diagram of how a muscle is thought to contract. With the release of energy from ATP, cross links are formed between actin (thin lines) and myosin (thick lines), and the actin molecules move toward each other, drawing the myosin units closer together.

Relaxed muscle

ATP

Contracted muscle

ADP + P_i

and the energy available is within the ATP molecule, the cross links are few in number; when ATP is split into ADP and P_i by an ATP-ase and energy is released, the amount of linkage is high and contraction takes place. The following relationship, greatly simplified, thereby holds:

1. Energy spent
2. Cross links formed between actin and myosin
3. Movement of actin units toward each other

$$\text{ATP} \rightleftharpoons \text{ADP} + \text{P}_i$$

1. Energy charged
2. Cross links broken between actin and myosin
3. Relaxation and separation of action units

The actual details of how these actions are accomplished are generally known, although the process is very complex and involves many individual chemical steps, but it appears that most types of cellular movement—beating of cilia and flagella, movements of microvilli and amoeba—occur in a comparable manner, that is, by contractile proteins acting in a coordinated way. The ability of muscles to contract quickly and frequently, and the numerous large mitochondria (Figure 6.15), would suggest a plentiful supply of ATP, ready at hand when needed. Based on information from frog muscles, this turns out not to be so; the amount of ATP immediately available is capable of giving only about eight twitches. If the ATP falls to low levels, the muscles go into rigor as happens when death sets in, but ATP is rapidly regenerated and its levels are kept adequate by drawing on another energy reserve, phosphocreatine (PC), under the control of an enzyme, creatine phosphotransferase (CPT):

$$\text{ADP} + \text{PC} \underset{\text{CPT}}{\rightleftharpoons} \text{ATP} + \text{C}$$

The phosphocreatine is sufficient in amount so that if used to exhaustion, it could produce about 100 contractions of a muscle, but a carbohydrate in muscles called glycogen can be drawn upon for prolonged muscular activity.

Smooth muscles, in contrast, are found in the walls of arteries, veins, uterus, and digestive tract (Figure 6.4); and they contract involuntarily and rhythmically rather than by voluntarily induced nerve impulses. Individual cells are smaller than those of striated muscle, are uninucleate, tapered at both ends, and lack the tough sarcolemma of a striated cell. The striation so characteristic of the other muscle cells is also missing. Both actin and myosin are present in the form of long, thin filaments, but they are not grouped into conspicuous structures. Whether contraction occurs in a manner comparable to that described for a striated cell is debatable, but since contraction is slow, gentle, and continuous, the energy requirements are less demanding. Mitochondria are consequently less conspicuous in size and number. Heart muscle is striated but behaves much like smooth muscle in its rhythmic contractions, which occur without the necessity of voluntarily induced nervous impulses.

If we now view muscle cells in terms of work performed, we can appraise their structure more meaningfully. The mitochondria are far more numerous in striated than in smooth cells, but in each type of cell their number and size are related to the cellular demands for energy. Once the muscle cells attain their differentiated character, the primary activity in a biochemical sense would be maintenance of the energy-manipulating system, that is, ATP, ADP, P_i, phosphocreatine, glycogen, and the enzymes involved in their interconversions. The mitochondria obviously would be essential, and they are large and numerous, but the endoplasmic reticulum, the Golgi apparatus, and the ribosomes are less conspicuous cytoplasmic elements, although ribosomes must be present to carry out the necessary protein synthesis. The myofibrils loom large as structural and functional elements in striated muscle, and as discussed they are peculiarly suited to perform the mechanical work of contraction.

Transporting cells Every cell has to engage in some sort of transport activity if it is to communicate with its environment, and the manner and kind of transport is suggested, at least to a degree, by the structure of the individual cells. In the vertebrate organism, there are, in addition, two systems whose cells are concerned principally with transport problems: (1) the cells of the lining of the digestive tract, which absorb digested and solubilized materials from the lumen of the intestine, move these materials across the cell, and then pass them on to the bloodstream for use elsewhere; and (2) the cells of the kidney, which extract wastes and fluid from the bloodstream and excrete them as urine. These two systems would seem to differ in that the first is basically an absorptive system and the other secretory, but since the cells lie between the bloodstream and the lumens of either the intestine or a kidney tubule, they must do both, that is, absorb on one side and excrete on the other. The cells of both systems are, therefore, comparable, at least in their grosser aspects, and we need consider only those of the kidney since the columnar epithelial absorbing cell of the intestine, with its numerous microvilli, has already been discussed (Figure 6.10).

Transport of substances across a living membrane is not generally a passive effort; it requires energy. We would expect, therefore, to find numerous mitochondria in cells engaged in active and constant transport; and indeed they are large in both number and size (Figure 6.11). Since the rate of movement of materials in and out of the cells is, in part, a function of the amount of cell surface, we would also expect to find various modifications of the plasma membrane. The microvilli on one side of the cell, and the deep indentations of the membrane at the opposite side, provide such an increase in surface area. In fact, it has been calculated that in one part of a kidney tubule, that of the proximal portion just below the renal capsule, there are about 6,500 microvilli per cell, increasing thereby the surface area approximately 40 times. They are similar in appearance to those of columnar epithelial cells (Figure 3.9 and 6.10).

The microvilli form a *brush border*, which faces the lumen of the tubule. Their function is to aid in the transformation of the initial kidney filtrate into urine by the reabsorption of water and food materials, such as dissolved sugars and salts, as the urinary fluid passes down the tubule. Passing through the cell and back into the bloodstream, the water and sugars leave the cell at its opposite end through the infolded membranes at the base where capillaries abound. Water and food are thus reclaimed in a highly selected manner, but wastes pass on. In addition, however, some of the cells lining the tubules secrete additional waste substances into the lumen. The molecular traffic in kidney cells, therefore, is an active, two-way flow. The structure of the cell is consistent with these activities, and since the degree of absorption and secretion varies along the tubule, the number of microvilli and the frequency of the deep folds in the basal membranes vary from cell to cell. Since the active transport of materials across these membranes requires a good deal of energy, the number and size of the mitochondria are also consistent with this need and can vary from cell to cell.

The kind of cell just described is not restricted to the lining of the intestine or the kidney but is characteristic of those in any tubular lining of an absorbing or secreting organ. Similar cells are found in such diverse organs as salivary glands, and secrete a variety of substances, or the salt-secreting gland of the herring gull. The Golgi apparatus, the endoplasmic reticulum, and ribosomes are present, but except for some cells that may secrete proteins, they are not conspicuous elements occupying large portions of the cell volume. In the kidney, for example, the materials passing through the cells are neither synthesized in any significant way nor repackaged in any form; the cellular contents reflect this.

The transport systems of higher plants—for example, the flowering plants, or angiosperms—are strikingly different from those in the vertebrate animal. The absence of digestive and waste disposal systems, and the rigid cellulose cell walls, mean that communication with the environment and internal transport of materials must be met by solutions other than those typical of animals. There are no microvilli and no highly folded plasma membranes to aid in the passage of materials, although cells may be irregularly contoured to increase their surface area. However, living plant cells do maintain active contact with each other (Figure 6.18). The cellulose wall is rather porous, at least to the passage of water, but in addition the cytoplasm of adjacent cells is continuous by means of *plasmadesmata,* strands of cytoplasm that connect one cell with another through tiny holes or pits in the cell wall. It is not necessary, therefore, for organic materials to cross plasma membranes to move from one cell to another, but active transport, with its energy demands, operates in plants as well as in animals.

The intake of materials into a plant from the environment consists of water and mineral salts from the soil and CO_2 from the atmosphere. The initial absorption in the soil is by root hairs, but the major task of transport of water and salts to all parts of the plant, in some instances to heights of several hundred feet, is by means of the xylem, whose conducting cells are dead. It is still uncertain how this is done. The tracheids and vessels of the xylem are so connected to each other by pits that the passage of water is readily accomplished (see Chapter 3), but to take water to great heights requires a great deal of force. One atmosphere of pressure can support a column of water 34 feet high, but in a tree the water must move rapidly and in a system where friction seems likely to be present. It has been calculated that it would take 30 atmospheres of pressure to raise water to the height of some tall trees, but such pressures of this magnitude seem not to be available. One can cite root pressure, transpiration pull, capillary action in narrow tubes, and the cohesive force of water itself as contributing to the movement of water, but the answer still eludes us. Transport, however, is a mechanical action, not seemingly dependent upon the action of living cells and not impeded or promoted by living membranes. In this sense, xylary transport is analogous to the arterial-venous transport of animals in being a hydraulic system, and it bears no resemblance to the active transport systems of intestines or kidneys.

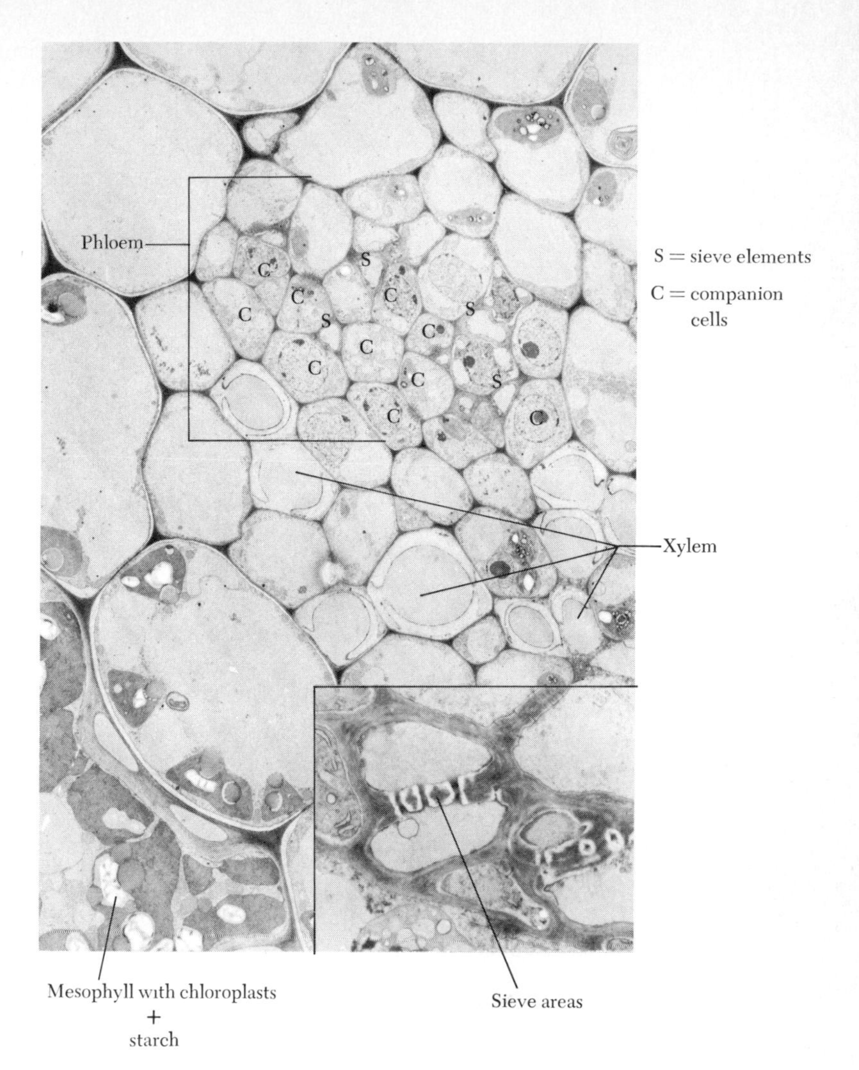

Figure 6.18
Electron micrograph of a cross section of a vascular bundle in the leaf of *Ficus elastica.* The xylem elements are identifiable by their heavy cell walls, while the sieve elements of the phloem are distinguishable from the companion cells by their general lack of cell contents. The latter possess large nuclei. The cells of the mesophyll show starch grains in their chloroplasts. The inset at the lower right shows a pair of sieve elements with pit areas. (Courtesy of Dr. J. D. Duckett.)

Organic materials, however, are moved from one cell to another by living processes. The phloem is the tissue in which this occurs, and the principal cells in it are sieve tube elements (Figures 6.7 and 6.18). The membranes of these cells do not seem to be modified in any obvious way, but the carbohydrates, elaborated during photosynthesis and primarily in the form of sucrose, can be transported from one area to another at a rate far faster than that possible through simple diffusion. The direction of flow also can be reversed in the phloem, suggesting that the physiological demands of the plant can play an active role in governing transport. The most probable way is through a flow under pressure. Sucrose is likely to be in highest concentration in the leaves as the result of photosynthesis; water would tend to flow in from the xylary cells of lesser solute concentration, thereby forcing the solute (sucrose) into a neighboring cell. As water moves in at one end because of high osmotic conditions, it must move out at the other end because the cell, with its heavy wall, cannot expand. The result is a flow downward from the leaves. Since animals cannot manufacture their own foods in a process analogous to photosynthesis, they exhibit no comparable process of transport. In fact, the cells of the kidney, withdrawing water from urine, do so against a gradient and not in response to a concentration gradient.

Assembly cells A great many cells are specialized for the biosynthesis of certain substances; pancreatic cells form and secrete digestive enzymes and insulin; outer skin cells, keratin; plasma cells, antibodies; and erythroblasts, hemoglobin. These substances are all proteins, and from what we know of protein synthesis, we would assume that substantial amounts of the several kinds of cytoplasmic RNA must be present to carry out this task. Figure 6.19 is an electromicrograph of a *plasma cell,* and the richness of the rough ER is evident. But contrast the cell in this figure, which is a secretory cell, with that in Figure 6.20, which depicts a

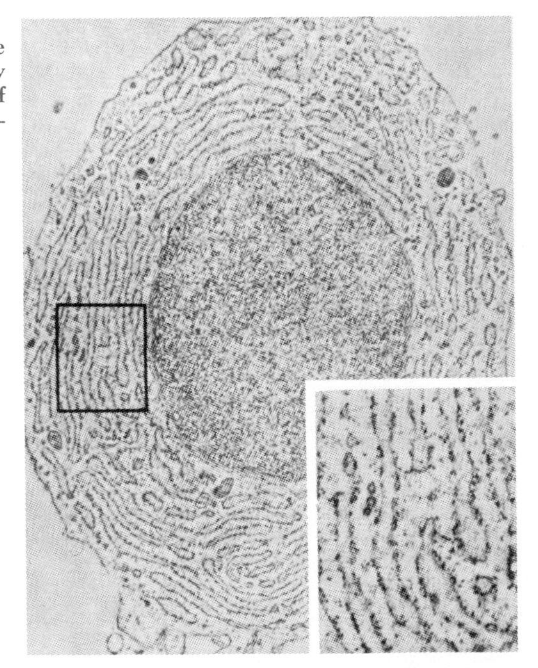

Figure 6.19
Electron micrograph of a plasma cell that produces antibodies; the antibodies are exported by the cell and enter the general circulatory system of the body. The inset is a higher magnification (× 23,000) of the ER. (Courtesy of A. Ham and J. Leeson, *Histology,* 4th ed. Philadelphia: J. B. Lippincott Co., 1961.)

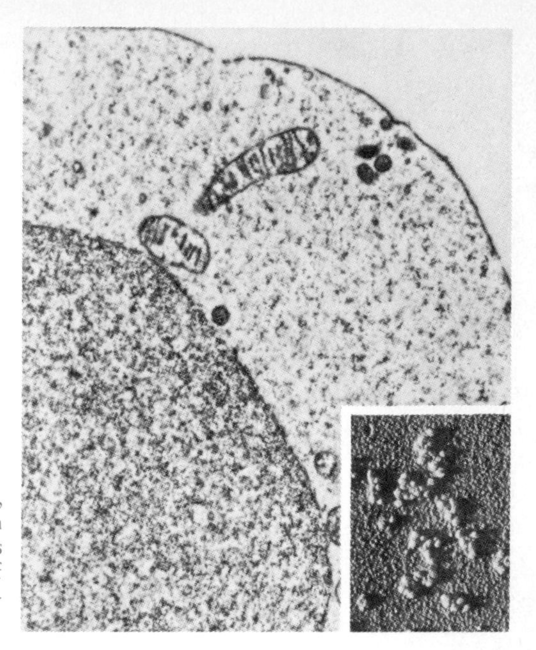

Figure 6.20
Electron micrographs of a proerythroblast, or immature red blood cell, and of the polyribosomes (inset, lower right) on which the protein portions of the hemoglobin molecule are manufactured. The nucleus of the proerythroblast is at the lower left of the image. (Courtesy of A. Ham and J. Leeson, *Histology,* 4th ed. Philadelphia: J. B. Lippincott Co., 1961.)

proerythroblast, an immature red blood cell. It, too, synthesizes a protein, hemoglobin, but its cytoplasm is strikingly free of the membranes of the ER, although rich in ribosomal particles containing RNA. The erythroblast also possesses a meager amount of Golgi membranes, whereas the plasma and pancreatic cells show a normal Golgi apparatus in addition to an abundant rough ER (Figure 6.21).

The contrasting differences here are probably related to the fate of the synthesized material. In the erythroblast, the hemoglobin remains within the cell; no further transformation or transportation of it is necessary. The plasma cells, however, and those depicted in Figure 5.2, Chapter 5, prepare their protein products for delivery outside the cell; these products are apparently transformed and packaged into an inactive form before being moved out of the cell for action elsewhere. From what we now know, it seems likely that the repackaging is a function of the membranous rough ER and the Golgi apparatus for ultimate discharge to the exterior of the cell (Figure 6.22). Once again, therefore, we find that the functions performed by the cell and its internal architecture are consistent, and this consistency permits us to interpret function in terms of structure, and vice versa.

Figures 5.4, Chapter 5, and 6.23 provide a further contrast. Figure 5.4 is a cell from the testis of an oposssum, the principal

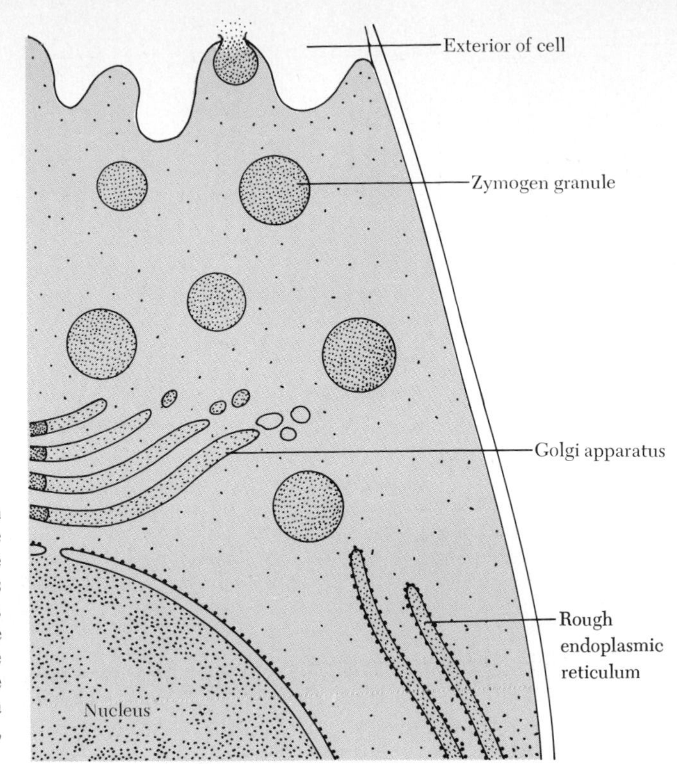

Figure 6.21
Diagram of a portion of a pancreatic cell, which assembles and packages zymogen granules. These are enzymes in an inactive form, and hence inert while within the cell. Once in the intestine, the enzymes are transformed into active forms to aid in digestion. The enzymes, after formation on ribosomes by the process of translation, pass into the cisternae of the rough ER, and then to the Golgi apparatus. They are then ready to move through the cell to the plasma membrane where they are discharged to the exterior, as indicated in Figure 6.22.

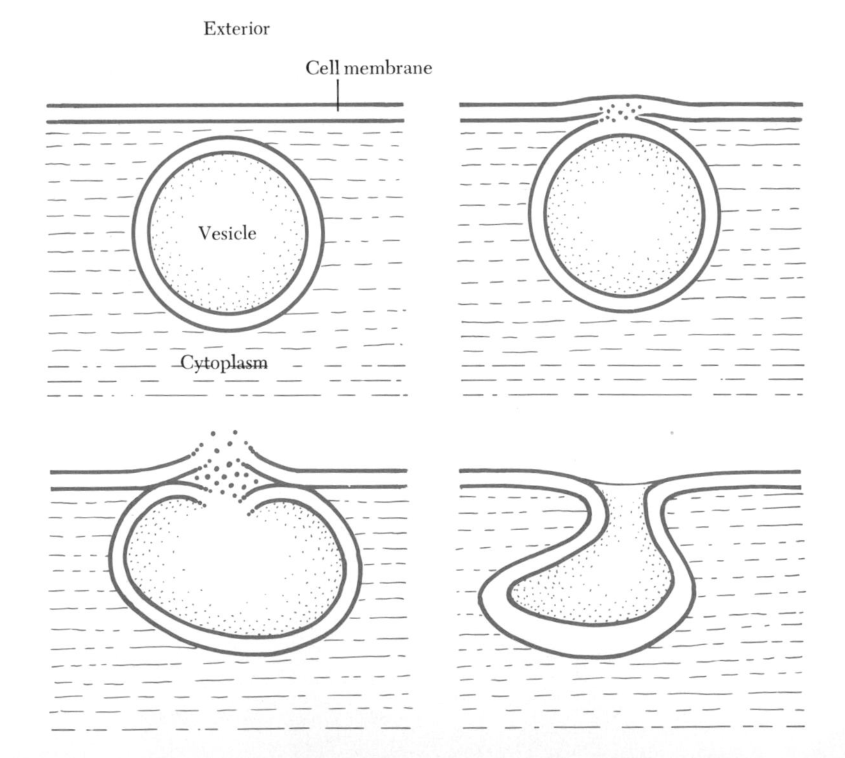

Figure 6.22
A diagrammatic representation of how a membrane-bound zymogen granule (see Figure 6.21) or vesicle containing dissolved substances can be discharged to the exterior of the cell without serious disruption of the plasma membrane. This is the process whereby mucus- and enzyme-forming cells of the digestive tract, or other kinds of secreting cells, discharge their contents so that the contents may be of use to other portions of the organisms.

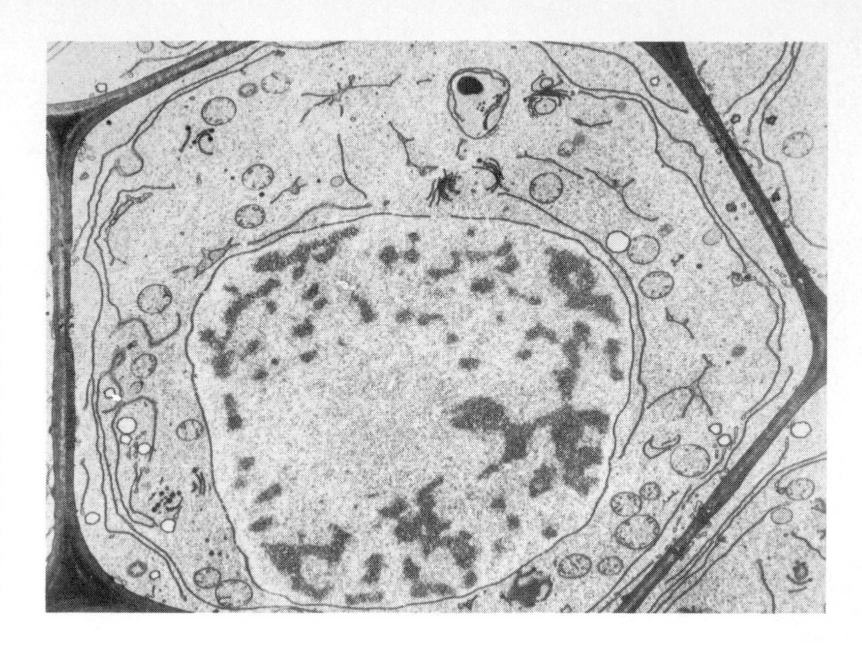

Figure 6.23
Electron micrograph of a cell in the actively dividing root-tip maize. Several clumps of Golgi membranes can be seen above the nucleus, with a plastid near them; the mitochondria are few and scattered; the ER consists of long, slender membranes with few, if any, attached ribosomes; pores are evident in the nuclear membrane, and the chromosomes show no distinctive structure. At this magnification, the free ribosomes in the cytoplasm are not clearly defined but are present in substantial amounts. The primary wall and middle lamella also can be seen, with the latter most evident at the angles as the darker material. (Courtesy of Dr. G. Whaley.)

function of which is the production of a steroid hormone, a compound similar in solubility to a fatty acid but differing from the lipids in chemical structure. There is little morphological evidence of either bound or free RNA, and none would be expected in any great amount since protein synthesis is not the principal cellular activity. The smooth ER, however, is believed to contain the enzymes for steroid synthesis, and it is consequently prominently developed. This is in contrast to cells of the pancreatic gland, which also produces a hormone (see Figure 5.3, Chapter 5). But insulin is a protein, and its prominent endoplasmic reticulum is rich with ribosomes. Figure 6.23, on the other hand, is an unspecialized cell from a rapidly dividing tissue of a plant root. Such a cell has all the organelles—the centrosome is missing since this is a plant cell—but no one element is emphasized at the expense of any other. However, since cells of this type are rapidly growing and dividing, and consequently manufacturing more nuclear and cytoplasmic materials, an abundance of ribosomal RNA in the cytoplasm is characteristic.

Bibliography

BLOOM, W., and D. W. FAWCETT. 1968. *A Textbook of Histology* (9th ed.). W. B. Saunders Co., Philadelphia. A well-known and widely used textbook, designed primarily for medical students, with a wealth of illustrative material on the many kinds of human cells as viewed with both light and electron microscopes.

BOURNE, G. H., and J. F. DANIELLI, eds. 1952–1975. *The International Review of Cytology*, Vols. **1–41.** Academic Press, New York. Many of the special articles in this review series deal with aspects of comparative cytology. The articles vary in excellence, and some are technically difficult and detailed, but their range spans the entire diversity of cells from prokaryotes to the most elaborate eukaryotic cell.

BRACHET, J., and A. E. MIRSKY. 1959–1964. *The Cell,* Vols. **I–VI.** Academic Press, New York. Each volume contains a number of articles by specialists, with many aspects of comparative cytology considered, making this an excellent reference source. Somewhat out of date, but still useful.

FAWCETT, D .W. 1966. *An Atlas of Fine Structure. The Cell: Its Organelles and Inclusions.* W. B. Saunders Co., Philadelphia. A truly outstanding collection of electron micrographs of a wide variety of cells, each accompanied by supporting descriptive material.

HAGGIS, G. H. 1970. *The Electron Microscope in Molecular Biology.* Longman, London. The emphasis is on aspects of molecular biology, and the manner by which electron microscopy complements the chemical approach. The illustrations of cells and viruses are excellent, and they are accompanied by additional line drawings, which aid in interpreting the electron micrographs. Some attention is given to instrumentation and the preparatory procedures employed in handling these cells.

LEDBETTER, M. C., and K. R. PORTER. 1970. *Introduction to the Fine Structure of Plant Cells.* Springer-Verlag, New York. The cells dealt with are those of the higher vascular plants, the angiosperms. The illustrations are uniformly superior, and the text is for purposes of interpreting fine structure.

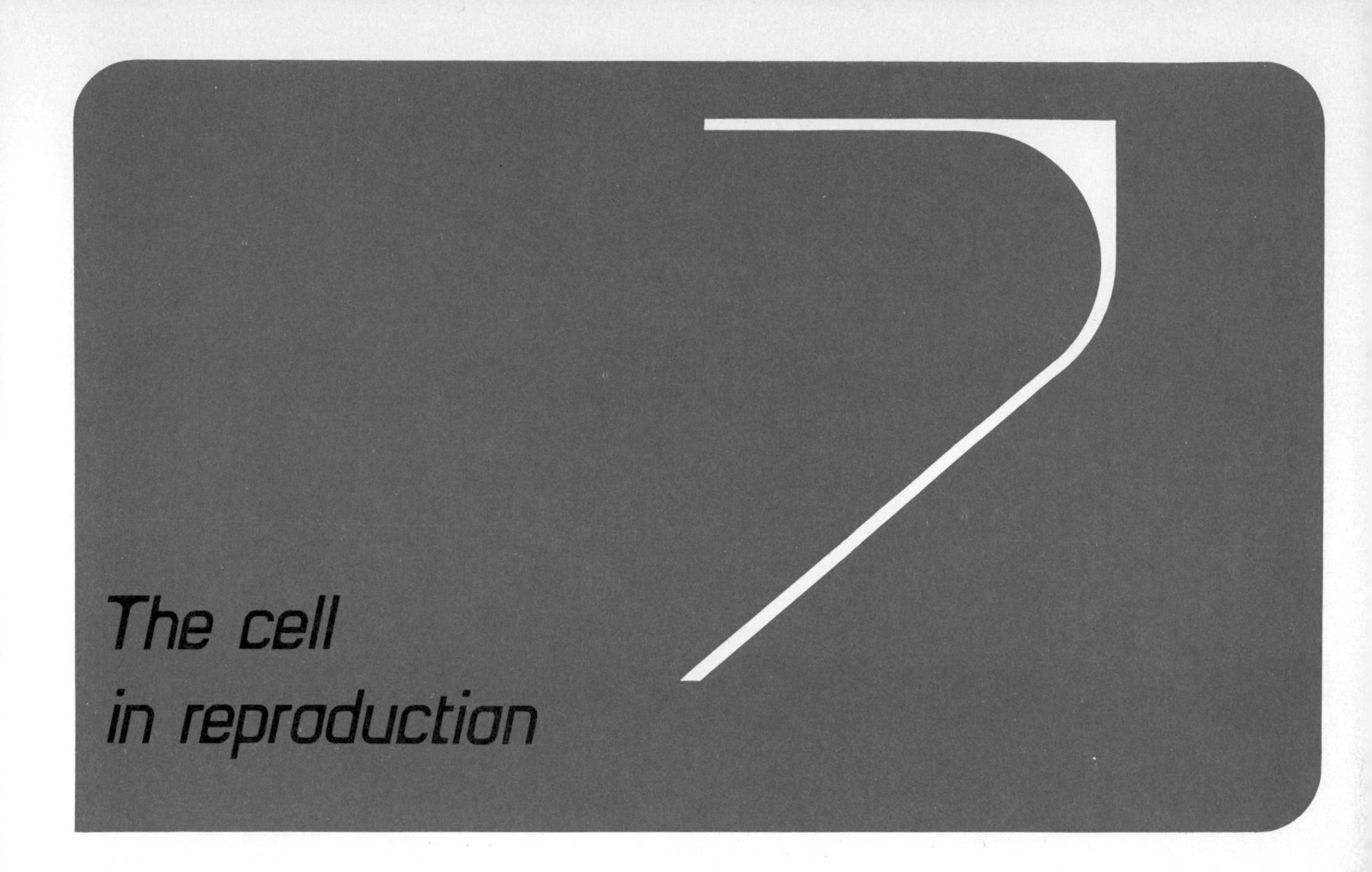

One of the important tenets of the cell theory is that cells do not arise de novo but are formed by reproduction of pre-existing cells. (Although the term "cell division" is frequently used to refer to the overall phenomenon of one cell giving rise to two daughter cells, we shall restrict the application of this term to those final stages of cell reproduction that do involve actual division of the cell). The billions of cells that comprise an adult human being are all derived by successive cycles of cell reproduction, starting with the single cell, the zygote, formed by fertilization of an egg by a sperm. The successive duplications that take place produce new cells, which become differentiated to form the recognizable structures of a multicellular embryo. This embryo continues to grow by cell reproduction and cell differentiation, these processes occurring in a coordinated manner and leading to the development of a mature adult.

Both the rates of cell reproduction occurring in an organism during its development and the localization of these cell reproductions are normally under strict control and determine the characteristic form of members of a species. Cell reproduction also continues throughout the lifetime of an organism at rates demanded by its inherent needs and in response to internal and external environmental conditions. The indeterminate growth of plants is a result of the continuous production of new cells in localized regions, such as the *apical meristems,* which contribute to the extension of roots and shoots, and the *cambium,* which contributes to their increasing girth. Most animals, on the other hand, exhibit determinate growth, and after attainment of mature proportions the active regions of cell duplication in bones, skin, intestinal epithelium, and in some glands are responsible for the replacement of cells that die or are lost rather than for addition of cells to the total number in the organism.

Most of the cells of plants and animals do not reproduce, however, although they may retain the capacity to do so under certain conditions. For example, the production of scar tissue in wound-healing responses in both plants and animals involves reproduction of cells that otherwise would not divide, while cancer is itself a consequence of the uncontrolled reproduction of cells that have become free of the restraints normally imposed upon them.

We know that cell reproduction does not only result in the formation of two cells from one, but also that it produces two cells whose information contents are identical to each other and to that of the original cell. We must look at the processes involved in cell duplication, therefore, as means of ensuring exact duplication of cellular information and its equal distribution to the two daughter cells. Since we also know that the nucleus is the control center of the cell, containing the informational DNA, we must concentrate primarily on the behavior of the nucleus during cell reproduction.

The overall process of cell reproduction is basically similar in all eukaryotic cells, although modifications do occur that reflect some degree of evolutionary divergence. The stages most obvious under the microscope involve the condensation of the nuclear material into visible chromosomes, followed by the equal distribution of the already duplicated chromosomal material into two daughter nuclei. The cytoplasm also participates in this process and is itself partitioned into the daughter cells. Prior to these visible stages of nuclear and cell division, a complex series of biochemical events occurs, resulting in the exact duplication of the chromosomal material to be segregated. The entire process of cell duplication, therefore, can be thought of as a cyclical affair, since the products of one duplication can themselves go on to divide. This *cell cycle* can be divided into two main stages: *mitosis,* which refers to those cytologically detectable events of chromosome behavior that re-

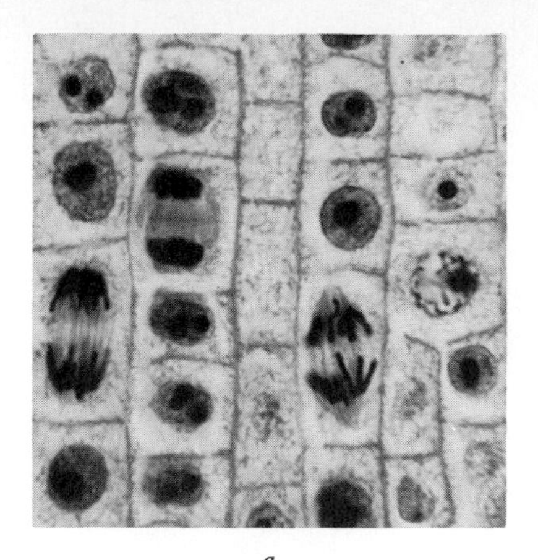

a

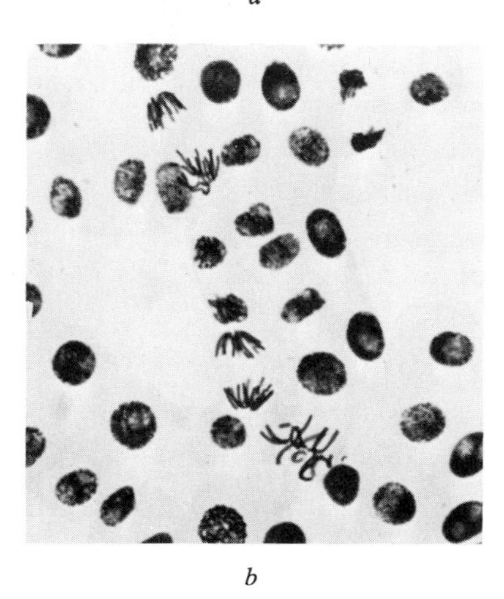

b

Figure 7.1
Panoramic view of sectioned and smeared root tip cells. (*a*) Sectioned view of dividing cells in the onion root, stained with iron hemotoxylin to show chromosomes, spindle, walls, and cytoplasm; the various stages of division range from interphase to telophase. (*b*) Smeared cells from the root of the broad bean, *Vicia faba*; smearing disrupts the arrangement of the cells, while the Feulgen stain used in this instance is specific for chromosomes and stains no other part of the cells. [(*a*) General Biological Supply House, Inc.; (*b*) courtesy of Dr. T. Merz.]

sult in equal distribution of the nuclear material; and *interphase,* the interval between two successive mitoses during which the cell is preparing to divide.

We shall describe mitosis and cell division as they typically occur in cells of higher plants and in animal cells, pointing out the differences that do exist between the two essentially similar processes.

Root tip cells in mitosis and division

As we know, cell reproduction in plants is confined to localized meristems. The apical meristems of young root tips are particularly suitable for studies of mitosis, since most of the cells of the meristem undergo duplication, and as many as 10 to 15 percent of the cells may be in mitosis at any one time.

The most commonly used technique for examining the behavior of the nucleus as it divides is to fix the cells, stain them with a DNA-specific stain (Feulgen reagent is commonly used), and "squash" them on a slide to obtain a monolayer of flattened cells. Root tips also can be embedded in paraffin wax after being fixed, and then sectioned on a microtome before staining. Root tip cells prepared in both ways are illustrated in Figure 7.1.

Most of the cells in a root tip are in interphase, because that stage usually lasts much longer than mitosis. The interphase nucleus is readily visible and is enclosed by the nuclear envelope. The chromatin appears fairly homogeneous, although some densely stained regions of heterochromatin are visible (Figure 7.2). One or more nucleoli can be seen within the nucleus, although they do not stain with DNA-specific dyes. During interphase, the nucleus and cytoplasm increase in volume, and, as we shall see later, active synthetic processes take place.

The first stage of mitosis is *prophase,* and its onset is marked by the chromatin becoming resolved into visibly distinct, long, thin threads, the chromosomes (Figures 7.3, 7.4). Each chromosome consists of two longitudinal subunits, the *chromatids,* which are in very close association with each other all along their length. These sister chromatids are relationally coiled around each other. As prophase continues, the chromosomes continue to condense, each chromatid becoming shorter and fatter (therefore, more distinct) by some controlled process of internal coiling. A lightly stained, less contracted region of the chromosome, the *centromere,* can be detected, each chromosome having its centromere located at a particular position (Figure 7.5).

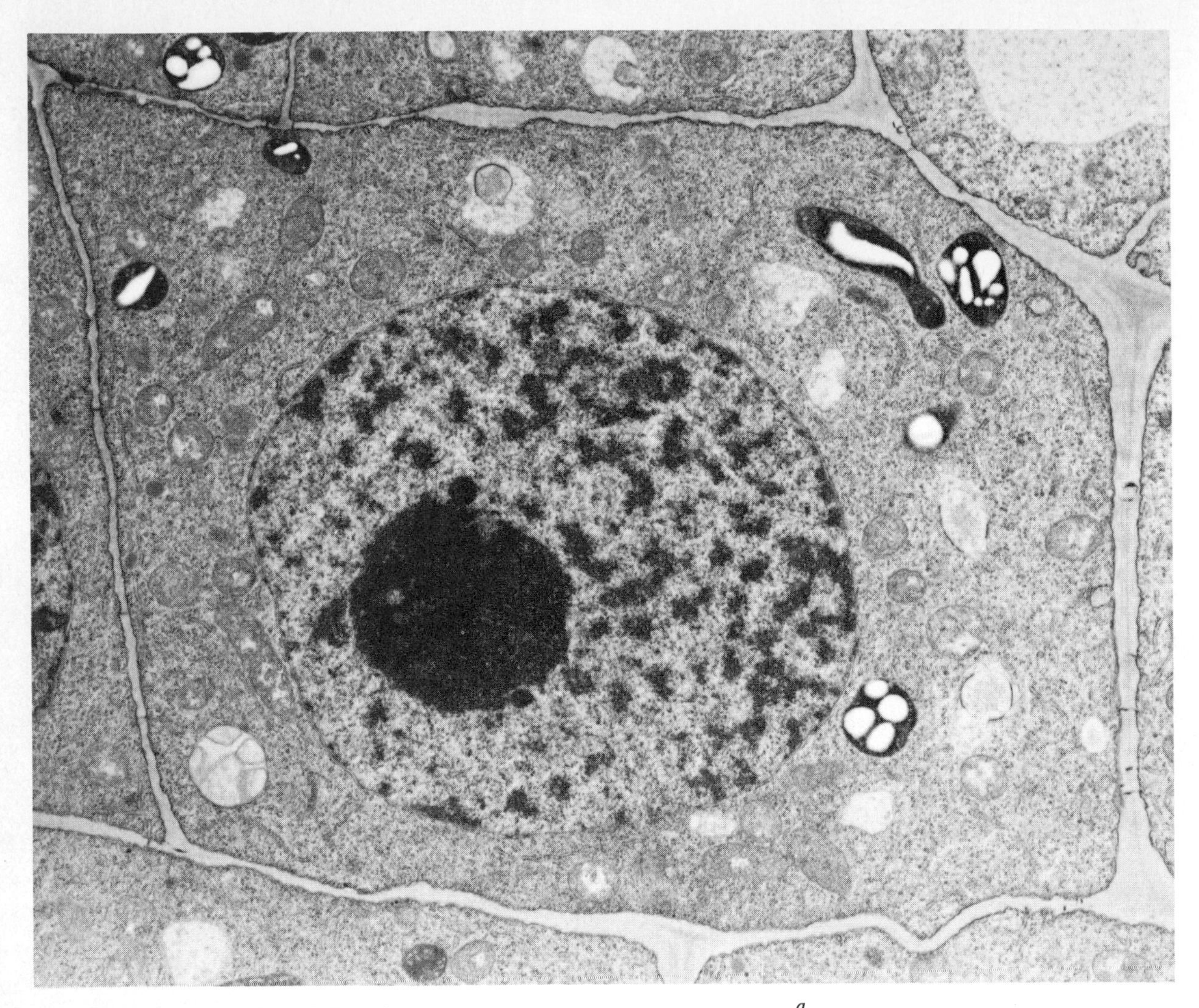

a

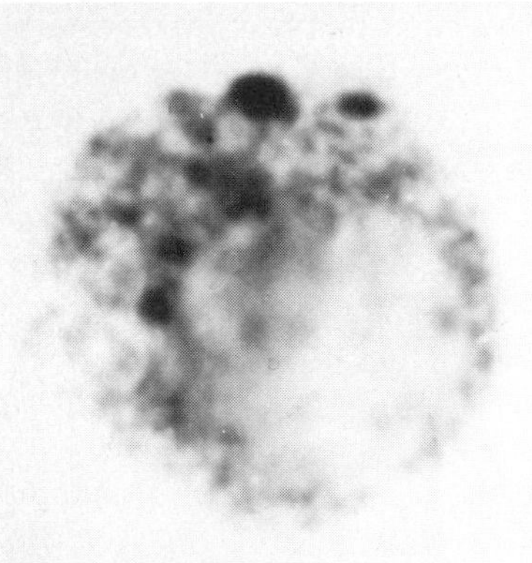

b

Figure 7.2
(*a*) Electron micrograph of root meristem cell in interphase. The darkly stained nucleolus is prominent, but little definite structure can be seen. (*b*) Light micrograph of interphase nucleus in a root meristem cell—the nucleolus is the central region that does not stain with DNA-specific stains and condensed regions of heterochromatin can be seen. (Courtesy of William McDaniel.)

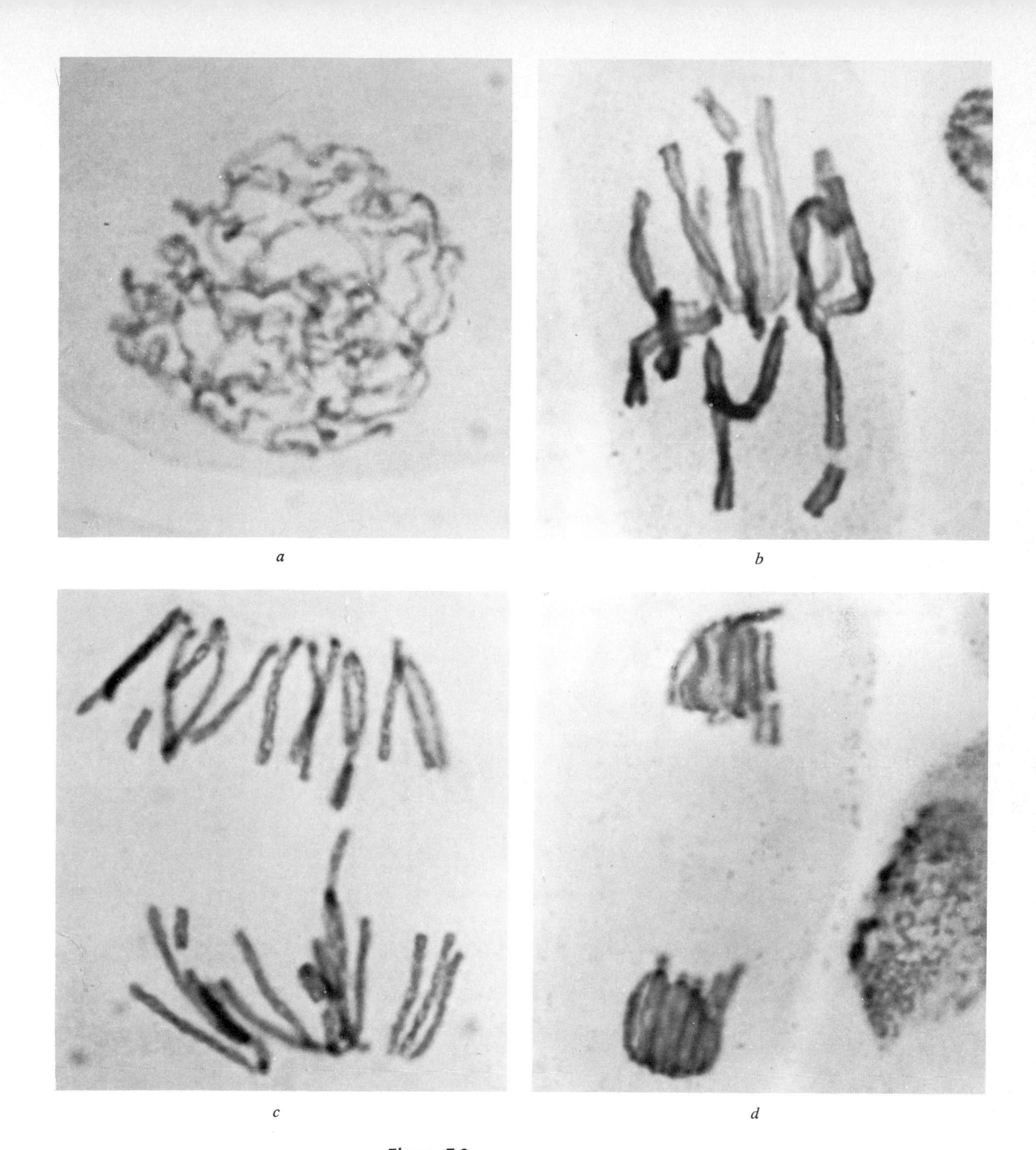

Figure 7.3
Stages of mitosis in root tip cells of the broad bean, *Vicia faba*: (*a*) prophase, (*b*) metaphase, (*c*) anaphase, (*d*) telophase.

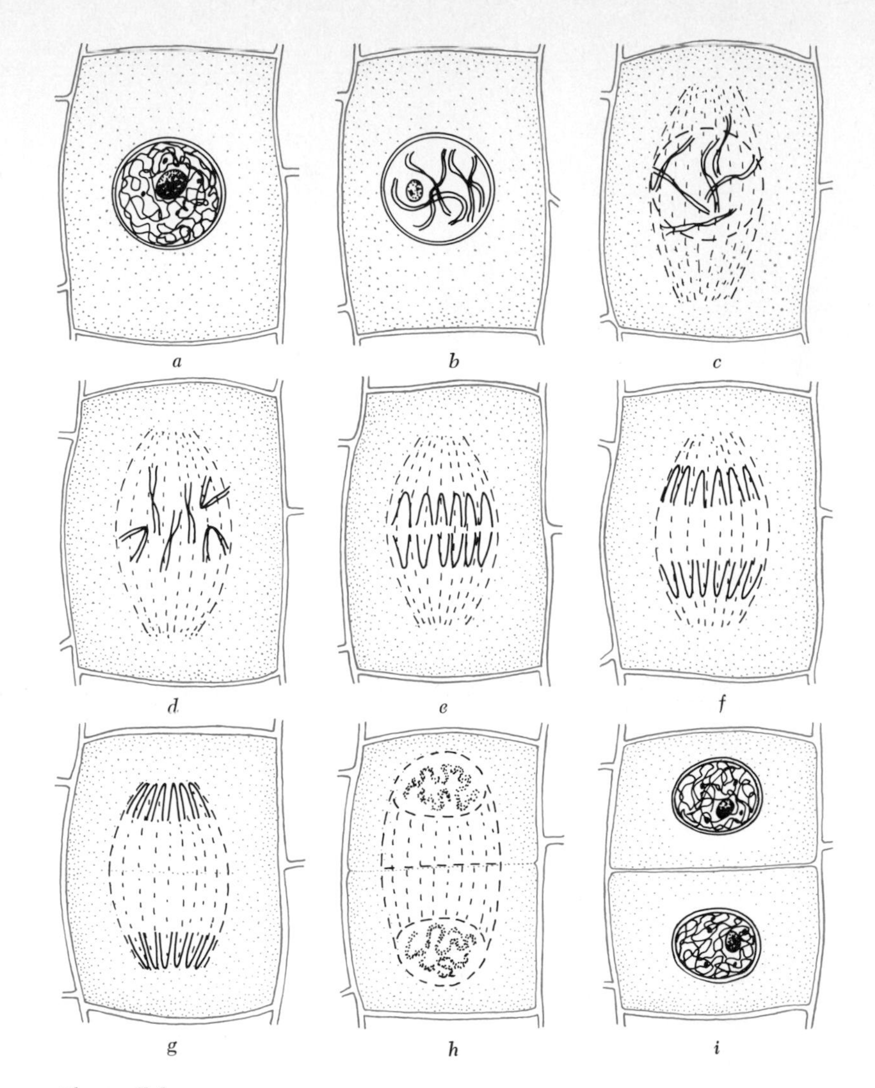

Figure 7.4

The progress of cell division, outlined in schematic form. As the cell prepares to divide, the chromosomes appear as distinct bodies in the nucleus, with a split along their length. The spindle appears at metaphase and separates the two chromatids of each chromosome at anaphase, after which the cell plate cuts the cell into two new cells. Karyokinesis, or mitosis, refers to the nuclear events of cell division; cytokinesis refers to the division of the cytoplasm by the cell plate.

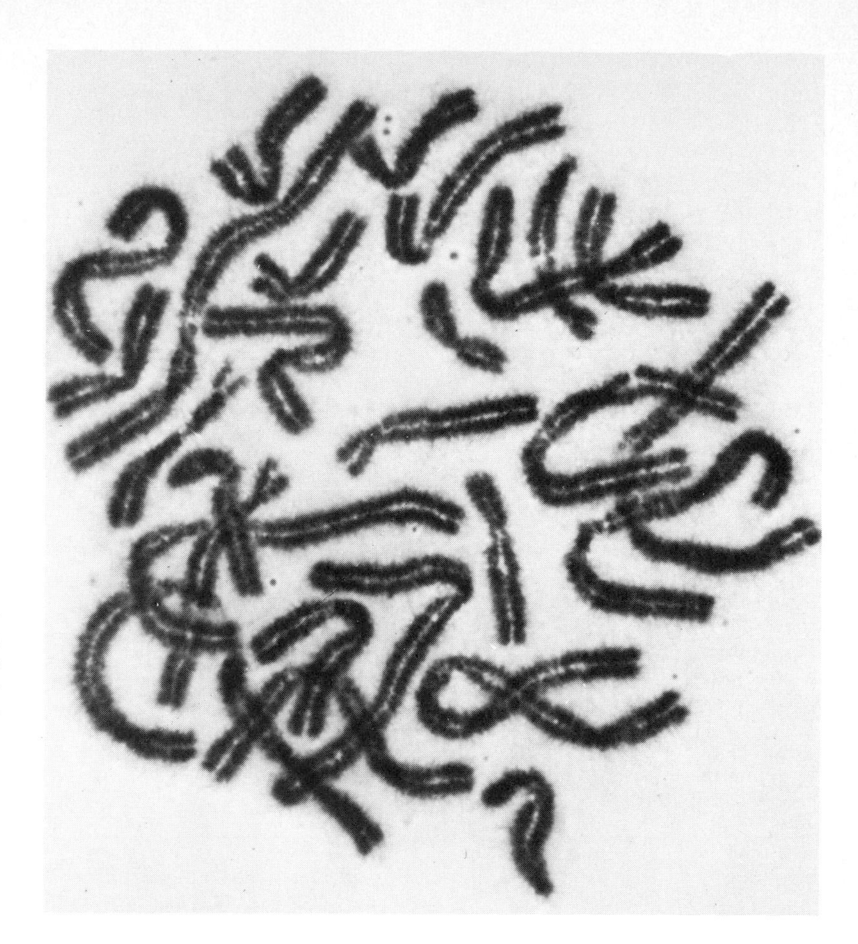

Figure 7.5
Late-prophase stage in a spermatogonial cell of the amphibian *Amphiuma*. Each chromosome is longitudinally split into two chromatids, and the centromeres are indicated by the constricted region in each chromosome. The fuzziness of the chromosomes is due to projecting loops of fine chromatin; these would be withdrawn into the body of the chromosome by full metaphase. (Courtesy of Dr. Grace Donnelly and Dr. A. H. Sparrow.)

During prophase, the nucleoli, which now can be seen to be associated with specific regions of specific chromosomes, the nucleolar organizers, gradually diminish in size, and eventually the nucleolar material becomes dispersed. The nuclear envelope also breaks down and the chromosomes are "released" into the cytoplasm. However, during prophase a *spindle* forms in the cytoplasm around the nucleus. The spindle is bipolar and consists of bundles of microtubules oriented longitudinally between the poles (Figure 7.6). Once the nuclear envelope disintegrates, each chromosome becomes attached to the spindle fibers by its centromere. The term "kinetochore" is often used to refer to the actual region of the centromere to which the spindle fibers are attached.

Following attachment of the chromosomes to the spindle, they become aligned in such a way that all of the centromeres lie in a plane equidistant from the spindle poles. This stage is *metaphase*,

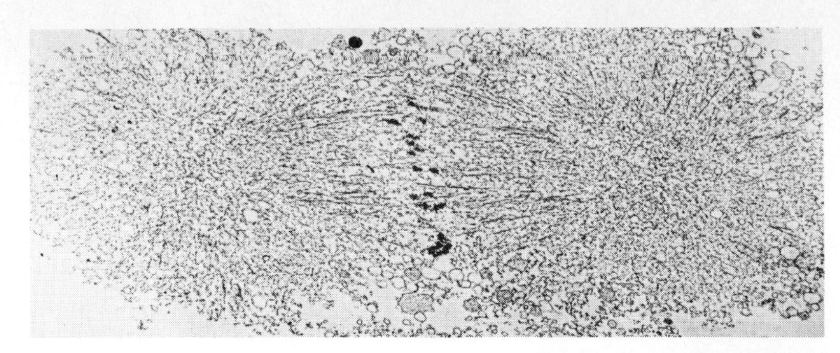

a

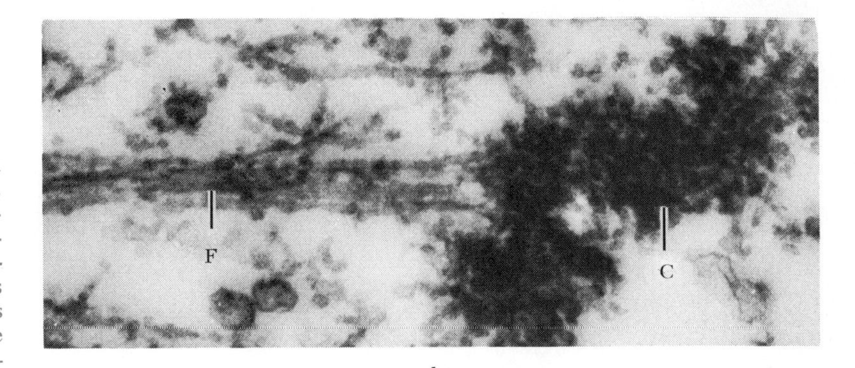

b

Figure 7.6

Electron micrographs of spindles and spindle structure in sea-urchin eggs. (*a*) Isolated spindle at low magnification (× 2,100), with the chromosomes appearing dark on the metaphase plate, a vague region across center of spindle; (*b*) spindle fibers (tubules) (F) attached to the chromosome (C) (× 53,000). (Courtesy of Dr. R. E. Kane.)

and it is also the stage at which the chromosomes have reached the point of maximum contraction (Figures 7.3, 7.4).

The next stage, *anaphase*, begins when the centromeres, each of which up until now has behaved as a functionally single unit, divide and begin to separate (Figure 7.4). The sister centromeres move apart on the spindle fibers, pulling sister chromatids apart after them. Thus, during anaphase two groups of chromatids, each chromatid representing half of an originally double chromosome, are moving toward opposite poles.

Telophase (Figures 7.3, 7.4) begins when the centromeres reach the poles of the spindle. During telophase the chromatids (which can now be considered as chromosomes consisting of a single unit) become decondensed and fuzzy in outline. A new nuclear envelope reforms around each daughter group, nucleoli reappear, and the daughter nuclei assume a typical interphase appearance.

Finally, true cell division, or *cytokinesis,* occurs, involving the formation of a cell wall between the daughter nuclei. This begins as a *cell plate,* or *phragmoplast,* formed by aggregation of vesicles from the Golgi complexes (Figure 7.7, see also Figure 5.6, Chapter

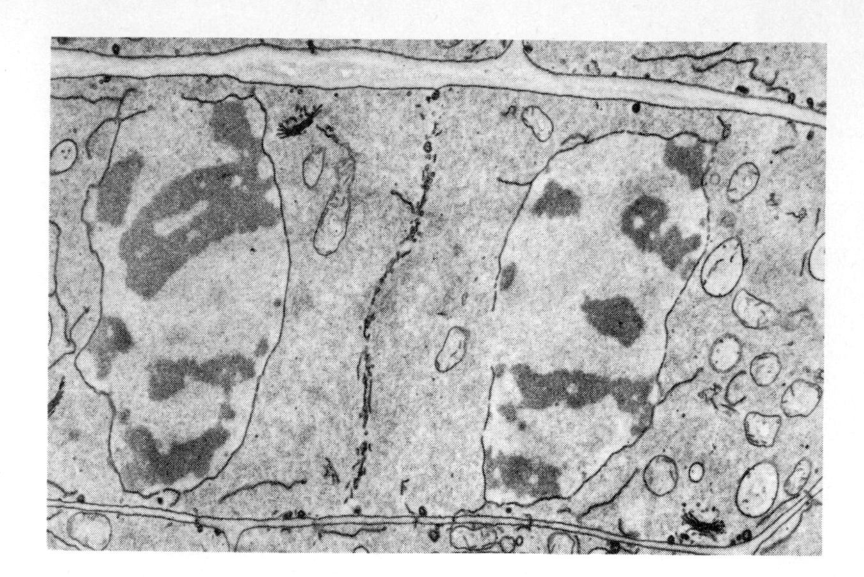

Figure 7.7
Electron micrograph of a maize cell in late telophase, with the cell plate forming across the center. (Courtesy of Dr. G. Whaley.)

5). These vesicles, which contain precursors of cell wall material, fuse with each other to form cell membranes and cell walls, thereby segmenting the cytoplasm into two parts. The spindle, which is also involved in cell plate formation, then disintegrates, and cell reproduction is completed. Thus, two new cells, each with a nucleus containing a full complement of information, have been formed from the original cell.

Mitosis and division in animal cells

The end result of cell reproduction is the same in both plant and animal cells: the formation of daughter cells of like genetic constitution. This stems from the fact that the chromosomes in each behave similarily. Differences do exist, however, and the division of cells in the embryo of the whitefish reveals these in striking fashion (Figure 7.8). The first major difference is in the appearance of the spindle apparatus. During interphase a structure known as the *aster* can be detected; this consists of microtubules radiating out from a central area, or *centrosome,* within which are two centrioles. The centrioles, each of which replicates during interphase, separate until they lie opposite each other outside the nuclear envelope. As the centriole pairs separate from each other, spindle microtubules form between them, until a bipolar spindle with an aster at each pole is formed (Figure 7.8, see also Figure 5.16, Chapter 5). As in the root tip cells, the chromosomes become

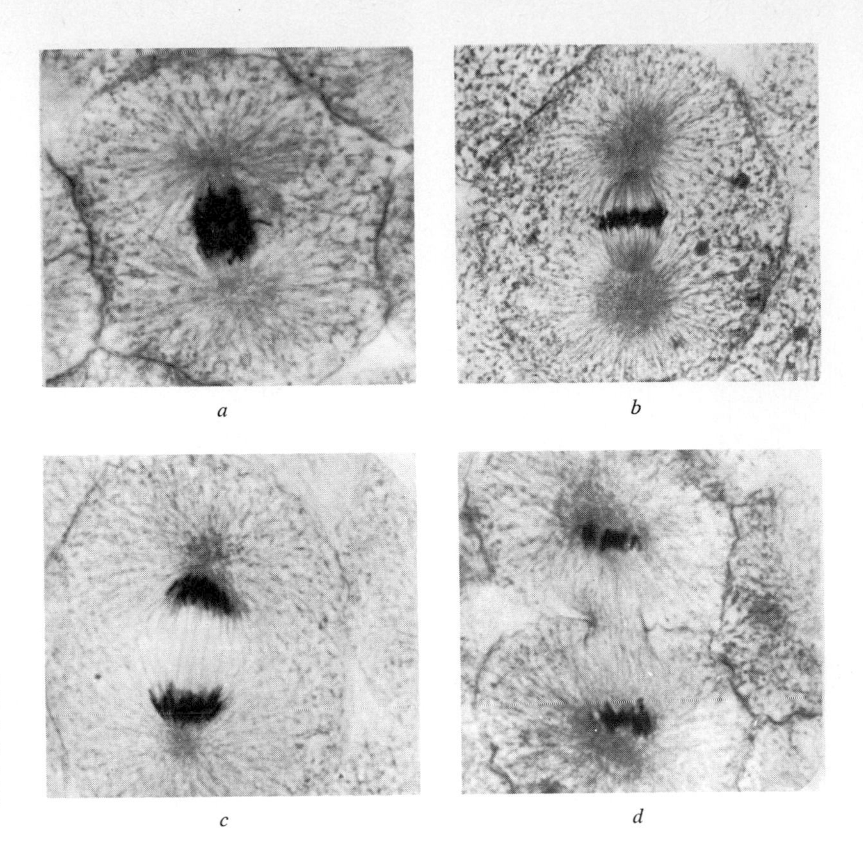

Figure 7.8
Stages in division in the whitefish. (*a*) prophase, with spindle beginning to form; (*b*) metaphase; (*c*) anaphase; (*d*) telophase, with the furrow cutting the cell into two new daughter cells. (General Biological Supply House, Inc.)

attached to the spindle fibers and are aligned with their centromeres in the equatorial plane; anaphase movement then segregates sister chromatids to opposite poles.

The actual division of the animal cell into two daughter cells differs from that of a root tip cell. Instead of a cell plate forming, a process of *furrowing*, beginning at the outer edges of the cell and midway between the poles, cleaves the cell in two. High concentrations of actin are present around the cleavage furrow, suggesting that microfilaments are important for the movements involved in cytokinesis in animal cells.

Colchicine

If we wish to examine the morphology of chromosomes, we can interrupt the normal sequence of events during mitosis by use of the drug colchicine. This plant alkaloid, which has been used to relieve the pain of gout in human beings, prevents the polymerization of tubulin subunits into microtubules, and hence inhibits spindle formation. The chromosomes can then lie free in the cell and undergo further contraction, making them more distinct.

Figure 7.9 shows the chromosomes of the broad bean, *Vicia faba,* as seen in a colchicine-treated cell. Each one has a distinct morphology that is characteristic. The location of the centromere is constant and is identified by the constriction it forms, dividing each chromosome into two *arms* of variable length. The centromere is the structure concerned with movement of the chromosome. Without it, a chromosome cannot orient on the spindle, and the chromatids cannot segregate from each other. In the longest chromosome found in the broad bean, another constriction is also present. This chromosome formed a nucleolus at that point, and the constriction, or gap, which is an uncoiled region of the chromosome, is the site occupied by the nucleolus before it disappeared. We can also see that the 12 chromosomes in the broad bean complement consist of one pair of long chromosomes with median centromeres and a secondary constriction plus 5 pairs of short chromosomes with subterminal centromeres. Colchicine is frequently used to analyze for chromosome abnormalities, such as those induced by radiation and certain chemicals, since any changes in the chromosome complement are readily recognizable and available for analysis.

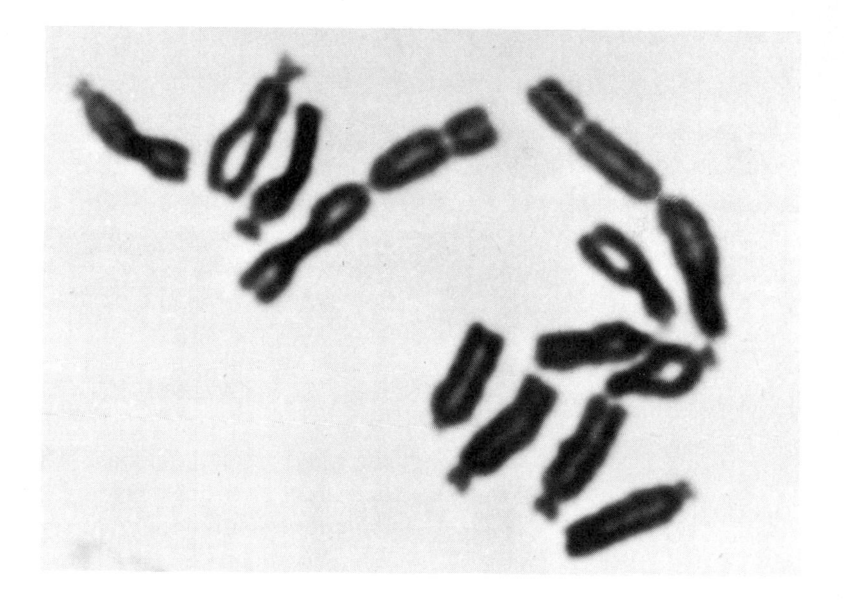

Figure 7.9
Colchicine metaphase in root tip cell of *Vicia faba.* The centromeres, which are not aligned in an equatorial plane, are evident, as are the secondary constrictions on the two long chromosomes.

The mitotic spindle

The spindle, which is such an essential component of the cell during mitosis and cell division, can be seen in the light microscope as an array of fibers that shows birefringence in polarized light, indicating a very regular molecular organization. The fibers are shown in electron micrographs to be bundles of microtubules; these become organized during prophase from a pool of subunits consisting of the protein tubulin. This pool may be at least partially derived from the breakdown of other cytoplasmic microtubules; in higher plants a band of microtubules forms around the perimeter of the cell just prior to prophase, and the appearance of the spindle microtubules is correlated with the disappearance of this "preprophase" band.

At least two classes of spindle microtubules can be recognized—one group runs the length of the spindle from pole to pole, and others reach from the poles to the centromeres. There are also (in many but not all cells) two components to anaphase separation of the sister chromatids; these are the shortening of the pole-to-chromosome fibers and the lengthening of the pole-to-pole fibers, the latter increasing the distance between the poles of the spindle. However, we still do not understand the mechanisms involved in the changes in spindle fiber length. One hypothesis is that depolymerization of the pole-to-chromosome microtubules into monomeric subunits occurs at the poles, somehow resulting in the shortened microtubules with their attached chromatids being pulled toward the poles. This hypothesis also considers the lengthening of the pole-to-pole fibers to be a consequence of addition of subunits by polymerization to the existing microtubules, resulting in further separation of the spindle poles. A second theory to account for chromatid separation suggests that somehow the two classes of microtubules can slide past one another in specific directions, the force being generated by the breakdown and formation of cross links between microtubules. There is some evidence for such bridges and for ATP-ase activity associated with these. Such a mechanism would be analagous to the sliding-filament model for muscle contraction (see Chapter 6). Of special interest with respect to a sliding-filament mechanism for anaphase movement is that actin molecules have been demonstrated to be associated with the pole-to-chromosome fibers of the spindle. It is possible, therefore, that actin-myosin interactions may be involved in movement on the spindle, but how they relate to the spindle microtubules is not clear.

Interphase

Although the fascinating behavior of the chromosomes during mitosis has attracted the attention of cytologists for many years, it is now clear that mitosis represents only a part (although certainly the most dramatic part) of the reproductive cycle of cells. It is during interphase, that period between the end of telophase and the onset of the subsequent prophase, that all of the preparation for mitosis and cytokinesis takes place. Growth of both nucleus and cytoplasm must occur between divisions if cell size is to be maintained at a reasonably constant level. This growth involves synthesis of all of the cellular constituents to be distributed to the daughter cells. Such synthetic activity is indicated by the fact that if radioactive precursors of nucleic acids, proteins, lipids, and so on are supplied to the cell, they are readily incorporated into those cellular macromolecules. The most significant events of interphase, however, are the exact replication of the DNA in the nucleus and, related to this, the duplication of the chromosomes from one chromatid to two identical chromatids.

Figure 7.10

Appearance of chromosomes in autoradiographs of cells in (*a*) first metaphase and (*b*) second metaphase following DNA synthesis in the presence of ^{3}H-thymidine. In (*a*) both chromatids are labeled, in (*b*) only one of the two chromatids of each chromosome is labeled.

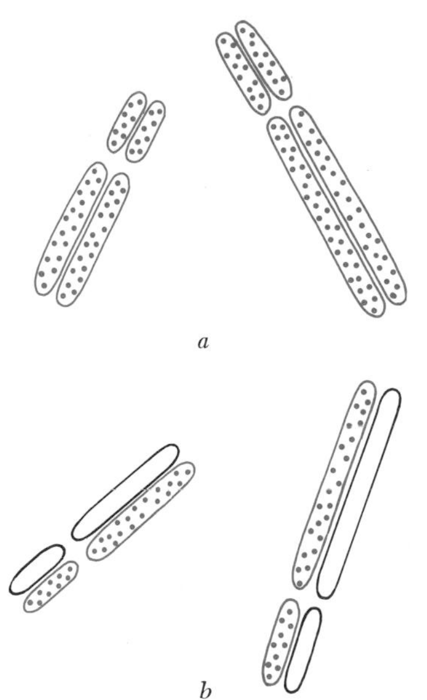

As discussed in Chapter 2, DNA replicates in a semiconservative manner, each double helix consisting of one "old" and one "new" polynucleotide strand. We also know that each chromatid of the two-chromatid chromosome consists of some "old" DNA, from the originally single chromatid, and some "new" DNA, synthesized during that interphase. This has been shown by supplying cells with tritiated thymidine (a radioactively labeled precursor of DNA) for a period during interphase, and then following by autoradiographic techniques the distribution of newly synthesized radioactive DNA during subsequent mitoses. At the first mitosis following synthesis, both chromatids of the chromosome are labeled, indicating that both chromatids contain newly synthesized DNA (Figure 7.10). If cells are allowed to complete another interphase in the absence of radioactive thymidine and enter a second mitosis, only one of the two chromatids will be labeled (Figure 7.10). These results are best explained (Figure 7.11) if we assume that a single unreplicated chromatid consists of two subunits. Replication in the presence of radioactive thymidine results in the formation of two chromatids, each consisting of one subunit of the original chromatid and one new subunit containing newly synthesized DNA (Figure 7.11). Thus, both chromatids will appear radioactive at mitosis. During the next interphase, the single chromatids, consisting of one radioactive and one nonradioactive subunit replicate again, this time in the absence of radioac-

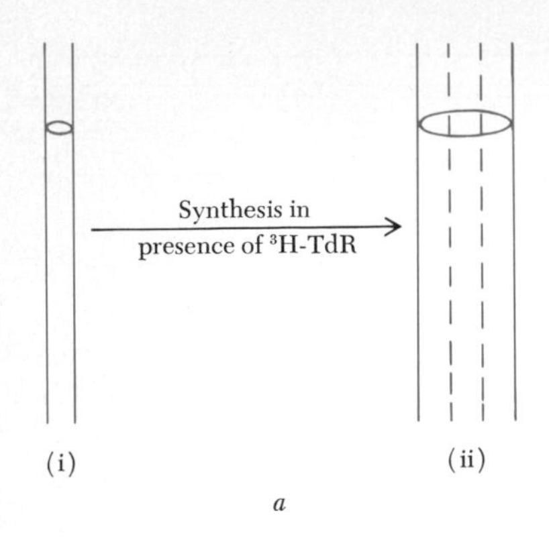

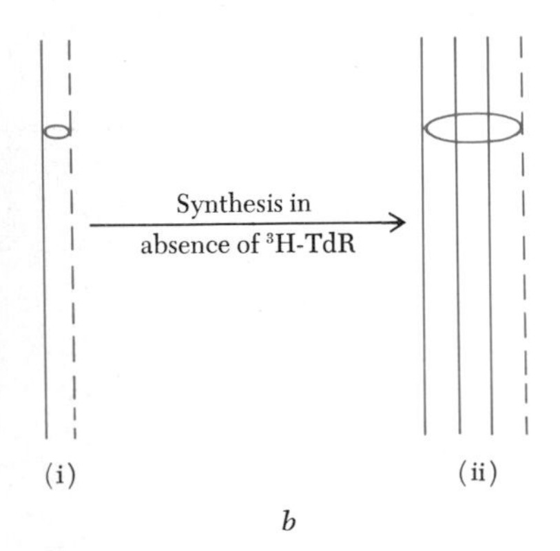

Figure 7.11

Explanation of distribution of radioactivity in chromosomes following DNA synthesis in the presence of tritiated thymidine (^{3}H-TdR). (*a*) (i) *G1* chromosome consisting of two intact subunits. Following DNA synthesis in the presence of ^{3}H-TdR, each chromatid consists of one radioactive and one nonradioactve subunit (ii). Therefore, both chromatids are labeled at first mitosis. (*b*) (i) *G1* chromosome following first mitosis, consisting of two subunits, one of which is radioactive. Following DNA synthesis in the absence of ^{3}H-TdR, only one of the two chromatids receives the radioactive subunit (ii). Therefore, only one of the two chromatids is labeled at the second mitosis.

tive thymidine. Thus, one of the two chromatids formed will consist of a radioactive subunit and a newly formed nonradioactive subunit; this chromatid will appear labeled. The other chromatid, however, will be unlabeled, since it consists of two nonradioactive subunits (Figure 7.11).

The obvious question to be asked about the replicating subunits of the chromatid is whether or not they correspond to the polynucleotide strands that make up the DNA double helix. In other words, does the unreplicated chromosome (or single chromatid) consist of one long double helix of DNA? This matter has long been disputed, but it now seems that indeed such is the case. By very carefully extracting DNA from cells, it can be shown that molecules of DNA are present of a sufficiently large size to account for all of the DNA in an unreplicated chromosome. Clearly, there can be only one double helix per chromosome in such cases. (Some multistranded chromosomes do exist, however, and we shall discuss these later.)

It can be shown by several methods that DNA replication is initiated at several points along the chromosome and proceeds from these starting points in both directions until the replicating forks meet. However, the newly replicated molecules must then untangle in such a way that sister chromatids can separate from each other at anaphase. For example, the nucleus of a human cell contains about 8 billion nucleotide pairs distributed among 46 chromosomes. Each chromosome, therefore, contains an average of about 175 million nucleotide pairs, with 10 nucleotide pairs every 3.4 nm (nanometers). Thus, the average length of the DNA of a human chromosome is 5.95 centimeters. How the separation of two such long molecules is accomplished is not fully understood but appears to involve breaks in one of the two strands. The chromosomal histone proteins also must be synthesized during interphase and somehow become complexed with the DNA and segregated with it into chromatids.

DNA synthesis does not occur throughout interphase, but occupies a discrete interval, the S (for synthesis) phase. The period between the end of telophase and the beginning of DNA synthesis is termed *G1* phase, the *G2* phase being that interval between the completion of DNA synthesis and the onset of prophase (Figure 7.12).

We can measure the duration of the cell cycle and its constituent phases by an ingenious technique involving the use of radioactive thymidine and autoradiography. If root meristems, for example, are supplied with tritiated thymidine for a short period, say 30 minutes, only those cells in S at the time will incorporate the pre-

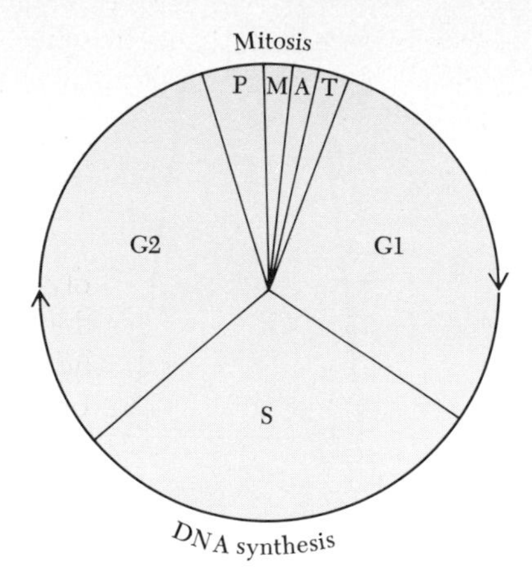

Figure 7.12

Diagrammatic representation of the mitotic cycle. *G1*—presynthetic interphase; *S*—DNA synthesis period; *G2*—postsynthetic interphase; P, M, A, and T—prophase, metaphase, anaphase, and telophase.

cursor into their DNA. Samples of root tips are then taken at different times following such a "pulse" label and prepared as autoradiographs. By counting the percentage of mitotic figures that are labeled at these different times, the progression through mitosis of the cells that were in S at the time of labeling can be followed (Figure 7.13). For a period equal to the duration of *G2* no labeled mitotic figures will be seen. The first labeled cells to reach mitosis will be those cells completing DNA synthesis (that is, cells that were at the end of S) during administration of the tritiated thymidine. These cells will reach mitosis after a time interval equal to the duration of *G2*, which can, therefore, be measured directly (Figures 7.13 and 7.14*a*). After this, the percentage of labeled mitotic figures will reach 100 percent, this level being maintained as long as the former S cells are reaching mitosis. Eventually, the cells that were at the beginning of S during the pulse label will be followed into mitosis by the unlabeled, former *G1* cells, and the percentage of labeled mitotic figures will drop (Figures 7.13 and 7.14*a*). Thus, the time period between the appearance of labeled mitoses and their subsequent disappearance is equivalent to the duration of S, which can, therefore, be determined. The next time labeled mitotic figures will be seen is when the daughters of the former S cells, which are themselves still labeled, complete interphase and enter mitosis (Figures 7.13 and 7.14*a*). Thus, the duration of the complete cell cycle can be estimated by measuring the time interval between successive appear-

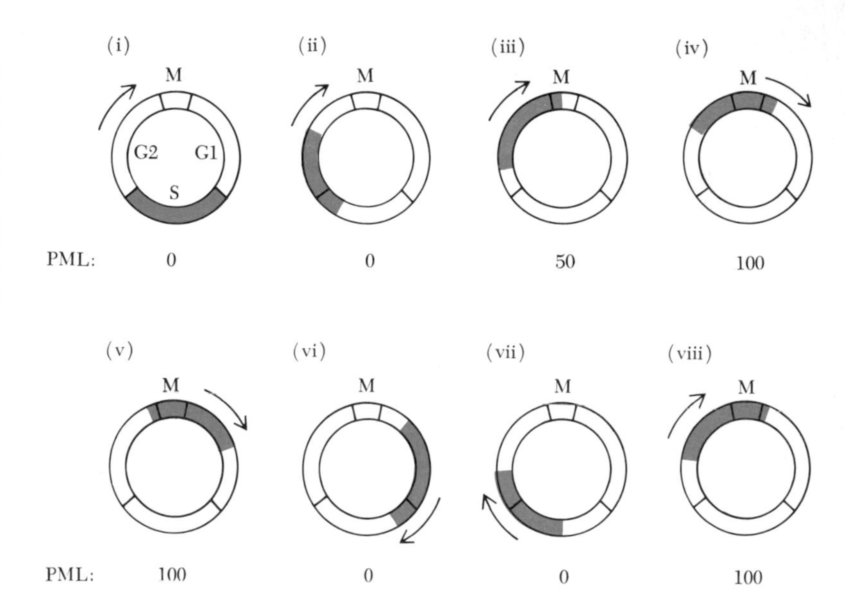

Figure 7.13

Diagrammatic representation of the progression through the mitotic cycle and past mitosis of a block of cells labeled while in S (shaded areas). (i) Position of cells at time of pulse label with tritiated thymidine (all in S). (ii) → (viii) Positions of cells at different times after administration of tritiated thymidine. *PML*—percentage of mitoses that will be labeled at these different times.

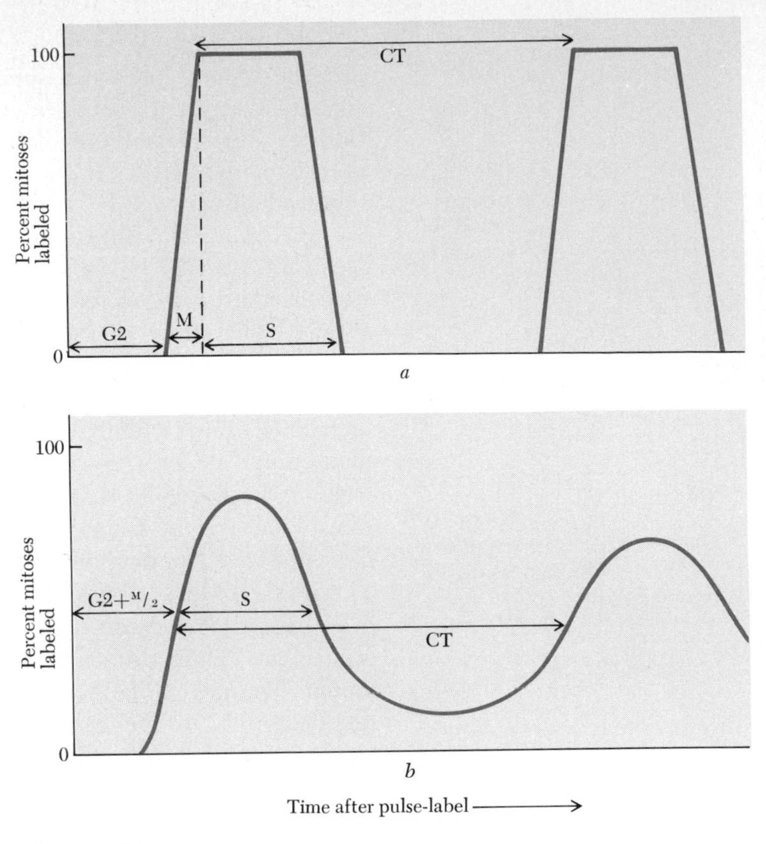

Figure 7.14
Percentage mitosis labeled curve to measure time parameters of the mitotic cycle. The percentages of mitoses which are labeled are determined at different times after administration of a pulse label of tritiated thymidine. The durations of some of the phases of the cycle can be measured from the plotted data. (See text and Figure 7.13). (*a*) Hypothetical ideal curve. (*b*) Actual curve obtained in practice and used for estimates. G2—Duration of *G2*. M—Duration of mitosis. S—Duration of *S* period. CT—Duration of mitotic cycle. G2 + M/2—Average duration of *G2* plus half of mitosis.

ances of labeled cells in mitosis (Figure 7.14*a*). The duration of mitosis itself can be estimated in one of several ways, and hence the duration of *G1* determined by subtraction. Figure 7.14*b* shows how average values of various phase durations are obtained in practice from such data.

The durations of the complete cell cycle and of its component phases, *G1, S, G2,* and *M* (mitosis) vary from cell type to cell type. For example, in lateral root tips of *Vicia faba,* the average cycle duration of most of the meristematic cells is about 14 hours at

22°C. About $2\frac{1}{2}$ hours are spent in *G1*, 6 hours in S, $3\frac{1}{2}$ hours in *G2*, and mitosis itself lasts 2 hours. Cells of the crypts of the intestinal epithelium of the mouse divide on average once every 19 hours, with *G1* lasting about $9\frac{1}{2}$ hours, S taking $7\frac{1}{2}$ hours, and *G2* plus mitosis together lasting 2 hours.

Even between cells of the same tissue, variation exists in the time parameters of the cycle. Most of this variation is in the *G1* phase, suggesting that once a commitment to DNA replication is made, cells proceed through the rest of the cycle at some fixed rate. This is not always the case, however, since cells can be arrested in *G2* during normal development.

Although DNA synthesis and mitosis are the most easily recognizable processes occurring during the cell cycle, as well as being the most significant, they themselves depend on other events that must occur. Progression through the cell cycle does require the synthesis of specific enzymes at certain times, and the production and utilization of these enzymes must be under strict temporal control. The cell cycle is a complicated affair; nucleus and cytoplasm interact in an orderly sequence of events in which all of the elements are in a state of readiness at the proper time. Each event depends on preceding events and is in turn necessary for the events to follow.

Presumably, the *G1* and *G2* phases represent stages in the cycle during which preparations for DNA synthesis and mitosis respectively are taking place, but we still are unfamiliar with the exact nature of these preparations. We do know, however, that during the cycle certain internal conditions that come to exist in the cell can signal the beginning of DNA synthesis and mitosis. This can be shown by fusing cells from different stages of the cycle. For example, if cells in early *G1* are fused with cells in S, the *G1* cells are stimulated to initiate DNA synthesis much earlier than normal. Furthermore, fusion of interphase cells with cells in mitosis results in the premature condensation of the chromosomes of the interphase cells; if the prematurely condensed chromosomes are from a *G2* cell, they consist of two chromatids, but those from a *G1* cell are single, not yet having undergone replication. Such experiments demonstrate clearly that specific internal environments during the cell cycle are responsible for determining progression of cells through the cycle.

Some of the important events we know to take place during the cell cycle are presented in Figure 7.15. Other changes do occur, including alterations in cell membrane configuration and permeability, modification of chromosomal proteins, and changes in sensitivity to radiation. However, the significance of these and other

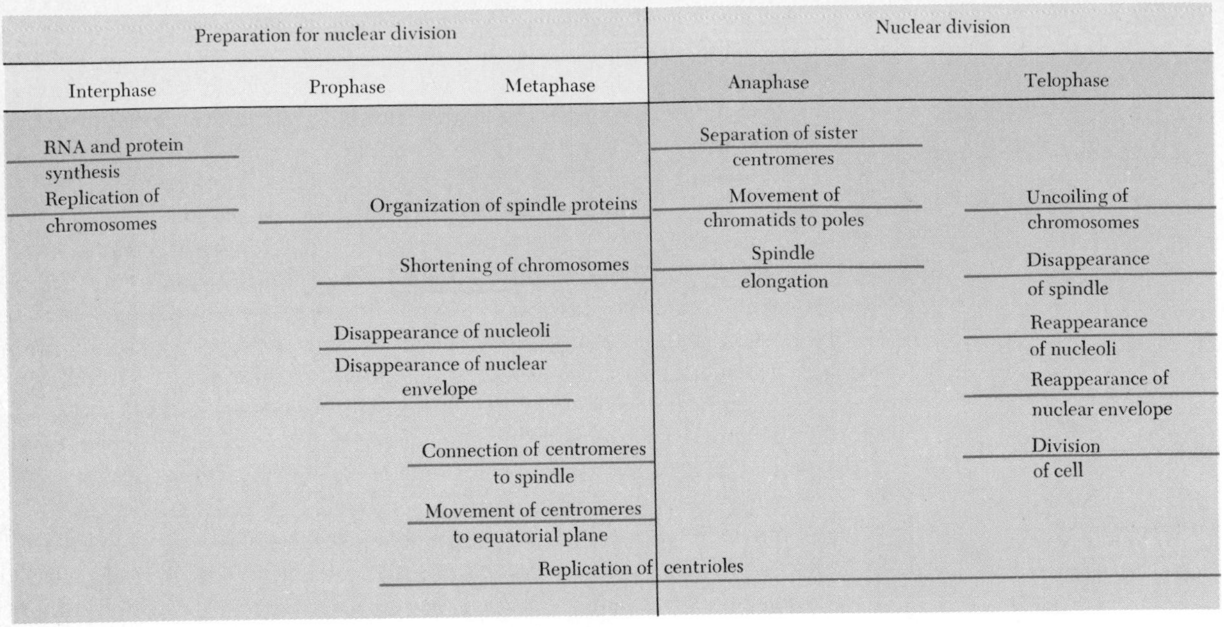

Preparation for nuclear division			Nuclear division	
Interphase	Prophase	Metaphase	Anaphase	Telophase
RNA and protein synthesis			Separation of sister centromeres	
Replication of chromosomes	Organization of spindle proteins		Movement of chromatids to poles	Uncoiling of chromosomes
	Shortening of chromosomes		Spindle elongation	Disappearance of spindle
	Disappearance of nucleoli			Reappearance of nucleoli
	Disappearance of nuclear envelope			Reappearance of nuclear envelope
		Connection of centromeres to spindle		Division of cell
		Movement of centromeres to equatorial plane		
		Replication of	centrioles	

Figure 7.15
Table of events taking place in preparation for, and during, nuclear division. (After D. Mazia.)

changes, and their relationship to chromosome replication and segregation, is still not fully understood.

Significance of cell reproduction

Cell reproduction is, of course, part of the process of growth. Although the dance of the chromosomes and the formation of the spindle and the daughter cells are the more obvious parts of the drama, it is also involved with the assimilation of materials from the outside, their transformation through breakdown and synthesis into new cellular parts, and the utilization of energy. Cell enlargement also takes place. We know of no cells, except the fertilized egg and a few of its derivative cells, that simply divide from one large cell into two others of half size and again into four cells of quarter size. This is not the usual way in which cell reproduction proceeds, for interspersed in the process are periods of growth; each division is a tumultuous affair, from which the cells must recover before proceeding again through the cycle.

The great significance of cell reproduction is the fact that it ensures a continuous succession of identically endowed cells. Chaos would result if only a random array of cells of varying qualities and capacities were to reproduce themselves; organized growth must proceed from cells of similar nature that subsequently can be molded according to the demands of the species. The species could not otherwise persist. We mentioned earlier that the chromosome is an intricate fabric composed of nucleic acids and proteins. Since the nucleus is the control center of the cell, and since the nucleus contains little else but chromosomes, the chromosomes must be the regulators of cellular metabolism and the structural characteristics of the cell. Therefore, if two cells are to behave similarly, they must have the same amount and type of nucleic acids and proteins. The longitudinal duplication of the chromosomes into identical chromatids and their segregation to the poles at anaphase must be exact to the minutest degree; the kind of cell reproduction described provides the mechanism needed. From the time a particular species was formed, this process has gone on with an exactitude that almost defies the imagination. Accidents and variations do occur—and indeed they must if evolution is to take place—but they are relatively few in number.

Cell reproduction in unicellular organisms is, of course, equivalent to exact reproduction of individuals and is responsible for increase in number of individuals of like type. Cell reproduction is, therefore, an act of survival, since cells that do not reproduce must eventually die. Reproduction and the growth associated with it bring fresh substance into the cell and effectively prevent aging, giving the cell potential immortality. In multicellular organisms, the production of new cells by cell duplication allows for division of labor, different cells and groups of cells becoming specialized to carry out different functions.

Viewed in this manner, cell division is, therefore, a first step toward cell differentiation. But this is the antithesis of survival, because differentiation is also a first step toward eventual death, since differentiated cells lose their capacity to divide. The significance of cell division, then, depends not only on the phenomenon itself but also on the kind of cell that is dividing and the consequences of division to an organism.

Although the exact process of cell reproduction ensures that all of the cells of a multicellular organism contain the same type and amount of information, different types of cells in an organism manipulate energy in different ways, and not all of the information they contain need be used. Exact duplication and segregation of information in cell reproduction, however, allows more flexibility to be retained by cells, since they always will have the potential to change their function should such be required. The kind of cell reproduction we have described, which does ensure that different cells do contain the information they need as well as some they may not, is also probably more efficient to the organism than an unequal but highly specific distribution of only those different types of information required by specific cells.

Bibliography

BASERGA, R., ed. 1969. *Biochemistry of Cell Division.* C. C. Thomas, Springfield, Ill. A collection of research papers describing biochemical aspects of the mammalian cell cycle.

MAZIA, D. 1974. The cell cycle. *Scientific American* **V. 230,** No. 1, 54–64. An account of the events of interphase and their possible regulation.

———. 1961. Mitosis and the physiology of cell division, in *The Cell,* Vol. **3,** eds. J. Brachet and A. E. Mirsky. Academic Press, New York and London. An extraordinarily comprehensive review of the mechanisms of mitosis, now only slightly out of date in parts.

MCLEISH, J., and B. SNOAD. 1958. *Looking at Chromosomes.* St. Martins Press, New York. Accounts of chromosome behavior during mitosis and meiosis in plant cells, illustrated with many beautiful photographs.

MITCHISON, J. M. 1971. *The Biology of the Cell Cycle.* Cambridge University Press, England. An excellent comprehensive survey of the cell cycle, including an account of methods of analysis of the cycle.

TAYLOR, J. H. 1974. Units of DNA replication in chromosomes of eukaryotes. *International Review of Cytology* **V. 37,** 1–20. A discussion of the evidence on which concepts of how the DNA of eukaryotic chromosomes are based.

TAYLOR, J. H., P. S. WOODS, and W. L. HUGHES. 1957. The organization and duplication of chromosomes as revealed by autoradiographic studies using tritium-labeled thymidine. *Proceedings of the National Academy of Sciences* **V. 43,** 122–28. The report of the first demonstration of semiconservative segregation of DNA in eukaryotic chromosomes.

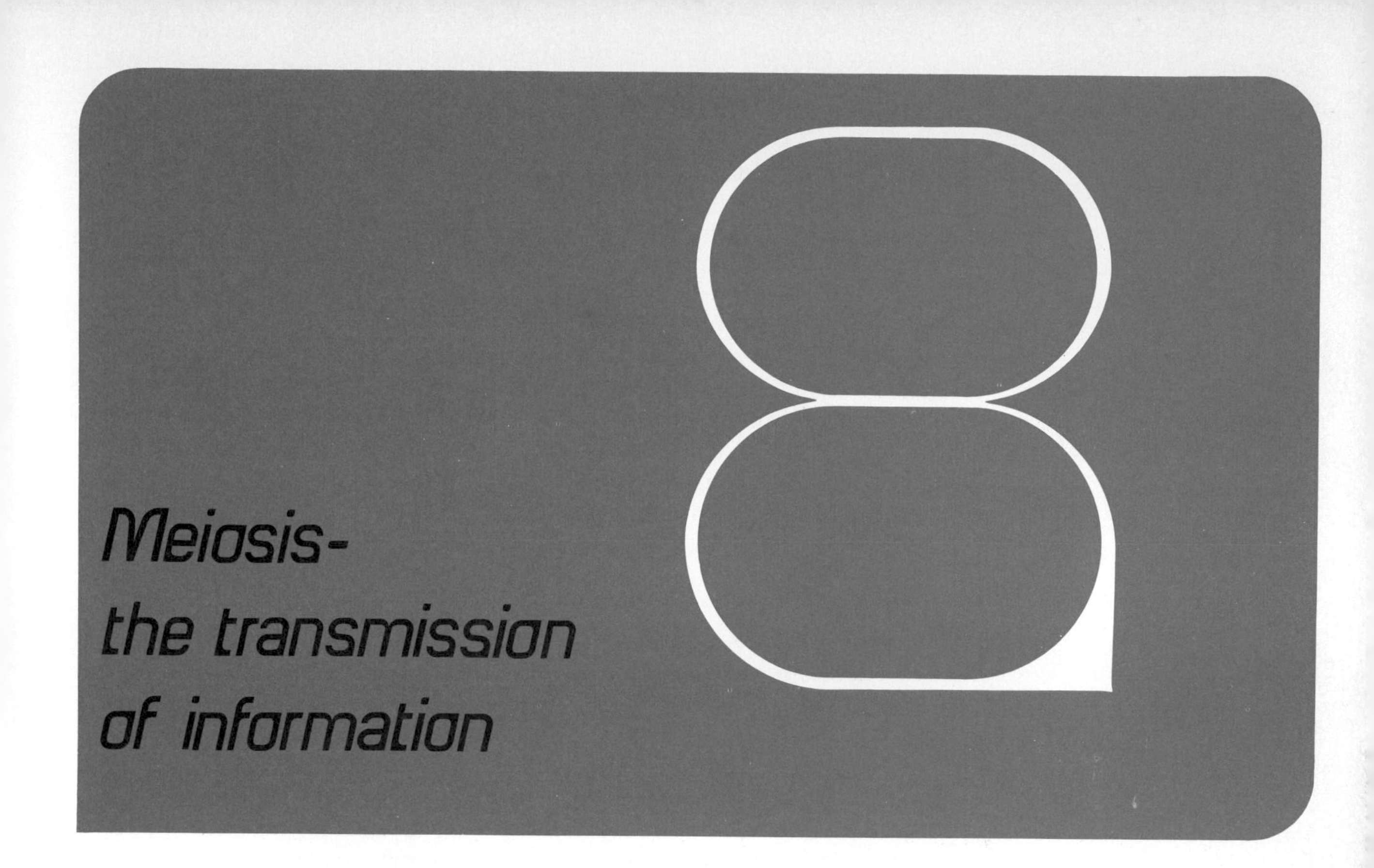

Meiosis-the transmission of information

If, over the course of several generations, we examine the individuals of any species—the human species is a good example—we are aware, very obviously, that all of the individuals fit within a frame of reference, morphological and behavioral, which we recognize as the general limits of the species. This distinction is readily made with human beings, for our nearest relatives, the chimpanzees and gorillas, are sufficiently different from us to eliminate the possibilities of mistaken identity. But such distinctions become more difficult when we are dealing with species of bacteria, moths, or oaks. Even here, however, close scrutiny reveals that these species produce individuals true to form in much the same way as does the human species. There is, consequently, from generation to generation, a *continuity* of the species, but we find on even closer scrutiny that the individuals do differ among themselves, a fact more readily observable among human beings but none the less equally detectable in other species.

The mechanisms that keep the species going, generation after generation, must be able to account for both the conservation of type (continuity) as well as the diversity among the members of the species. Since we recognize that the cell is the basic unit of organization, that eggs and sperm, among sexually reproducing species, are the slender cellular bridges connecting one generation to the next, and that DNA is the molecular basis of heritable information, we must turn to the cellular level for an understanding of these phenomena.

The continuation of any species, man or amoeba, oak tree or bacteria, depends on an unending succession of individuals. No individual is immortal, so the members of a population must reproduce if the species is to escape extinction. In unicellular organisms such as the amoeba, somatic cell division, as described in the preceding chapter, serves this function; it is a reproduction device that leads to the continued formation of new individuals. And, since somatic cell division and the mitotic cycle are mechanisms by which DNA replicates itself and chromosomes are distributed in a precise manner to daughter cells, all offspring arising through mitosis have the same number of chromosomes as the original parent—barring, of course, any accident of mitotic segregation.

The amoeba, however, like many unicellular and some multicellular organisms, is asexual; it does not produce sexual cells—*eggs* and *sperm*—and all descendant individuals have a unilateral or uniparental inheritance. But other unicellular and most multicellular organisms reproduce by sexual means; at some time during their life cycles they produce *gametes* (a general term applied to any type of sexual cell), which unite in pairs to form a single new cell called a *zygote*. From this cell a new individual develops, the product of biparental inheritance. The union of gametes is called *fertilization* or *syngamy*.

It is important to recognize that when two gametes unite through fertilization, the principal event is the fusion of gametic nuclei. Let us consider what this means in terms of chromosome number. The cells of the human being, for example, contain 46 chromosomes (Figure 8.1). If we assume for the moment that mitosis is the only type of nuclear division, the human egg and sperm would each contain 46 chromosomes since they would be mitotic descendants of the original zygote. The zygote formed by their union would then contain 92 chromosomes, and so, too, would the eggs and sperm produced by the individual developing from the new zygote. The grandparents of this individual would have had 23 chromosomes in each cell; the individuals of the next generation would possess 184 chromosomes. Starting with parents each possessing 46 chromosomes, each individual, by the end of the tenth generation, would have cells containing 23,552 chromosomes.

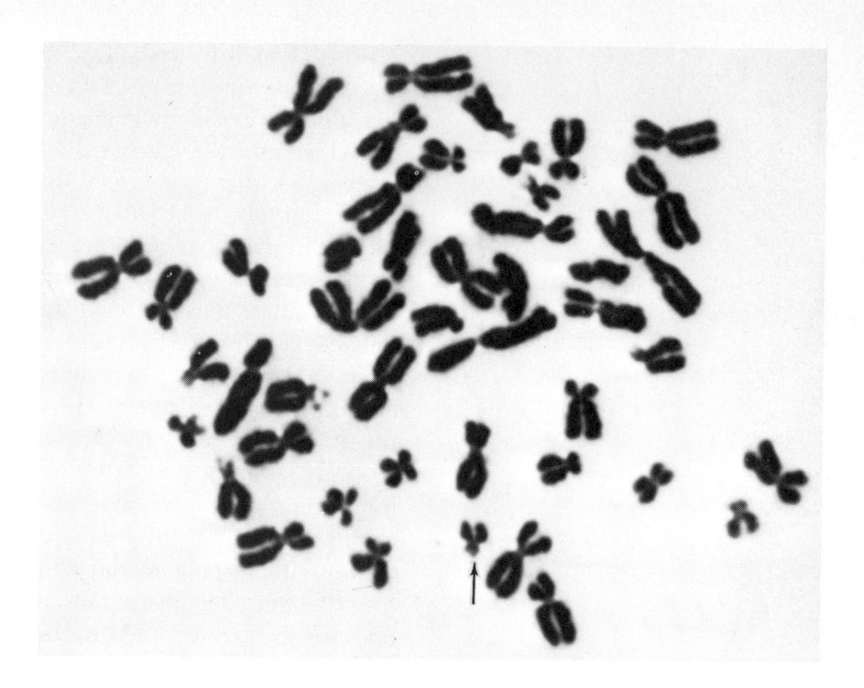

Figure 8.1
The somatic chromosomes of the human male, derived from cultured leucocytes. The Y chromosome is probably the small one identified by an arrow; the X is a medium-sized chromosome and difficult to identify with certainty. (Courtesy of Dr. Barbara Migeon.)

Obviously, this would be a ridiculous state. We are merely emphasizing that, in a sexually breeding population, the increase in chromosome number resulting from fertilization cannot and does not go on indefinitely. At some time during the life cycle of an individual, a compensatory mechanism must reduce this number, for we know that the cells of individuals belonging to the same species have a striking constancy of chromosome number. Thus, normal human cells have 46 chromosomes; those of maize, 20; of the mouse, 40; of the rat, 42; and so on. The lowest diploid number known among eukaryotes is 4, found in *Haplopappus,* a plant in the family Compositae, whereas some plants and animals have numbers as high as several hundred, and the ancient fern *Ophioglossum* exceeds 1,000 chromosomes per cell. (The tapeworm *Ascaris* has only one pair of meiotic chromosomes, but these fragment into many smaller chromosomes in somatic cells, making the nature of the meiotic chromosomes peculiar as to structure. No prokaryote has more than a single chromosome per nuclear area, but the prokaryotic and *Ascaris* situations are quite dissimilar.)

This numerical constancy for each species is repeated generation after generation. The gametes, therefore, of sexually reproducing species must have one-half the number of chromosomes found in the zygote and in the other cells of the body (since the latter arise from the zygote by mitosis). The reduction in number of chromosomes is accomplished by a special type of cell division called *meiosis*, which in its barest essentials consists of *two nuclear and cytoplasmic divisions but only one replication of chromosomes.* It is a type of division peculiar to sexually reproducing eukaryotes.

Before considering the details of meiosis and the features that distinguish this type of cell division from mitosis, we need to recognize certain terms that conveniently describe the chromosomal states of eukaryotes (Figure 8.2). The chromosomes in the nuclei of gametes are variously said to be of a *reduced, gametic, haploid,* or n number; those in the zygote and all cells derived from it by mitosis are termed *unreduced, zygotic, diploid,* or $2n$ number. Thus, a human egg or sperm, prior to fertilization, possesses 23 chromosomes, in contrast to the 46 in the zygote. Furthermore, the 46 chromosomes are not all individually different; they exist as 23 pairs, as indicated in Figure 8.3, the members of each pair being similar in shape, size, and genetic content. The members of each pair are said to be *homologous* to each other and *nonhomologous* with respect to the other chromosomes. In a zygote every pair of homologous chromosomes, or *homologues,* thus consists of one member contributed by the sperm and one by the egg.

If there is an alternation of haploid and diploid chromosome numbers during the life cycle of an individual (Figure 8.2), there similarly must be a comparable alternation in the amounts of DNA per nucleus. These amounts are expressed as C values. If C represents the amount of DNA in a haploid sperm or egg, a diploid cell such as a zygote or a cell derived from the zygote by mitosis would possess a $2C$ value. As the cell prepares to divide, the DNA would have replicated to a $4C$ value during the S period of interphase, and then be reduced to $2C$ during the anaphase separation of chromosomes. The cyclic events are represented in Figure 8.4.

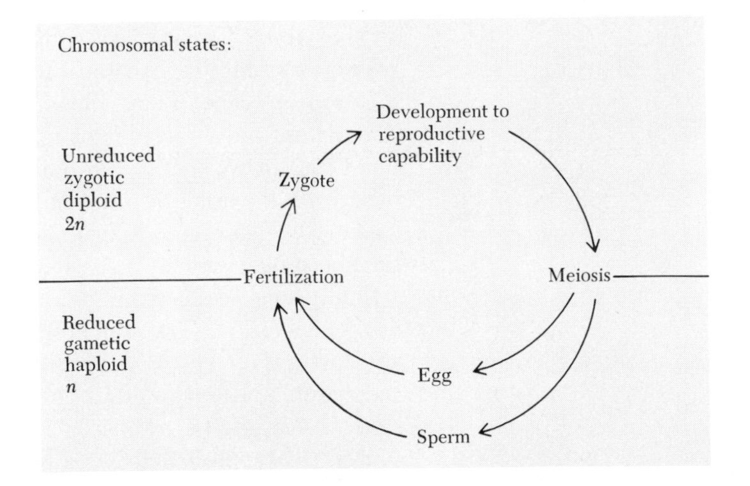

Figure 8.2
The life cycle of sexually breeding animals, with meiosis and fertilization being the events that govern chromosomal states. In higher plants, meiosis would produce haploid spores, which germinate to form the gametophytic generation, which, in its turn, would produce the eggs and sperm.

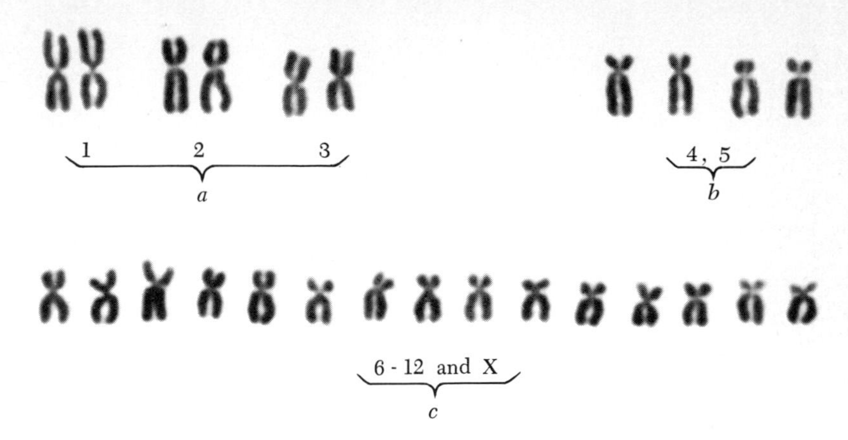

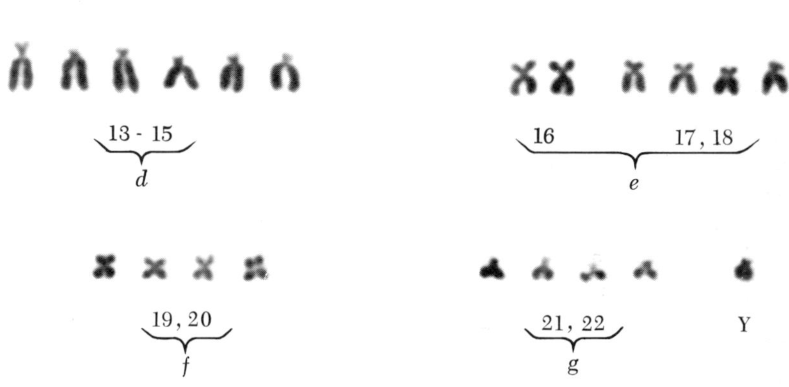

Figure 8.3
The chromosomes of a normal human male, with the chromosomes arranged in homologous pairs, numbered according to size, and also designated by letter groups. The male has an XY sex-determining system; the small Y chromosome is indicated at the lower right in group g, while the X, which is difficult to identify positively, is one of those included in group c. (Courtesy of Dr. Barbara Migeon.)

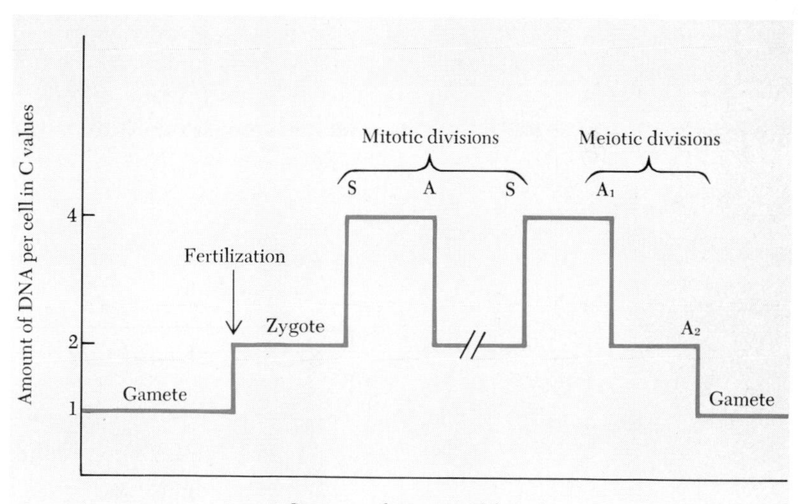

Figure 8.4
Sequences of stages in the life cycle of a sexually reproducing animal correlated with the changes in the amount of DNA per cell. S = period of DNA synthesis in interphase; A = mitotic anaphase; A_1 = first meiotic anaphase; A_2 = second meiotic anaphase.

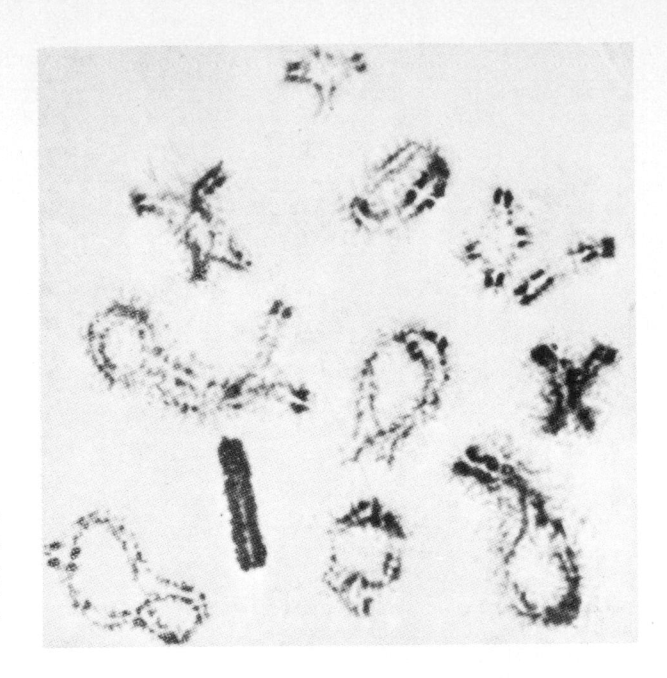

Figure 8.5
The chromosomes in meiosis in the spermatocyte of the grasshopper, *Schistocerca gregaria.* This insect has an XO sex-determining system, and the unpaired X chromosome is seen as the deeply stained rod. (J. H. Tjio, and A. Levan. 1954. *Ana. Est. Exp. Aula Dei* 3(2): 225–228.)

Since meiosis is two cell divisions, but with only a single period of DNA replication, gametes would possess only a 1*C* value prior to fertilization.

One exception to the similarity of paired homologues in shape and size is the pair of chromosomes characterizing the two sexes. Figure 8.3 shows this pair of chromosomes in a human male, although the size and shape of the X makes identification difficult. The human female is XX, and consequently her paired chromosomes are similar and homologous; the male is XY, but the two chromosomes, while differing in size and genetic content, are sufficiently homologous to pair in the prophase of meiosis (see meiosis stages below). Not all sexually reproducing species possess recognizable sex chromosomes, and the XX-XY system is only one variation among many that are known (Figure 8.5). All function in some manner to bring about the expression and functioning of the sexual state.

Stages of meiosis

Meiosis is a rather complicated type of cell division, yet, remarkably, like somatic cell division and mitosis, the crucial nuclear events and end results are essentially the same wherever encountered. Consequently, a single account of it applies equally well to a fungus, an insect, a flowering plant, or a human. Except for the type of cell resulting from meiosis, the process is basically similar in both sexes as well.

We can separate meiosis into a sequence of steps similar to those in mitosis (Figures 8.6 and 8.7). Prophase, however, is a more leisurely process and hence longer in duration, and the modifications introduced affect the character of the resultant cells. Five separate prophase stages are recognizable, even though the prophase progression is continuous.

The *leptotene* stage initiates the first visible steps of meiosis (Figures 8.8 and 8.9). Meiotic cells and their nuclei are generally

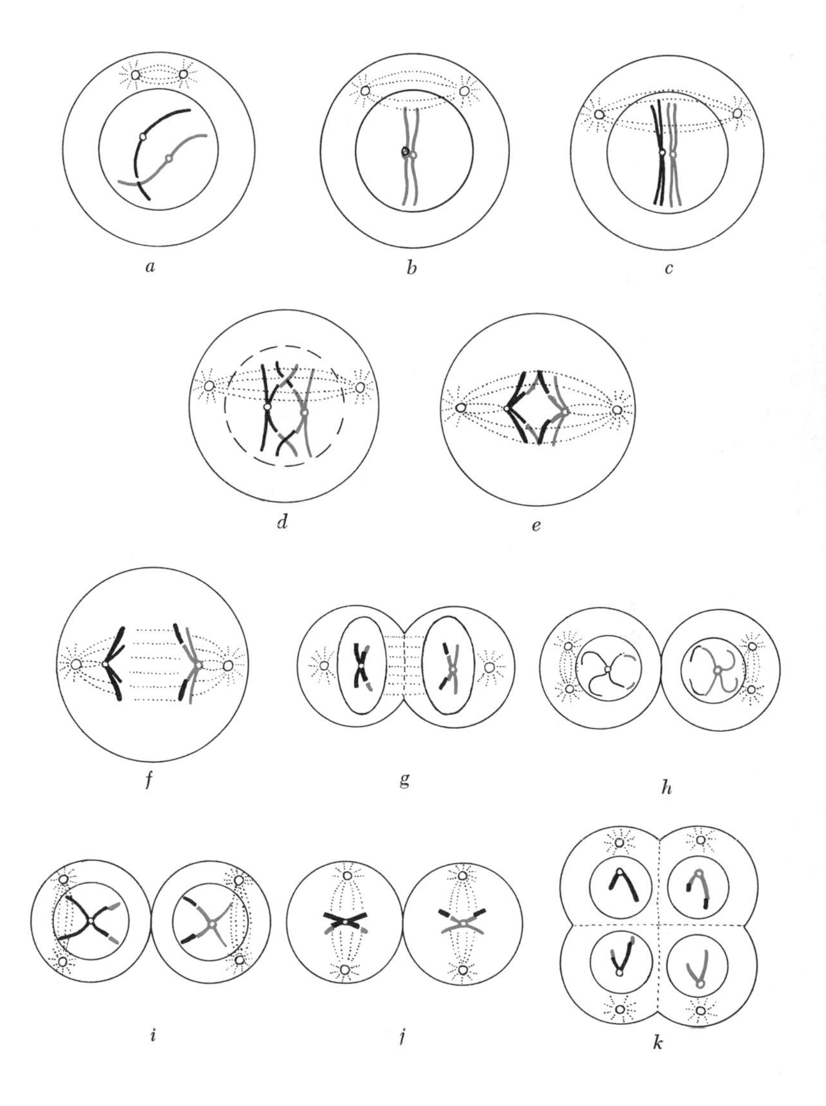

Figure 8.6

Diagrammatic representation of the stages of division in meiosis I and II: (*a*) leptotene, prior to synapsis; (*b*) beginnings of synapsis at zygotene; (*c*) pachytene; (*d*) diplotene; (*e*) metaphase I; (*f*) anaphase I; (*g*) telophase I; (*h*) interphase between the two meiotic divisions; (*i*) prophase II; (*j*) metaphase II; (*k*) telophase II. For simplification, only one pair of homologues has been included in this figure.

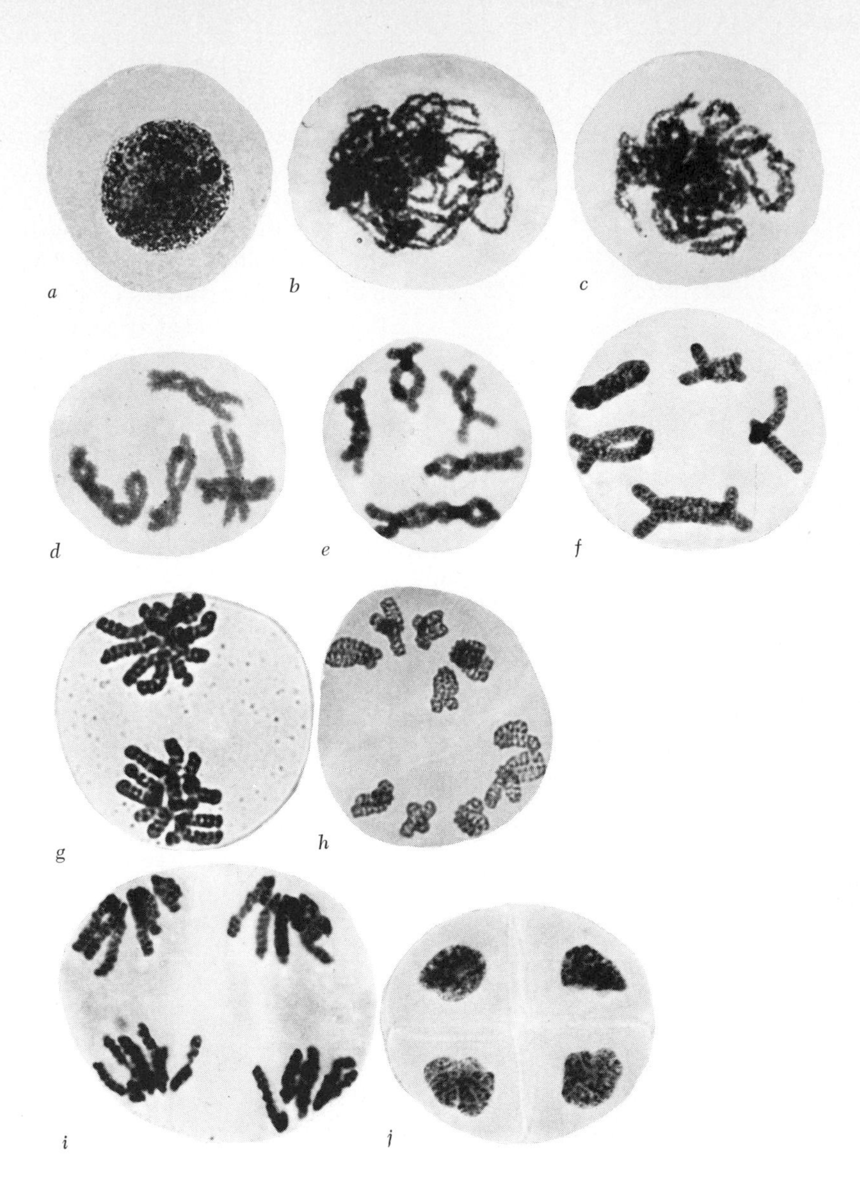

Figure 8.7

Stages of meiosis in the microsporocytes of the wakerobin, *Trillium*: (*a*) zygotene; (*b*) pachytene; (*c*) early diplotene; (*d*) late diplotene; (*e*) diakinesis; (*f*) metaphase from a polar view; (*g*) late anaphase I; (*h*) metaphase II; (*i*) anaphase II; (*j*) quartet stage, with four microspores. (Courtesy of Dr. A. H. Sparrow.)

larger than those of the surrounding tissues. The chromosomes, present in the diploid number, are thinner and longer than in mitosis and are, therefore, difficult to distinguish individually. Leptotene chromosomes, however, differ from those in ordinary mitotic prophase in two ways: (1) they *appear* to be longitudinally single rather than double, although DNA synthesis has already occurred, indicating that they are in fact double; and (2) their structure is more definite, with a series of dense granules, or *chromomeres*, occurring at irregular intervals along their length. These

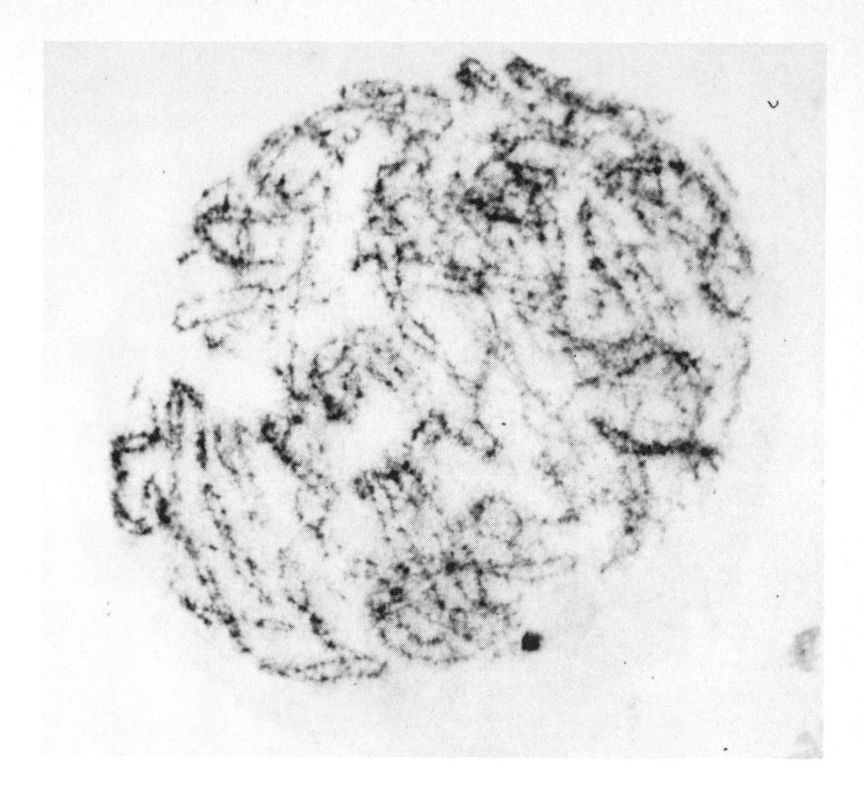

Figure 8.8

A spermatocyte nucleus of the amphibian, *Amphiuma means tridactylus*, in the leptotene-diplotene stage of meiosis. Some regions of homologous chromosomes are synapsed, others are not. The chromomeres at this stage of meiosis are quite small and regularly spaced; they would become larger and more irregularly spaced in the following stage of pachytene. (Courtesy of Dr. Grace M. Donnelly, and A. H. Sparrow, 1963. *J. Heredity* LVI: 91–98.)

Figure 8.9

Electron micrographs of spermatocyte nuclei in the milkweed bug, *Oncopeltus*, prepared by being first floated onto a water surface, lifted off, and then dried by the critical point methods, a technique that dehydrates the specimen without serious distortion. Left, leptotene stage; middle, zygotene stage; and right, pachytene stage. The denser clumps of chromatin are sex chromosomes, which condense earlier than the autosomes. (Courtesy of Dr. S. Wolfe.)

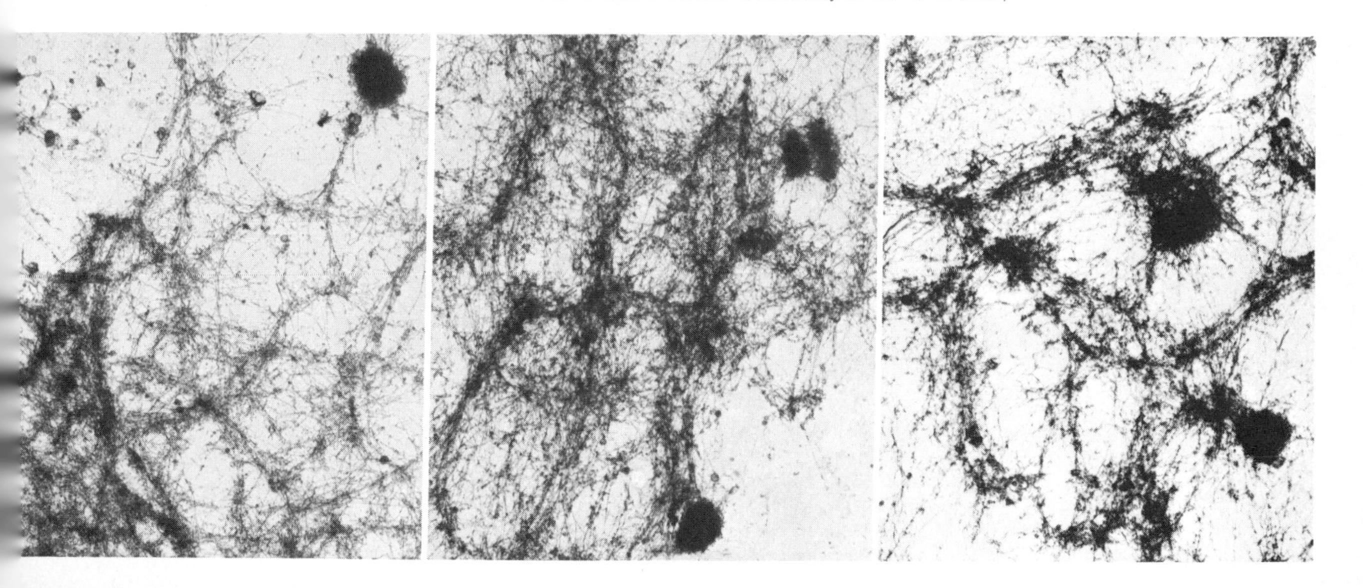

bodies are not to be confused with the beaded structure of chromatin seen in Figure 2.22, Chapter 2; the level of magnification is quite different, with chromomeres of meiosis being hundreds of times larger than the chromatin beads.

The chromomeres of any given organism are characteristic in number, size, and position; they can, as a result, be used as landmarks for the identification of particular chromosomes, especially at the succeeding zygotene and pachytene stages where they are larger, fewer in number, and more readily identified. Chromomeres are regions of chromatin that have been compacted through localized contraction, possible by association with the proteins of the chromosome, and in this state the DNA contained within them is thought to be metabolically inactive and nontranscribing. At one time it was believed that the chromomeres were visible manifestations of single genes, but since there are only about 2,000 chromomeres among the 24 chromosomes of the garden lily, and far fewer in the woodrush *Luzula* or tomato (Figures 8.10 and 8.11), and

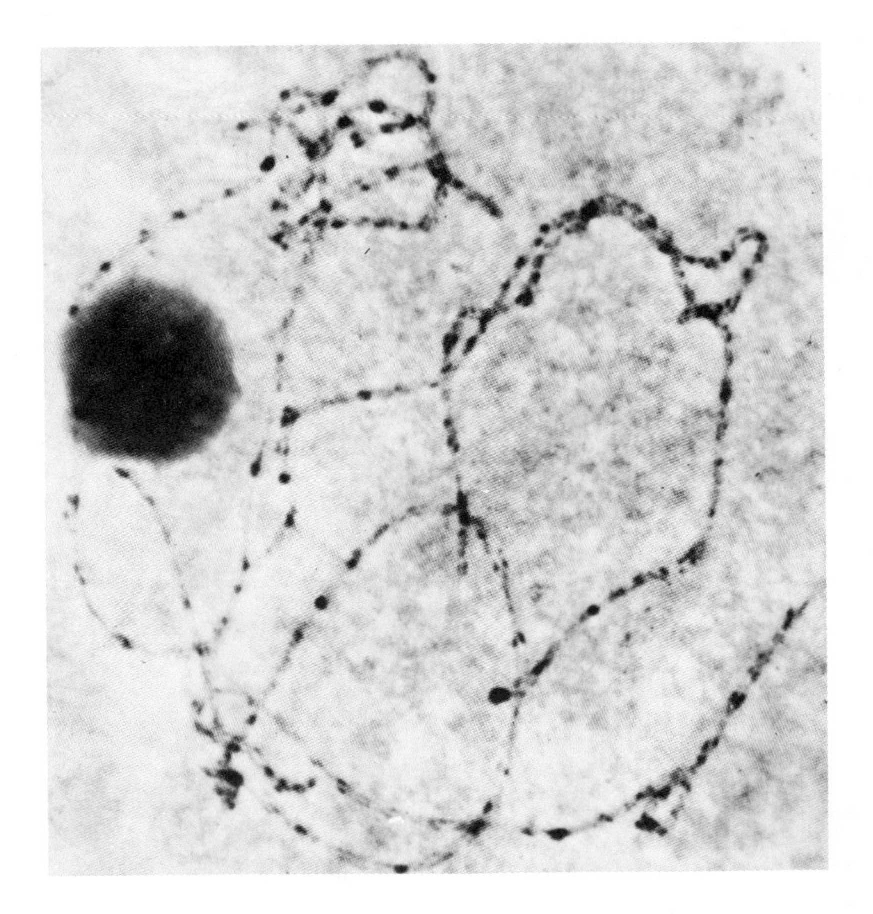

Figure 8.10
A microsporocyte of *Luzula* in the pachytene stage of meiosis, showing the chromomeres irregularly spaced along the length of the paired homologues. It can be seen that the pairing is chromomere-by-chromomere in a highly accurate fashion. The dark body at the left is the nucleolus. (Courtesy of Dr. S. Brown.)

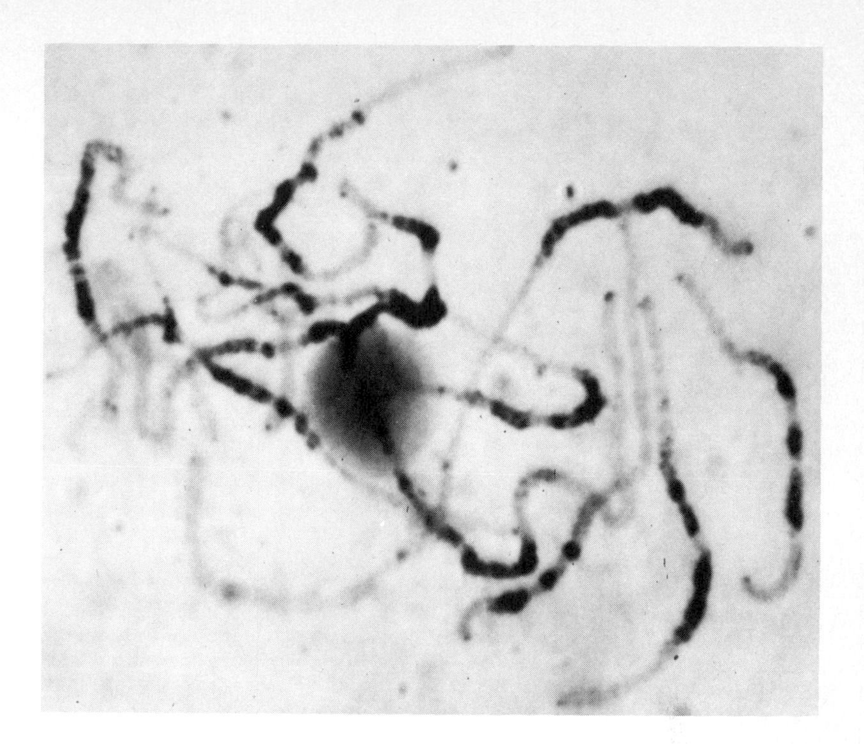

Figure 8.11
A microsporocyte of the tomato, *Lycopersicum esculentum,* in the pachytene stage of meiosis, and showing the large chromomeres concentrated around the centromeres of the paired homologues, with smaller chromomeres located at the ends of the chromosomes. (Courtesy of Dr. C. M. Rick.)

since the pattern of the chromomeres changes with the stages of meiosis, it no longer seems reasonable to equate meiotic chromomeres with single genes.

Movement of the chromosomes initiates the *zygotene* stage, and this movement results from an attracting force of unknown nature that brings together homologous chromosomes. The pairing of homologues, known as *synapsis,* begins at one or more points along the length of the chromosomes and then proceeds, much as a zipper would, to unite the homologues along their entire length. This is an exact, not a random, process, for the chromomeres in one homologue synapse exactly with their counterparts in the other (Figure 8.12). When synapsis is complete, the nucleus will appear as if only the haploid number of chromosomes is present. Each, however, is a pair of homologous chromosomes, and these are now referred to as *bivalents.*

At the level of resolution of the electron microscope, the synapsing homologues form a *synaptinemal complex* (Figure 8.13). The paired chromosomes seem to lie on either side of the complex, which consists of two denser outer boundaries and a less dense, ladderlike central core. There are various speculations as to the meaning of this structure, but as yet no certain information is known except that it seems always correlated with synapsis fol-

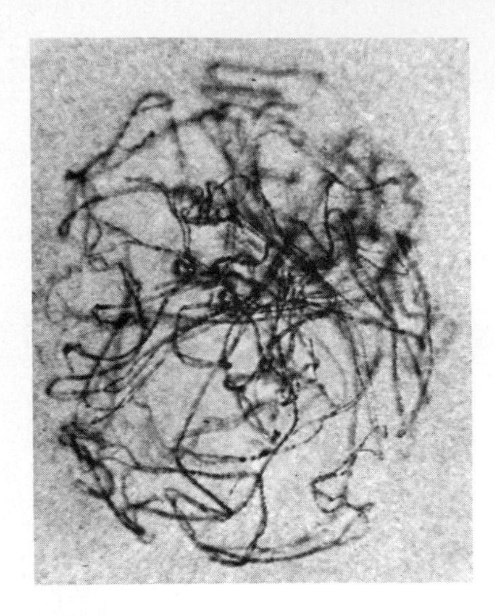

Figure 8.12
Zygotene stage of meiosis in a microsporocyte of the lily, *Lilium regale.* In the lower right-hand corner, both paired and unpaired regions of homologues can be seen. (Courtesy of Dr. J. McLeish.)

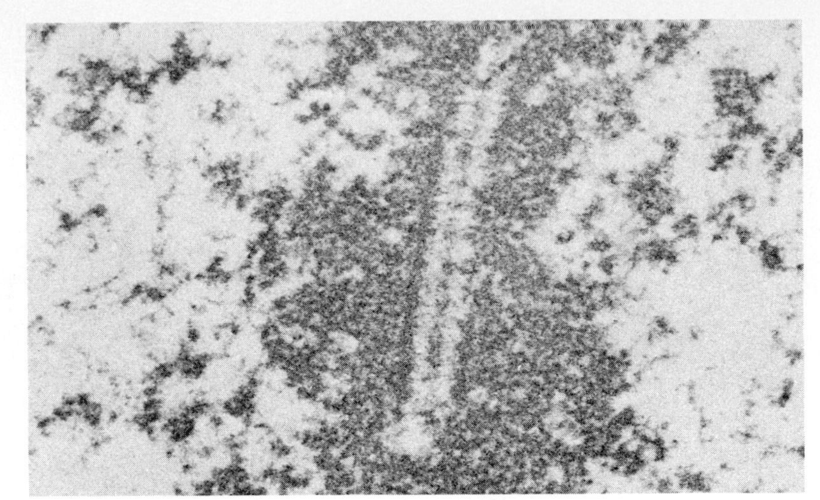

Figure 8.13
Electron micrograph of a synaptinemal complex in a microsporocyte of *Tradescantia paludosa.* The dark masses are the chromatin of paired homologues, but an interpretation of the light and dark areas of the complex is not entirely clear at the present time even though synaptinemal complexes are obviously indicative of pairing relationships.

lowed by crossing over. In cells where synapsis has been interfered with, or where synapsis occurs but crossing over does not, as in *Drosophila* spermatocytes, no synaptinemal complex is found.

Zygonema is the period of active synapsis. The next, or *pachytene,* stage is distinguishable by the fact that, in some species, the paired chromosomes of each bivalent are easily seen (Figure 8.10), and since the chromosomes have continued to shorten and thicken by coiling, they are more readily identified one from the other. In other species the homologues are so tightly paired as to be indistinguishable one from the other (Figure 8.14). The chromomeres and the attachment of the nucleolus to a particular chromosome in maize may be visible with high magnification (Figure 8.15 and 8.16).

The pachytene stage ends when the synaptic forces of attraction lapse and the homologous chromosomes separate from each other (Figure 8.17). This is the *diplotene* stage, and, as Figure 8.18 indicates, each chromosome now consists of two clearly visible chromatids. Each bivalent, therefore, is composed of four chromatids. Longitudinal replication of each chromosome took place prior to this stage, but it did not become obviously evident until the attraction between homologues ceased (Figure 8.19).

Figure 8.14
Pachytene stage in a spermatocyte of a salamander. The chromomeres are quite regularly spaced along the length of the paired homologues, and pairing seems to be complete for all of the chromosomes. (Courtesy of Dr. J. Kezer.)

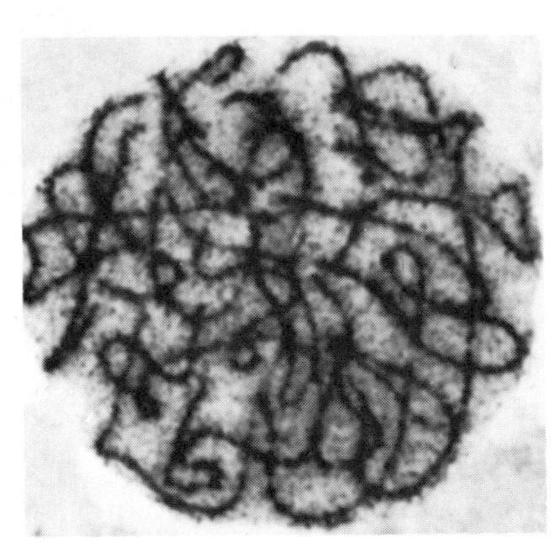

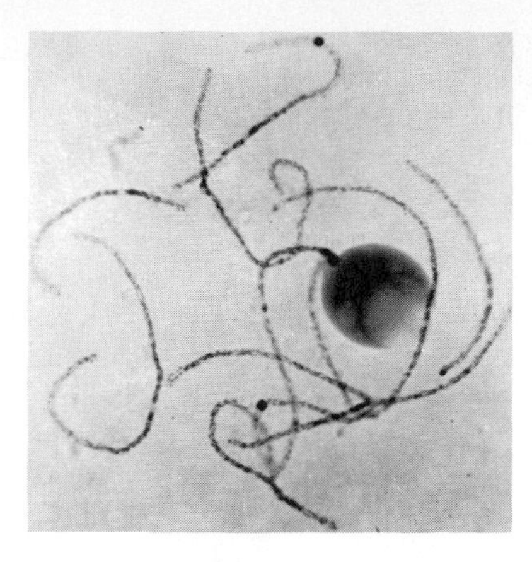

Figure 8.15
Pachytene stage in a microsporocyte of maize, with chromosome 6, in a paired condition, attached to the nucleolus by its nucleolar organizer. Maize chromosomes in meiosis are often characterized by prominent chromomeres callel knobs, two of which are readily visible at the top and bottom of the figure, while a larger knob attaches chromosome 6 to the nucleolus. (Courtesy of Dr. M. M. Rhoades.)

Figure 8.16
Chromosome 6 of maize magnified somewhat greater than that seen in Figure 8.15. The nucleolar organizer is indicated by the figure 3, while other recognizable regions of the paired homologues are also labeled with numbers. (Courtesy of Dr. B. McClintock.)

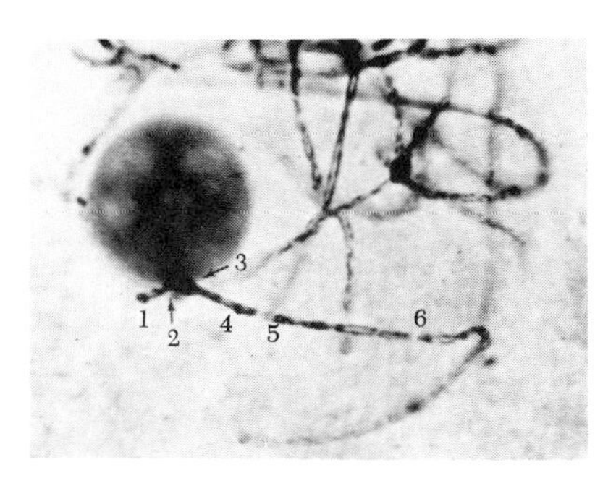

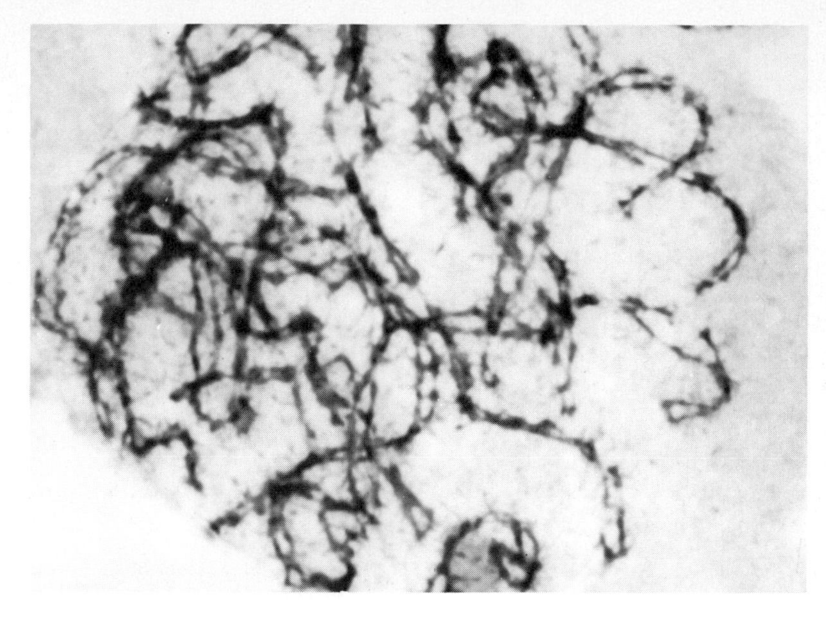

Figure 8.17
Early diplotene stage in a spermatocyte of a salamander. The paired homologues are beginning to fall apart as the synaptic attraction between them lapses. (Courtesy of Dr. J. Kezer.)

Figure 8.18
A middiplotene state in a salamander, with the position of chiasmata visible in a few of the homologous pairs. The four chromatids in each bivalent are readily apparent. (Courtesy of Dr. J. Kezer.)

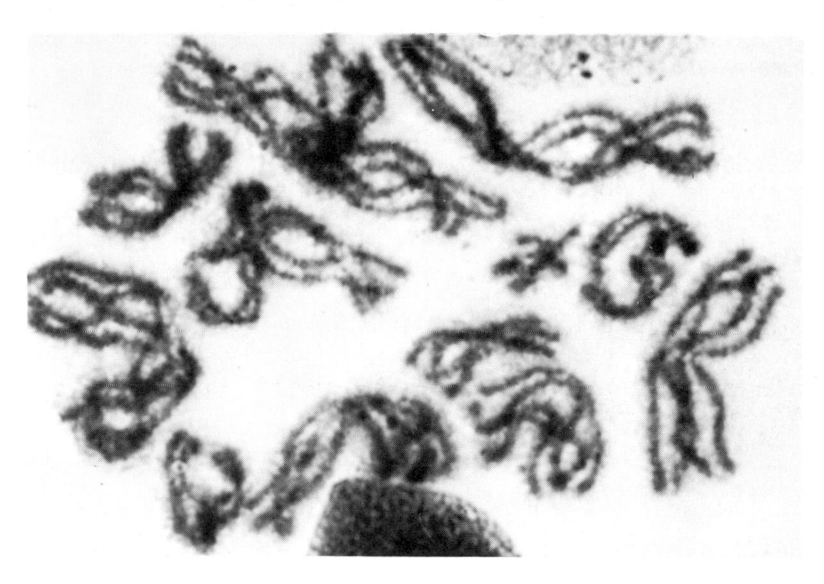

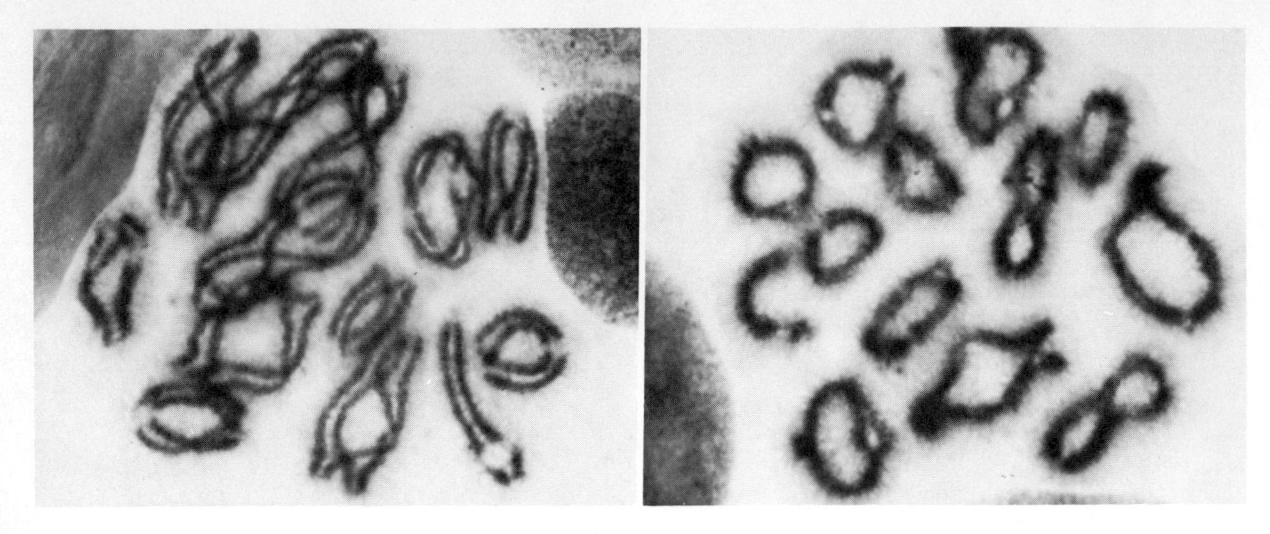

Figure 8.19
Somewhat later stages of diplotene than indicated in Figures 8.17 and 8.18. The right-hand figure shows the characteristic fuzziness of the paired chromosomes at this stage, a fuzziness due to loops of chromatin extending beyond the main contracted portion of the chromosomes. (Courtesy of Dr. J. Kezer.)

Separation of the homologues, however, is not complete. At one or more points along their length, contact is retained by means of *chiasmata* (singular, *chiasma*). Each chiasma results from an exchange of chromatin between the chromatids of two homologues (we shall discuss later in this chapter the significance of this phenomenon as it relates to heredity), but the relationship of the chromatids to a chiasma is clearly indicated in Figure 8.20. To produce such an exchange of chromatin between two chromatids, the double helices of both chromatids must be broken and then reunited to effect the exchange. The repair of such broken ends

Figure 8.20
A single bivalent at the diplotene stage, which illustrates the position of the two chiasmata formed previously, and the position of the centromeres. The accompanying diagram traces all four chromatids throughout the length of the bivalent. (Courtesy of Dr. J. Kezer.)

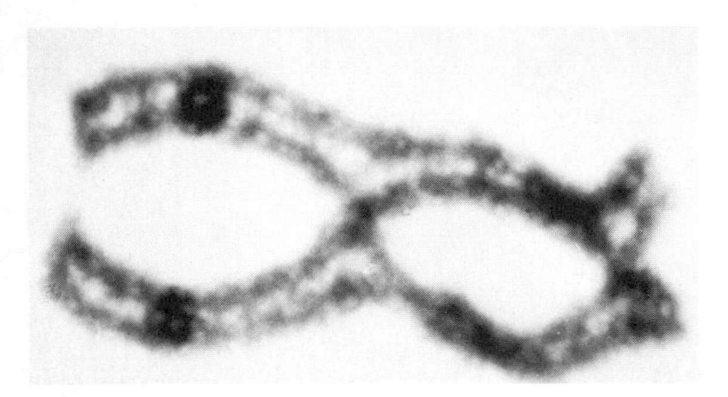

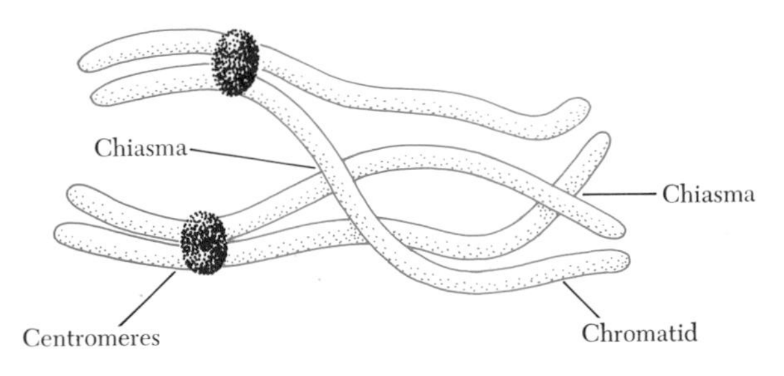

requires the formation of new DNA; through the use of radioactive thymidine, such repair has been detected.

When only one chiasma has formed, the bivalent in the diplotene stage appears as a cross. If two are formed, the bivalent is generally ring-shaped; if three or more form, the homologues develop a series of loops (Figures 8.19 and 8.20). In different cells, the number and approximate positions of the chiasmata vary, even for the same bivalent, but as a rule long chromosomes have more chiasmata than short ones, although even the shortest seem able to form at least one chiasma.

The next prophase stage is called *diakinesis,* but the distinction is not sharp between it and the diplotene stage. During diakinesis the nucleolus becomes detached from its special bivalent and disappears, and the bivalents become considerably more contracted (Figure 8.21). Also as contraction proceeds (Figure 8.19), the chiasmata tend to lose their original position and move toward the ends of the chromosomes.

We have mentioned that the chromosomes shorten as they progress from the leptotene stage onward through prophase. This is accomplished by the development of a series of coils in each chromatid, which gradually decrease in number as their diameters increase. The process is no different from the shortening of chromosomes in mitosis; the coils here, however, are more easily observed, particularly when the cells have been pretreated with ammonia vapors or dilute cyanide solution before staining. Figure 8.22 illustrates the coils as they appear in the spiderwort, *Tradescantia.*

The breakdown of the nuclear membranes and the appearance of the spindle terminate prophase and initiate the *first metaphase*

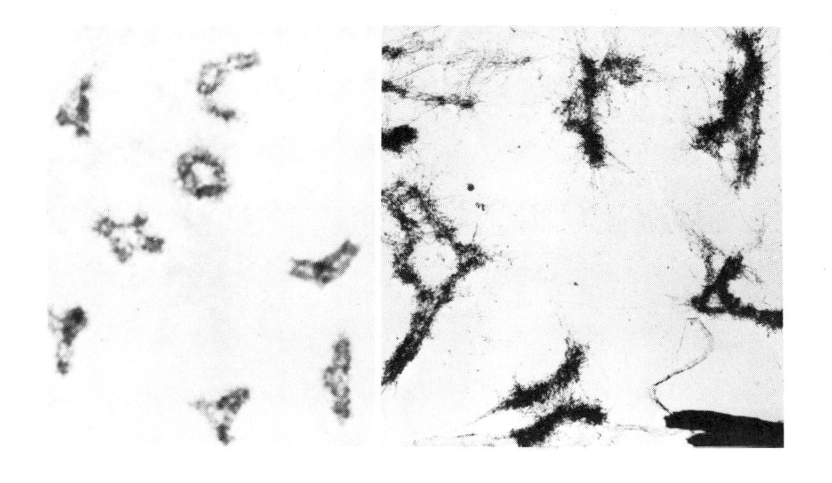

Figure 8.21
Diakinesis in the milkweed bug, *Oncopeltus.* The figure at the left was prepared by the usual fixing and staining methods for light microscopy; that on the right was floated onto a water surface before being lifted off and dried by the critical point method, a technique that dehydrates the specimen without serious distortion. It was then photographed in the electron microscope. (Courtesy of Dr. S. Wolfe.)

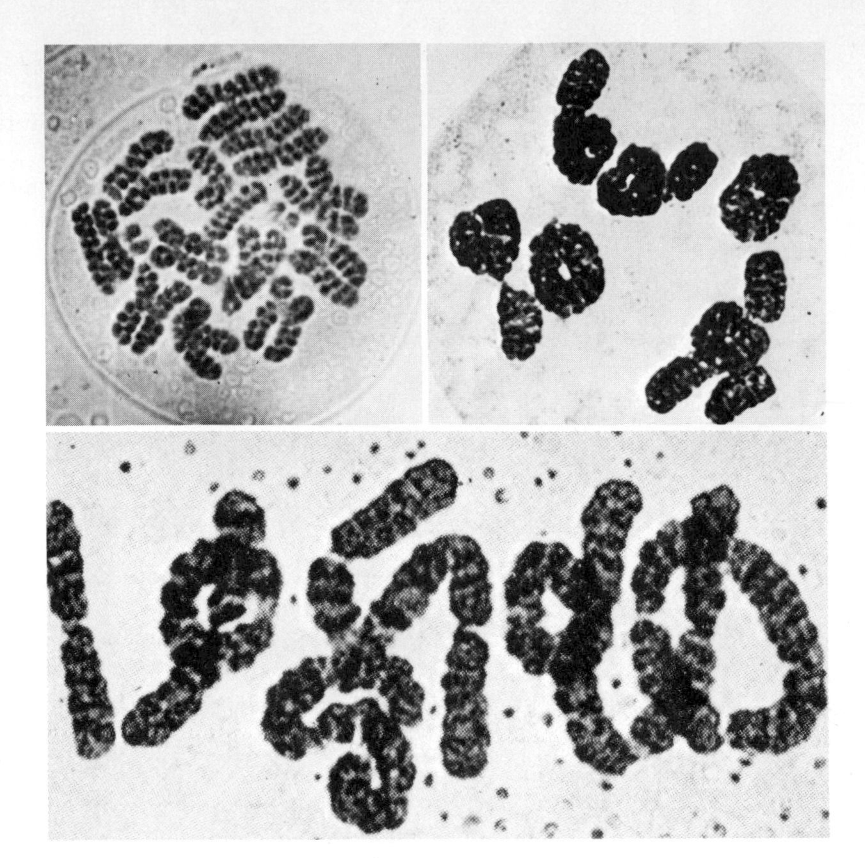

Figure 8.22
Coiling of the chromosomes of *Tradescantia.* Top left: a flattened anaphase I of a diploid form with the coils clearly evident in each chromatid; top right: a triploid form at metaphase I; bottom: a tetraploid form at metaphase I.

of meiosis (Figures 8.6 and 8.7). The bivalents then orient themselves on the spindle, but instead of all centromeres being on the equatorial plate, as in mitosis, each bivalent is so located that its centromeres lie on either side of, and equidistant from, the plate (Figures 8.23 and 8.24). This seems to be a position of equilibrium.

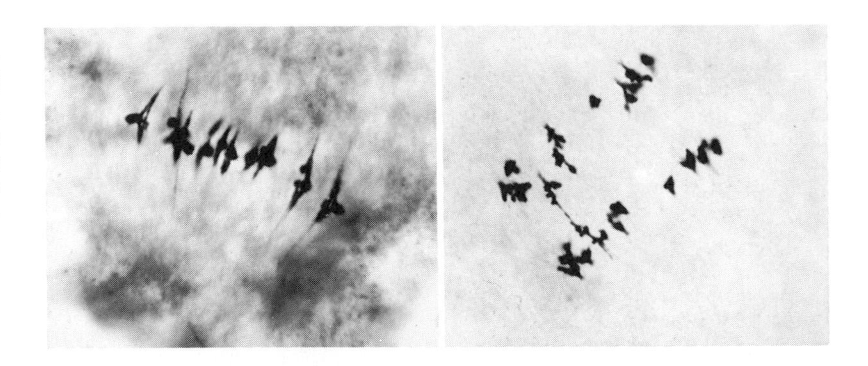

Figure 8.23
Metaphase I (left) and anaphase I (right) in maize. The stretching of the chromosomes to the poles can be seen in the left figure; the right one indicates that at times the homologues may have difficulty separating from each other. (Courtesy of Dr. M. M. Rhoades.)

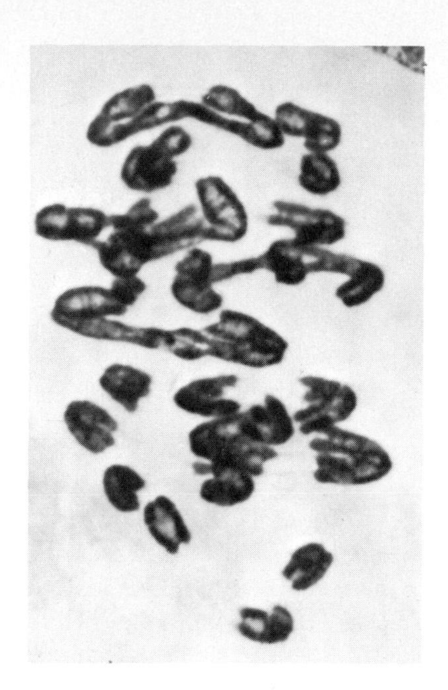

Figure 8.24
Anaphase I in a salamander. (Courtesy of Dr. J. Kezer.)

The *first anaphase* of meiosis begins with the movement of the chromosomes to the poles (Figure 8.25, see also Figures 8.23 and 8.24). The two centromeres of each bivalent remain undivided, and their movement to the opposite poles of the spindle causes the remaining chiasmata to slip off and free the homologues from each other. When movement ceases, a reduced, or haploid, number of chromosomes will be located at each pole. Unlike mitotic anaphase, in which the chromosomes appear longitudinally single, each chromosome now consists of two distinctly separated chromatids united only at their sister centromeres. The nucleus then forms, the chromosomes uncoil, and the meiotic cell may be bisected by a membrane wall. This is the *first telophase of meiosis* (Figure 8.6).

After an interphase that, depending on the species involved, may be short or long—or even absent altogether—the chromosomes in each of the two haploid cells enter the *second meiotic division* (Figures 8.6 and 8.7). If an interphase is absent, the chromosomes

Figure 8.25
Metaphase I and anaphase I in the milkweed bug, *Oncopeltus*. Top left, a side view of metaphase I, with the sex chromosomes only loosely paired with each other; bottom left, a polar view of metaphase I, with the sex chromosomes occupying the center of the metaphase plate, and with the X chromosome to the right and the lighter Y chromosome to its left; right, anaphase I. (Courtesy Dr. S. Wolfe.)

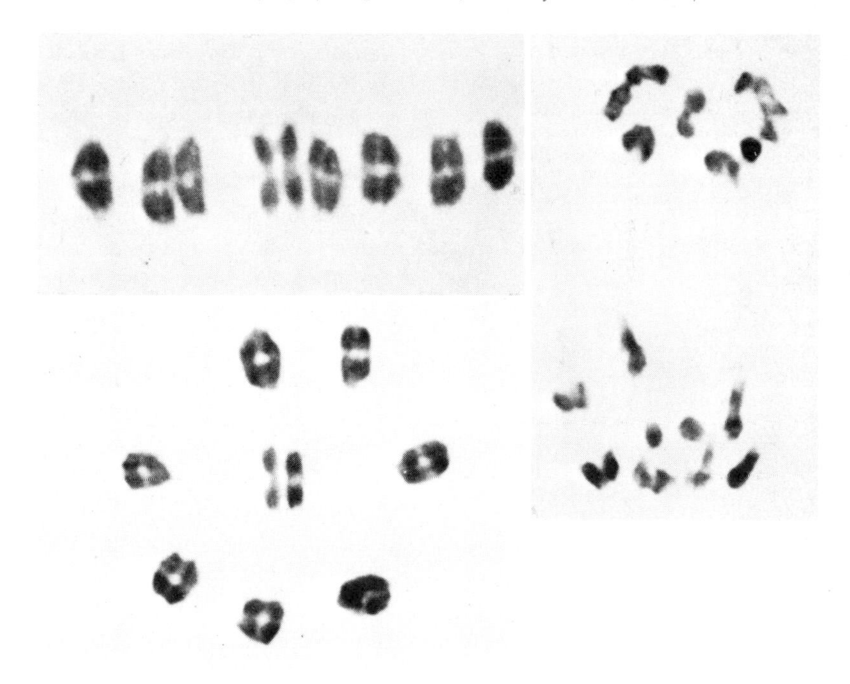

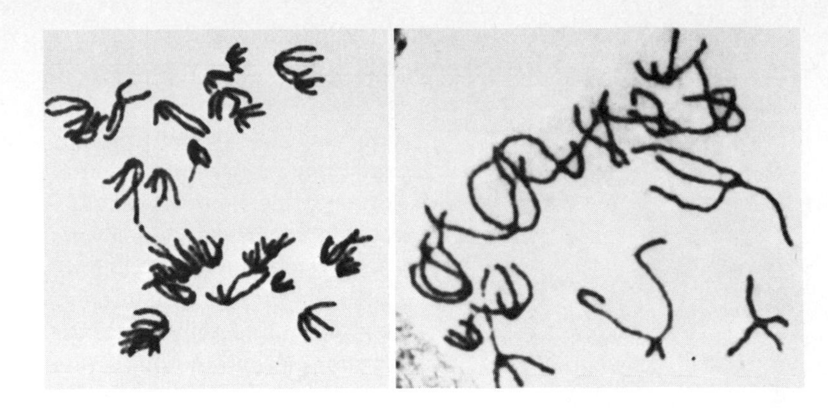

Figure 8.26
Anaphase I and prophase II in a salamander to show that there is relatively little change in the character of the chromosomes during the interval between the two meiotic stages. (Courtesy of Dr. J. Kezer.)

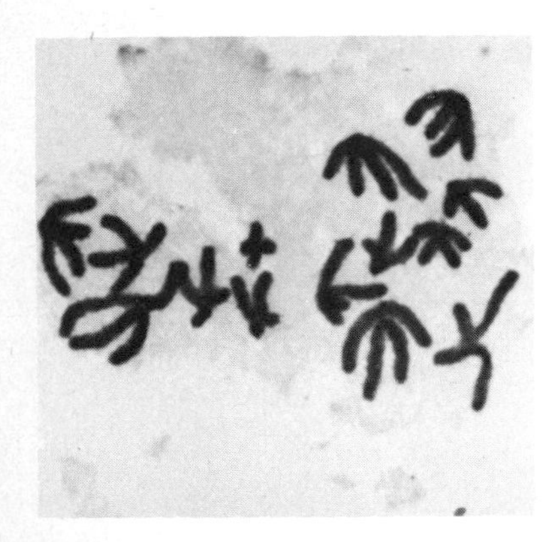

Figure 8.27
Anaphase I in a salamander to indicate that the chromatids are held together only at the centromere regions. These chromosomes are somewhat more compacted than those in Figure 8.26. (Courtesy of Dr. J. Kezer.)

pass directly from the first telophase to the *second prophase* without any great change in appearance (Figure 8.26). If an interphase is present, a nuclear membrane forms in telophase, the chromosomes uncoil, and a somewhat more prolonged second prophase is found. But whatever the case, the chromosomes reaching the *second metaphase* (Figure 8.27) are essentially unchanged from what they were in the previous anaphase; that is, *no chromosomal replication occurs during interphase,* and the centromere of each chromosome remains functionally undivided. A spindle forms in each of the two cells, and at the *second anaphase,* the centromeres separate and the chromosomes move to the poles (Figure 8.28). The nuclei are reorganized during the *second telophase,* giving four haploid nuclei that become segregated into individual cells by segmentation of the cytoplasm.

Looking back over the events of meiosis, we find that the chromosomes remained unchanged in longitudinal structure from the diplotene stage to the end of the second meiotic division. The replication of each chromosome occurred during the premeiotic interphase, but this was followed by two divisions; in the first, homologous centromeres and hence the homologues separated from each other at an anaphase I to reduce chromosome number, an event made possible because synapsis joined them and chiasmata held them together until metaphase I; in the second, sister centromeres and hence the two chromatids of each chromosome separated.

At this point you may well ask why the reduction in chromosome number could not be accomplished just as efficiently with a single division instead of two. Where only a single meiotic division is found, as happens during sperm formation in the normally haploid male honeybee, and where a reduction in chromosome number is not a necessary feature in the life cycle, it is essentially

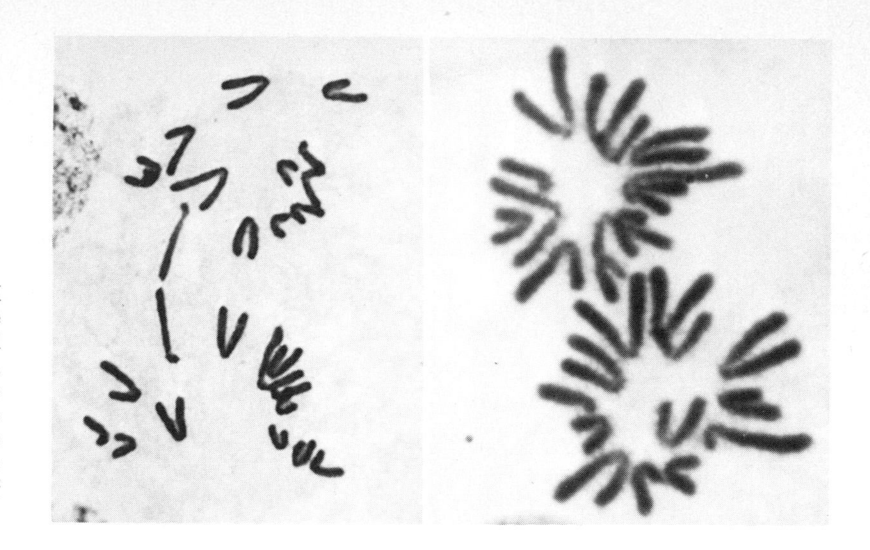

Figure 8.28
Side and polar views of anaphase II in a salamander. The chromatids are now separating, and each nucleus will round up and become, in this instance, the nucleus of a sperm. Notice that the distinctiveness of the chromosomes can be seen in that they are not of the same size and their centromeres are not similarly located. (Courtesy of Dr. J. Kezer.)

like the second rather than the first division and, therefore, more mitotic in character except for the nature of the resultant cells. In organisms that have a diploid chromosome number, the reduction could take place in one division if no prior replication of the chromosomes had occurred, but since DNA replication appears to be part of the initiating mechanism of division, the second meiotic division, without another round of replication, is necessary to bring about the reduction in chromosome number. The occurrence of crossing over between homologous chromosomes also indicates that one meiotic division would be insufficient to accomplish the segregation of genes when in a heterozygous condition (see page 207).

In the animal kingdom, meiosis leads to the formation of sexual gametes, the egg and sperm usually being the only cells carrying a haploid complement of chromosomes. In the plant kingdom, however, meiosis can occur at various times during the life cycle, and the haploid products may be sexual gametes or asexual spores, depending on the particular group of plants being studied. We shall consider here only the formation of the products of vertebrate meiosis, that is, the egg and sperm, and meiosis as it occurs in the flowering plant *Zea mays*.

Meiosis in the vertebrate animal The primordial germ cells of the human embryo make their appearance approximately 20 days after fertilization; they migrate from their origin in the wall of the yolk sac to the developing gonads during the fifth week of development. Once located in the female gonad, these cells become the source of the female germ cells. They divide rapidly to form clusters of *oogonia* near the outer wall of the ovary, and each cluster is transformed into a layer of flat epithelial, or nurse, cells surrounding a central cell, which, at the end of the third month of development, becomes the *primary oocyte*. The cluster is known as the *primordial follicle* (Figure 8.29). The primary oocyte enters meiosis as soon as it is formed, and by the seventh month of prenatal development, all of the oogonia have stopped dividing, and the oocytes have reached the *dictyotene* stage, which follows pachynema but which differs from the typical diplotene stage in that the chromatin is quite diffuse in appearance (Figure 8.30). The oocytes remain in this state until sexual maturity.

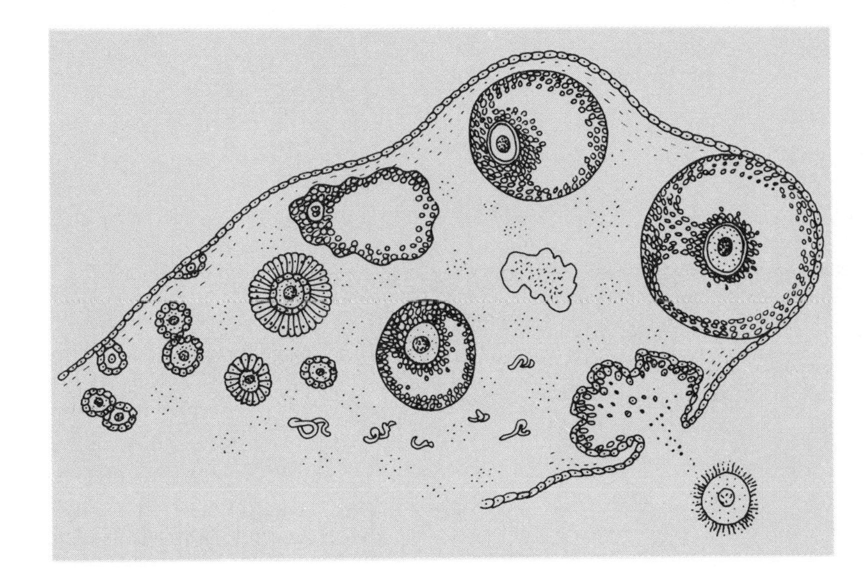

Figure 8.29

Diagrammatic view of a section of a mammalian oocyte, showing the progressive development of the oocytes as they arise from the germinal epithelium at the left, form a layer of nurse cells around them, increase in size, sink into the interior of the ovarian tissue, and finally escape to the outside by rupture of the wall of the Graafian follicle.

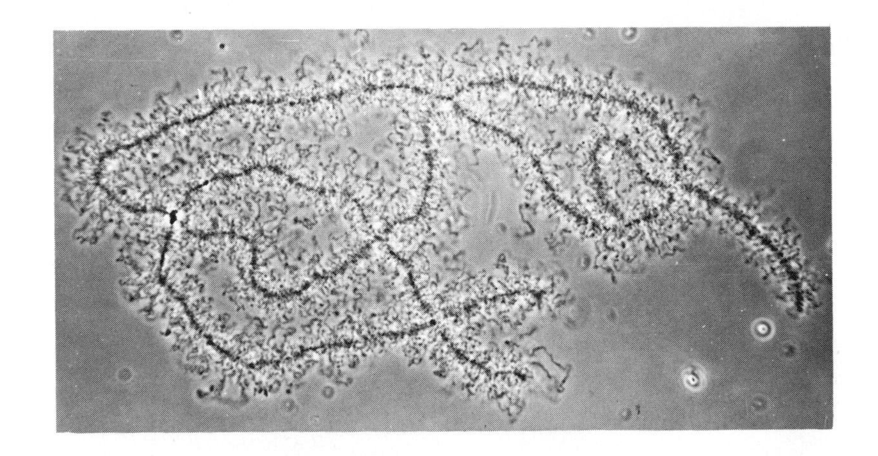

Figure 8.30

A meiotic bivalent of the newt, *Triturus*, consisting of two homologues held together at four points by chiasmata, and in the diffuse diplotene stage. The fuzziness results from the projection of loops of chromatin away from the linear body of the chromosomes, and these loops are in active states of synthesis, somewhat comparable to that depicted in Figure 2.28. (Courtesy of Dr. J. Gall.)

By birth, therefore, a human female will presumably have formed all of the oocytes she will have, and they will have progressed well into the prophase of meiosis. There is some uncertainty about this, however, for the number of oocytes has been estimated to range for 40,000 in a new-born infant to over 400,000 in a woman 22 years of age. During each ovarian cycle, several oocytes begin development, but usually only one achieves maturity; the remainder disintegrate. As a rule, therefore, only a single functional oocyte produces a fertilizable egg during each ovarian cycle, and if the childbearing years are assumed to cover the period of 12 to 50, only 400 or so of the oocytes reach maturity. The youngest of them will have spent no less than 12 years in meiosis; the oldest may have spent as many as 50 years. The entire course of events from primordial germ cell to Graafian follicle is depicted in Figure 8.29.

Initially, the *primary oocytes* lie close to the germinal epithelium, but later they increase in size and sink into the interior of the ovary where they become surrounded by *follicle cells,* which probably have both a protective and nutritive function. The whole structure is now known as a *Graafian follicle.* During this process of enlargement and encapsulation, the oocyte is building up reserve food material, the yolk. This food, which may be protein or fat in mammals, is generally distributed throughout the cytoplasm as yolk spheres or granules. In the frog, however, the yolk so completely fills the cell that the cytoplasm is restricted to a small fraction of the cell surrounding the nucleus; the well-known yolk in the hen's egg is also enormous compared to the amount of cytoplasm.

Eventually, the Graafian follicle ruptures and the egg (Figure 8.29), or *ovum,* is released from the ovary and passes into the *oviduct,* or Fallopian tube, where it can be fertilized by a sperm. By this time, however, meiosis has been resumed and has reached metaphase of the second meiotic division. Meiosis will be completed only if the egg is fertilized, the sperm acting as initiating agent. Only a *single* functional cell results, however. The other three cells, or *polar bodies,* are cast off and will degenerate, *but the process has effectively reduced the chromosome number without depriving the egg of the cytoplasm and yolk the embryo will need when it begins to develop.*

The first meiotic division in the primary oocyte takes place close to the cell membrane, and the outermost nucleus, together with a small amount of cytoplasm, is pinched off as a polar body (Figure 8.31). The second meiotic division results in the pinching-off of a second polar body; the first polar body, meanwhile, has also un-

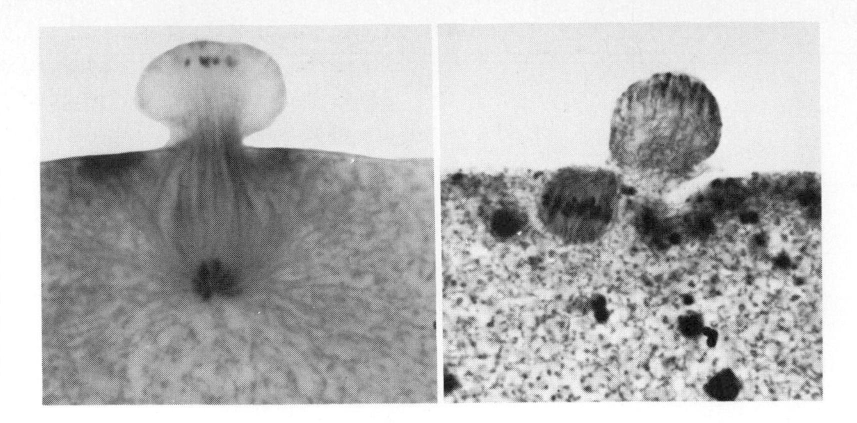

Figure 8.31
Polar body formation in the eggs of the whitefish, *Coregonus.* Left, anaphase of the first meiotic division, with the first polar body being pinched off. Right, metaphase of the second meiotic division, which will lead to the pinching off of a second polar body. The first polar body may or may not divide again. (Courtesy General Biological Supply House, Inc.)

dergone a second meiotic division, thus giving a total of three polar bodies. The haploid nucleus remaining in the egg is now known as the *female pronucleus.* It sinks into the center of the cytoplasm and is ready for union with a similar haploid nucleus brought in by the sperm during fertilization.

The primitive germ cells of the human male enter the developing gonad during the fifth week of development. They become incorporated into the *sex cords,* which at first are solid structures but which after birth develop a lumen and become the *seminiferous tubules.* These make up about 90 percent of the bulk of the testes.

The germinal epithelium contains *spermatogonia,* cells that continue to increase their number by mitotic division throughout the sexual life of the male (see Figure 7.5, Chapter 7). These derivative cells mature into *primary spermatocytes,* which undergo a first meiotic division to produce *secondary spermatocytes;* the latter pass through a second meiotic division, giving four cells called *spermatids.* These become motile sperm by a remarkable transformation of the entire cell. The human male differs from the female in that spermatogonia initiate the production of primary spermatocytes, and hence viable sperm, at the beginning of sexual maturity and continue to do so until old age.

The mature sperm consists essentially of a head and a tail. The head is a highly compacted nucleus, capped by a structure known as the *acrosome* (Figure 8.32). It is derived from the Golgi materials of the spermatid and apparently functions as a device for penetrating the egg during fertilization. Just behind the compacted nucleus is the *middle piece,* formed by an aggregation of the mitochondria. It develops as a sheath around the filament, or tail, and provides the tail with energy for locomotion. The filament, in turn, has developed as the result of a tremendous growth from one of the centrioles; the other centriole remains just beneath the nucleus and at the time of fertilization enters the egg along with the male nucleus. Virtually no cytoplasm except particulate structures is used to form the mature sperm. Each spermatid, therefore, has been transformed from a rather undifferentiated cell into a highly specialized cell capable of reaching the egg under its own power and of penetrating it once it has made contact.

The egg and sperm differ not only in shape; they are vastly different in size. The egg is one of the largest cells of the human body, exceeded in volume only by some neurons. It is just barely visible to the naked eye, having a diameter of about 0.1 mm, but its volume is 2,000,000 μm^3. The sperm, on the other hand, is one of the smallest cells, having a volume of only 30 μm^3. The fact that their genetic contribution to the next generation is equal once again points to the nucleus, the chromosomes, and to DNA as the crucial elements of inheritance.

The mature egg and sperm must unite within a limited period, for neither has an indefinite life span. The critical period may be a few minutes or it may be spread over several hours or days. In mammals, fertilization can occur as the egg leaves the ovary and passes down the oviduct on its way to the *uterus*. Insects, however, mate only once, and sperm are stored in the female and used throughout the entire egg-laying period; in the honeybee, for example, this period may last a year or more. It is now possible to store mammalian sperm for an indefinite period by freezing them, and by means of *artificial insemination* the sperm of a single sire may be used to fertilize the eggs of many females. This practice has been widely used in animal breeding programs, thus passing on the superior qualities of one sire to many offspring, and it has

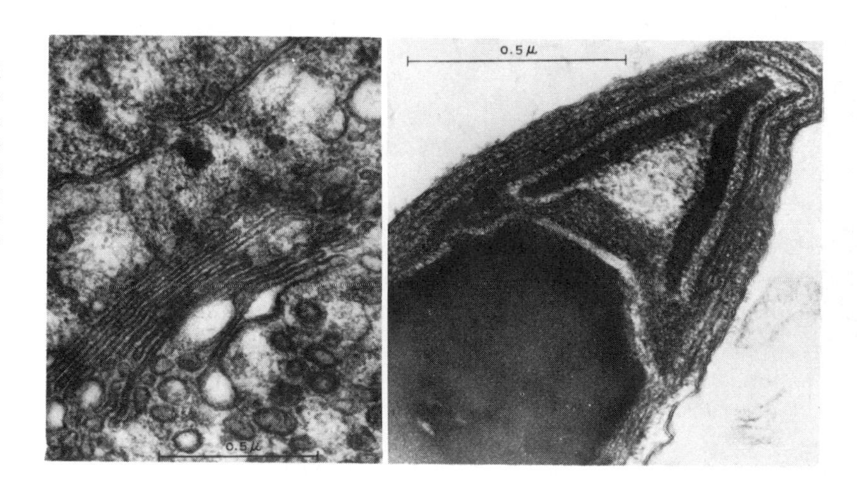

Figure 8.32
Electron micrographs of the Golgi complex (left) and the acrosome (right) of the sperm head of the house cricket, *Acheta domestica*. The cone-shaped acrosome results from the transformation of the Golgi complex during the development of the spermatozoan from the spermatid. (Courtesy Dr. J. Kaye.)

been successfuly carried out in humans when, for one reason or another, normal conception fails.

The essential process of fertilization is the union of male and female pronuclei, but the sperm also functions as an activating agent. That is, nature has ensured against the egg beginning its embryonic development in an unfertilized state; if it did, haploid embryos would result, and even if such embryos were viable and developed into sexually mature adults, the process of meiosis would be hopelessly complicated. Unfertilized eggs of mammals and other related vertebrates can be induced to initiate development by various artificial means, but this rarely occurs naturally.

Fertilization is also a specific process in that the sperm of one species will not, as a rule, fertilize the egg of another species. It now appears that several chemicals are present to ensure proper fertilization and to prevent the penetration of foreign sperm. The egg produces a protein substance called fertilizin, which reacts with an antifertilizin on the surface of the sperm; fertilizin may act to attract sperm of its own kind, but once the two substances interact, the sperm becomes firmly attached to the egg membrane and is then drawn into the interior of the egg. Other sperm are barred from entry by the changes that then take place in the *vitelline membrane* of the egg, an outer coating found on most eggs.

Only the nucleus and one centriole of the sperm enter the egg. The nucleus fuses with the female pronucleus, the centriole divides and begins formation of the first division spindle. In summary, therefore, the entry of the sperm into an egg contributes: (1) a stimulus to development; (2) a set of haploid chromosomes, which is the paternal hereditary contribution to the newly formed zygote; and (3) a centriole, which is involved in the machinery of cell division.

Meiosis in a flowering plant Meiosis in a flowering plant such as maize, or Indian corn, gives rise to asexual *spores* instead of gametes. These spores give rise to male and female gametophytes, or haploid plants, which produce gametes by mitosis (Figure 8.33).

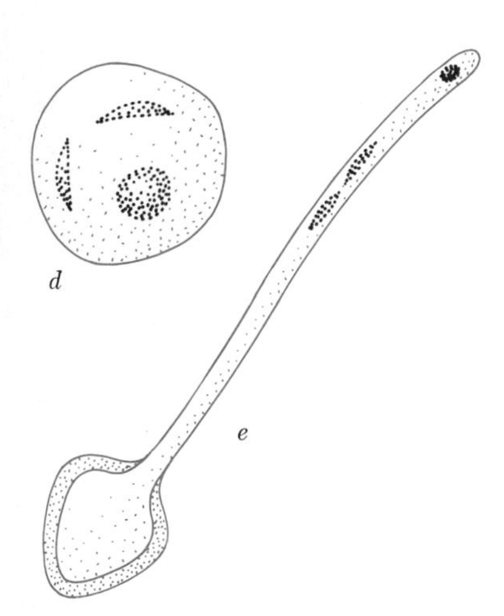

Figure 8.33
Development of the pollen grain in maize. (*a*) Quartet of microspores resulting from the two meiotic divisions of the microsporocyte. (*b*) A single microspore. (*c*) A binucleate microspore with tube and generative nuclei resulting from the first microspore division. (*d*) Mature pollen grain with the tube nucleus and the two sperm, which resulted from a mitotic division of the generative nucleus (the pollen grain is shed in this state). (*e*) A germinating pollen grain with the tube nucleus at the end of the pollen tube and the two sperm that will be carried to the embryo sac for purposes of fertilization.

A comparison with the life cycles of vertebrates can be gained from Figure 8.2.

The flowers of maize are unisexual although borne on the same plant. The male flowers are grouped into an inflorescence or tassel, and meiosis takes place in the sporogenous tissue of the anthers. The *pollen mother cells,* or *microsporocytes,* undergo the two divisions of meiosis to produce four haploid *microspores* (Figure 8.7). The single nucleus of each microspore divides mitotically again to produce two nuclei within its wall: a generative nucleus and a tube nucleus. The former will divide mitotically again to produce two sperm nuclei, after which the microspore matures and is shed as a three-nucleate pollen grain (Figure 8.33).

The female inflorescence, or cob, is made up of individual flowers, each consisting of an ovary and a long silk that is both style and stigma. The ovary bears a single ovule having the structure shown in Figure 8.34. The *megasporocyte* is the equivalent of

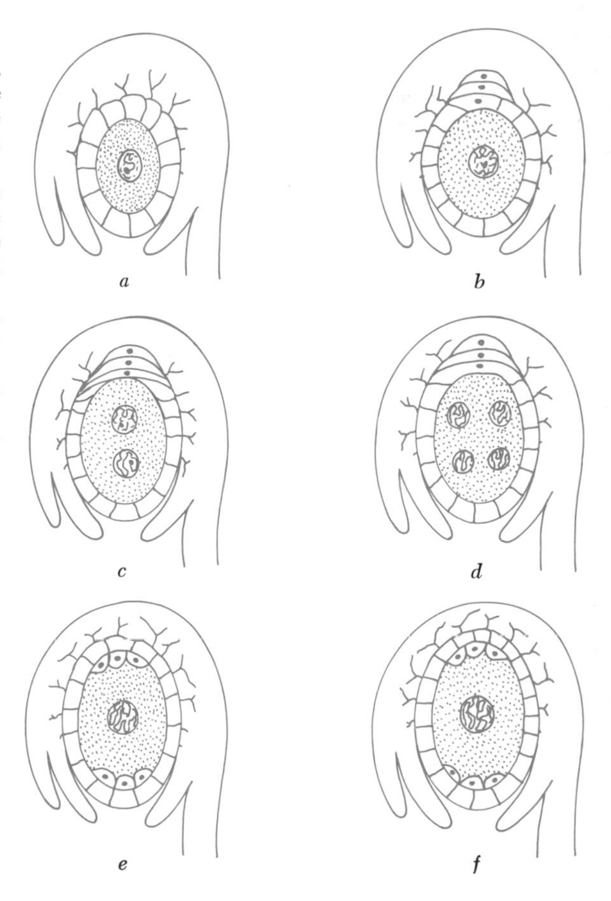

Figure 8.34

Formation of the embryo sac of maize within the ovule. (*a*) The megaspore, which will undergo meiosis to produce a linear tetrad of cells (*b*), three of which will disintegrate while the remaining one enlarges to form the embryo sac. (*c*), (*d*), and (*e*). Successive divisions of the haploid megaspore produces eight nuclei, which in E arrange themselves as follows: three antipodals are at the innermost side of the embryo sac; three nuclei arrange themselves at the opposite end and consist of two synergids and one egg; the remaining two nuclei unite in the center of the embryo sac to form the fusion nucleus, now diploid. (*f*) Fertilization occurs by one of the sperm (solid body) uniting with the egg to form the zygote, while the other fuses with the fusion nucleus to form a triploid endosperm nucleus. As the ovule develops, the endosperm nucleus will divide repeatedly to form a nutritive tissue upon which the developing zygote will feed, particularly during the process of germination.

the animal oocyte, but instead of producing an egg directly through meiosis, it forms a *linear tetrad* of four haploid *megaspores,* three of which abort. The remaining one develops into a *megagametophyte,* and three mitotic divisions convert the single nucleus into eight nuclei. At the end opposite the *micropyle,* three of the nuclei form a group called the *antipodals;* three remain at the micropylar end of the ovule, with one egg being flanked by two *synergids;* and the remaining two nuclei fuse in the center of the megagametophyte to produce the diploid *polar nucleus.* This is the state of the ovule when ready to be fertilized.

The mature pollen grain lands on the silk and germinates by producing a pollen tube, which carries the two sperm to the micropylar end of the ovule. Both sperm enter. One will fertilize the egg nucleus to give rise to a *diploid zygote;* this will eventually form a new plant or sporophyte. The other sperm fuses with the polar nucleus to form a *triploid endosperm nucleus;* this, by repeated mitotic divisions and wall formations, will give rise to the endosperm, a rich nutritive tissue that nourishes the growing embryo in its early stages. Flowering plants, therefore, have a double fertilization process, and the seed is a mosaic of tissues: the diploid integuments, which are maternal in origin and which became the hardened seed coats; the diploid zygote and the resultant embryo coming from the fused egg and sperm; and the triploid endosperm, consisting of two maternal nuclei and one male nucleus. Only the embryo, of course, continues on into the next generation.

Chromosome theory of inheritance

Meiosis has been discussed as a logical and necessary part of the life cycle of a sexually reproducing organism, and it is the antithesis of fertilization as regards the number of chromosomes (Figure 8.2). DNA also has been discussed as the molecular basis of inheritance. It is the crucial portion of the chromosome that maintains its linear integrity, and genes are segments of DNA in a chromosome. It follows, then, that the behavior of genes and chromosomes should parallel each other in a most exact manner, and that the phenomena of mitosis, meiosis, and fertilization can be interpreted in genetic as well as cytological or molecular terms. The merging of the genetical and cytological aspects has come to be known as the *chromosome theory of inheritance;* the molecular phenomena simply confirm and extend the theory.

Mendelian laws of inheritance Gregor Mendel, the Austrian monk and botanist, knew nothing of chromosomes or of haploid and diploid stages of the life cycle when he published his basic laws of inheritance in 1865, but by following the passage of identifiable and mutually contrasting characters in the garden pea (tall versus dwarf, round versus wrinkled seeds, and so on) he was able to make sense of what was taking place. Figure 8.35 illustrates Mendel's first law of inheritance, or the law of segregation. The parental (P_1) plants would breed true if self-pollinated but if crossed would yield seed that would produce F_1 (first filial) individuals, all of which would be tall. Self-pollinated F_1 plants would produce the next generation (F_2 individuals) in which the ratio of tall to dwarf would be approximately 3:1. The same kind of result was obtained when he used other pairs of contrasting characters, so it was apparent that what he was observing was a general phenomenon and not a special case of inheritance.

Mendel reasoned that each parent contributed a *factor* to the offspring, and that the two factors would be responsible for the expression of a character. There would, of course, be as many factors as there are characters. In the F_1 individuals, dwarfism was not expressed, so that tall was *dominant* to dwarf, which was labeled *recessive*. In the F_1 generation, tall segregated from dwarf when eggs or pollen were being formed and then recombined in random fashion to produce the F_2 individuals in the ratios indicated. Figure 8.36 also interprets these events in terms of meiosis, mitosis, and fertilization, using a single pair of homologous chromosomes.

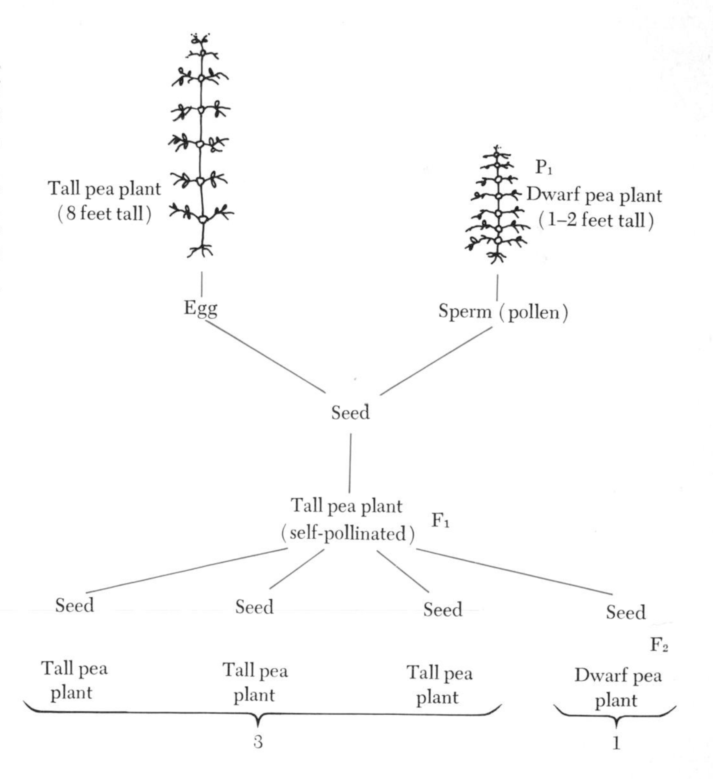

Figure 8.35

A diagrammatic representation of Mendel's experiment in which he crossed two distinct strains of peas—one tall, the other dwarf. These two strains would breed true to each other when self-pollinated. When crossed with each other, the resultant F_1 generation consisted of tall plants only; when these plants were self-pollinated, the seeds developed into an F_2 generation containing both tall and dwarf plants in a ratio of 3:1. The actual numbers of individuals in the F_2 generation were 787 tall and 277 dwarf.

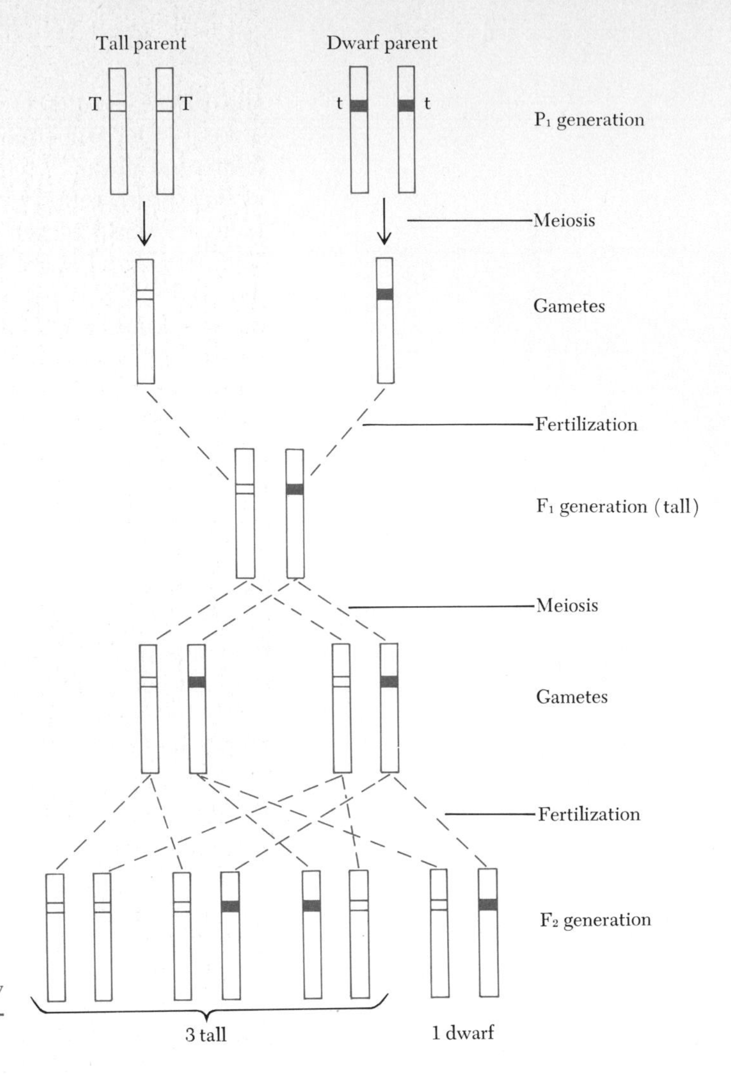

Figure 8.36

Diagrammatic representation of Mendel's first law of segregation, with the dominant (*T*) and recessive (*t*) alleles on particular chromosomes.

Mendel's law was a brilliant piece of abstract reasoning from carefully kept quantitative data; our present knowledge of chromosomes permits us to place the same information on a physical and molecular basis.

Mendel's second law, of the independent assortment of factors, is illustrated in Figure 8.37. If it is assumed that each pair of genes is related to a different pair of homologues, and if it is assumed further that the pairs of homologues segregate independently of each other, then the behavior of the genes and the chromosomes is exactly parallel. This situation is made more complicated by considering what happens when four pairs of homologues segregate (Figure 8.38). If the bivalents are randomly oriented on the meta-

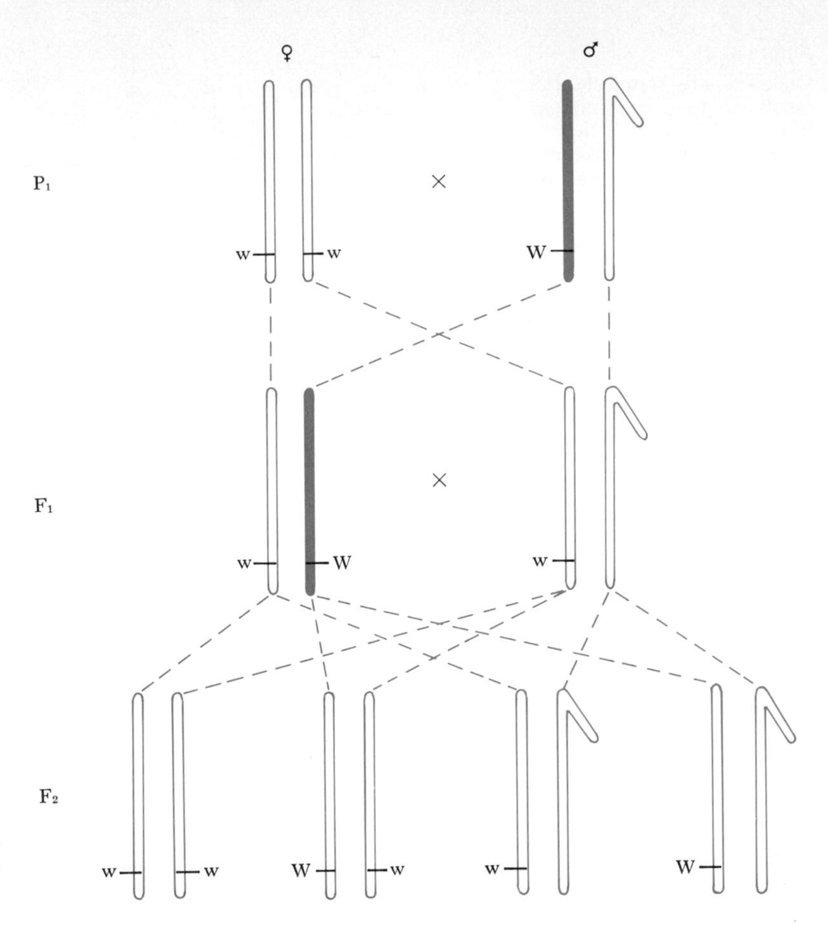

Figure 8.39
Diagrammatic representation of a sex-linked cross involving a red-eyed male (W) and a white-eyed (w) female. Such a cross yields white-eyed males and red-eyed females in the F_1 generation, and an F_2 generation of 1 w female to 1 W female to 1 w male to 1 W male. The Y (hooked) chromosome is devoid of genes.

phase I spindle of meiosis, each gamete should contain four haploid chromosomes, with 16 different possible combinations of maternal and paternal chromosomes. The number of gametic combinations possible can be readily obtained by calculating the value of 2^n, where n equals the number of pairs of chromosomes. In the human, having 23 pairs, the number of possible gametic chromosome combinations is 2^{23}, or 8,388,608. The chance of any single human egg or sperm containing only paternal or maternal chromosomes of the previous generation is, therefore, small indeed. Furthermore, since the same number of gametic combinations is also true for the other sex, the possibility of siblings other than identical twins being genetically identical is remote.

Fertilization, then, is the means whereby, through the union of the nuclei of egg and sperm, the genetic contributions of each parent are combined within a single cell. Mitosis will ensure that

Figure 8.37

Diagrammatic representation of a Mendelian cross involving two independent traits or characters. Round (R) and Yellow (Y) are dominant to wrinkled (r) and green (y), and all are seed characters. The F_2 ratio is theoretically 9 R-Y-, 3 R-yy, 3rrY-, 1 $rryy$, and the actual numbers found by Mendel for this particular cross were, respectively, 315, 101, 108, 32.

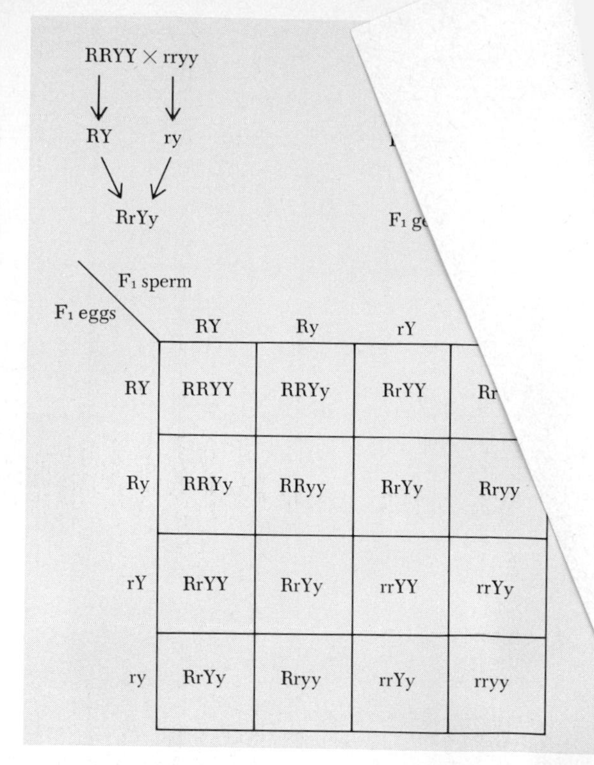

Figure 8.38

The random segregation of paternal (black) and maternal (white) chromosomes during meiosis to give 16 different gametic combinations when four pairs of homologues are involved.

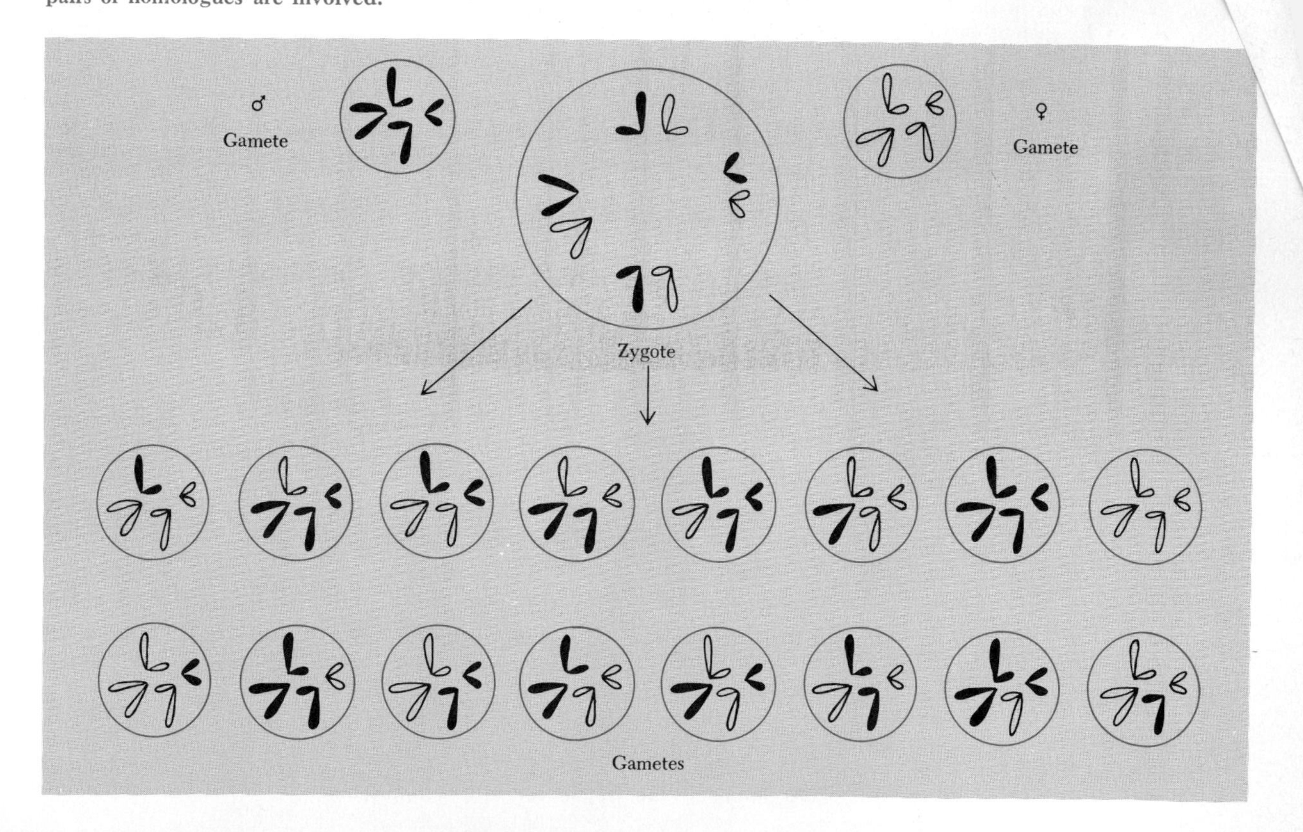

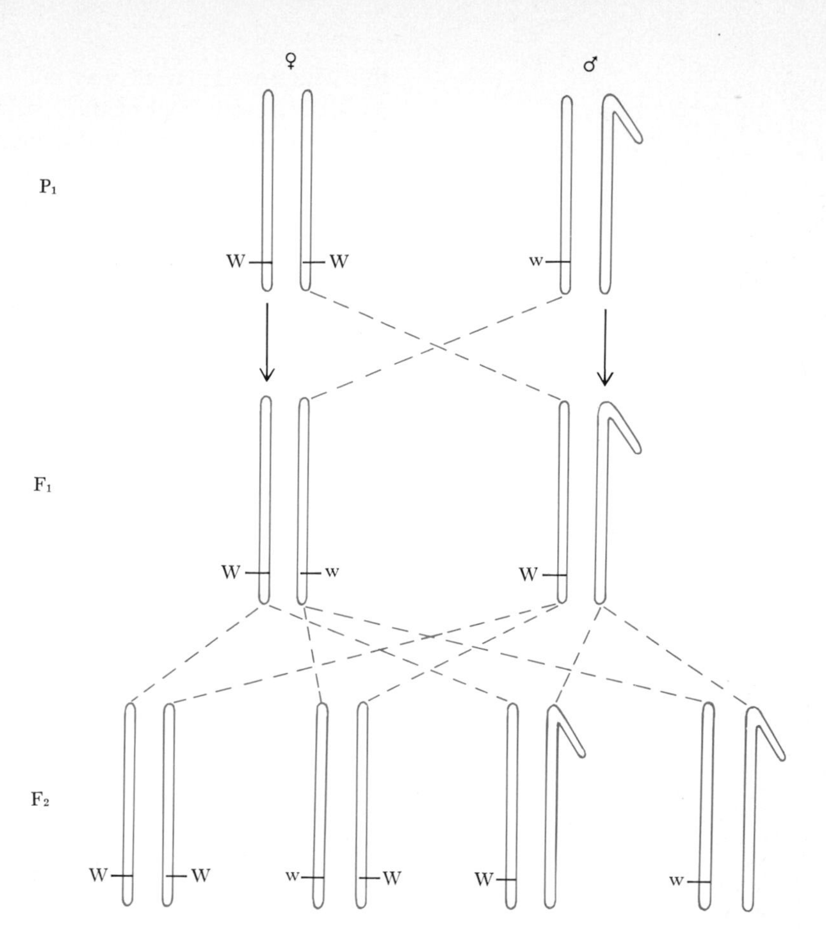

Figure 8.40

Diagrammatic representation of a sex-linked cross involving a white-eyed (w) male and a red-eyed (W) female. All F_1 individuals will be red-eyed; all F_2 females will be red-eyed; half of the males will be red-eyed, the other half white-eyed.

each cell of the body has the same genetic constitution. Meiosis, with synapsis bringing together homologous chromosomes, provides for the segregation of genes to the individual gametes. We may then ask what proof do we have that a particular gene is on a particular chromosome, and that pairs of homologues segregate independently from each other? The answers, coming long before DNA was known to be the crucial genetic substance, involved the relation of a particular chromosome with sex determination.

The X chromosome, so named like an algebraic expression because its value or function was once unknown, is associated with sex determination, and in *Drosophila melanogaster*, the fruit fly, as in human beings, females are XX and males XY. It was also suspected, around 1910, that the first mutation known in this organism, *white eye* (w), was located on the X chromosome. As Figures 8.39 and 8.40 indicate, the inheritance pattern depended upon whether the white-eyed parent was male or female. The inheritance of *w* parallels exactly that of the X chromosome, provided it is assumed that *w* is recessive to the dominant red-eyed varient (*W*).

Occasionally, however, in crosses involving white-eyed females and red-eyed males, an exceptional white-eyed female or an exceptional red-eyed male appears in a culture (Figure 8.41). This would seem contrary to Mendelian segregation, unless it is assumed that an accident occurred in meiosis in the female such that either both X chromosomes or neither entered the egg nucleus. The exceptional females should, therefore, have two X chromosomes and a Y chromosome, the exceptional male one X chromosome and no Y chromosomes. Cytological examination of these flies showed this to be the case, plus the fact that the exceptional males were always sterile due to the absence of the Y chromosome. The exception, therefore, proved the rule in that the *w* gene is on the X chromosome and hence is sex-linked, with the deviant behavior of the genes due to the deviant behavior of the chromosomes. This

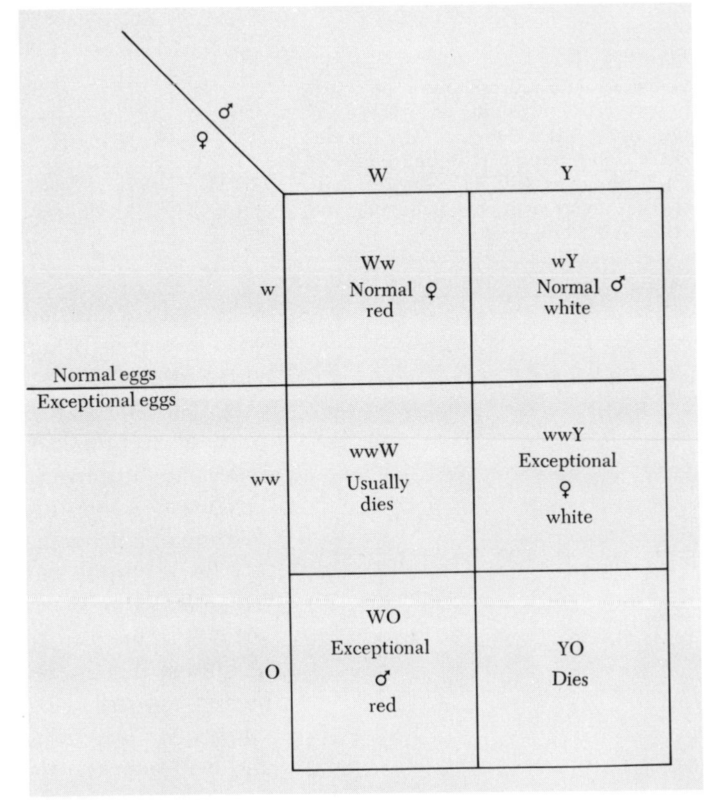

Figure 8.41
Diagrammatic representation of C. B. Bridges' proof that a particular gene is on a particular chromosome. The exceptional eggs—those with two X chromosomes or those with none—arise as a result of nondisjunction, that is, in anaphase I of meiosis the two X chromosomes failed to disjoin, and both went to one pole and none to the other. The exceptional white-eyed females would always possess a Y chromosome in addition to their two X chromosomes; the exceptional red-eyed males would be sterile because they lacked Y chromosomes. In *Drosophila*, the organism used in this instance, the Y chromosome only governs patterns of fertility or sterility; it possesses no other genetic function.

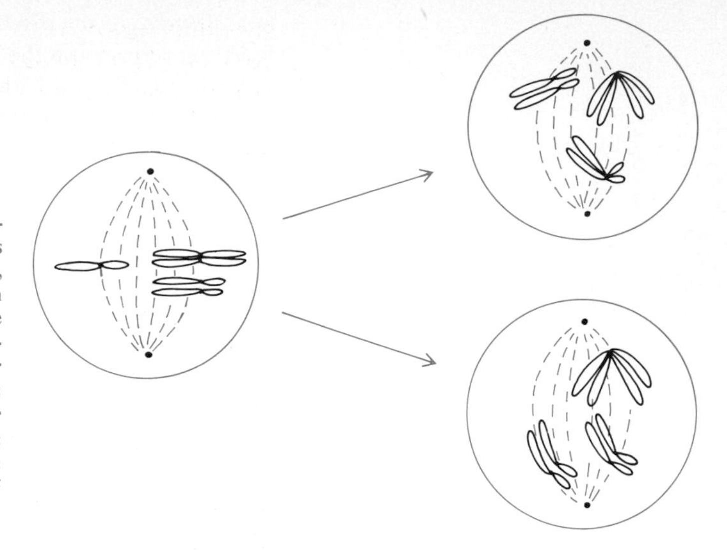

Figure 8.42

Grasshoppers generally have an XO sex-determining mechanism in males, and this is sometimes found with a pair of heteromorphic homologues, that is, two homologues that differ sufficiently in morphology to be distinguishable one from the other. In the above diagram, the single X chromosome in the spermatocytes moves to one pole or the other, and it is a matter of random chance as to which of the homologues it segregates with or from. By such observations, Eleanor Carothers showed that the independent segregation of genes is paralleled by an independent segregation of chromosomes.

phenomenon of meiotic aberrancy was termed *nondisjunction,* and it is the cause in humans of many of the discovered aberrant sex-determining chromosomal situations.

Some organisms such as grasshoppers lack the Y chromosome, and the males, therefore, are XO instead of XY. Also, in some individuals the homologous chromosomes are morphologically different and distinguishable from each other. Taking advantage of these two aspects (Figure 8.42), it is possible to show that the single X chromosome goes to one pole as often with one homologue as it does with the other, supporting, through chromosomal observation, Mendel's contention that genes segregate independently of each other.

Crossing over and chiasma formation

The number of haploid chromosomes in an organism is limited; four in *Drosophila melanogaster,* 23 in humans, 10 in maize. But as the recognized number of mutant genes increased, it became evident that some of the genes must occupy sites on the same chromosome. Each gene, tested individually as in Figure 8.35, would yield F_2 ratios in conformity with Mendelian expectations, but when studied in pairs or in groups of three, the F_2 ratios would often depart significantly from the expected. When retested, the same discrepancies would appear. These findings tended to delay acceptance of the Chromosome Theory of Inheritance until the phenomenon of *linkage* was discovered and understood. Mendelian law may hold, therefore, for the segregation of genes individually, but it is not universally applicable under all circumstances.

Figure 8.43 illustrates the kind of data that might be obtained from a cross when the genes are linked. Notice that the procedures differ from that in a typical Mendelian cross in that rather than obtaining F_2 ratios by the selfing or crossing of F_1 individuals, the F_1s are bred to double recessive individuals by what is termed a *testcross*. The purpose behind this procedure is to test for gametic ratios produced by the F_1s, and by the combining of all F_1 gametes with those from the double recessive, the gametic ratios will be revealed directly. The data reveal that all possible combinations of genes are recovered among the testcross progeny, but it is also evident that the parental types $A\ B$ and $a\ b$ are far more numerous than the new recombinants, $A\ b$ and $a\ B$. The degree of linkage, or the percent of recombination, is determined by adding all new types and dividing by the total number: $(149 + 151) / 3000$. This gives .10 or 10.0 percent, a figure which is a genetic measure of distance between the two genes on the same chromosome. It would have made no difference in quantitative results if the parental combinations had been $A\ b$ and $a\ B$, indicating that the frequency of recombination is a function of chromosomal distance between genes and not of particular genetic combinations.

What has been described is an instance of *incomplete linkage;* the genes are linked, but not absolutely. And the genes are recombined by a process of crossing over. The chromosome, then, which is a molecule of DNA extending from one end to the other, is a series of linked genes. If genes can be recombined by crossing over, homologous chromosomes must be able to exchange chromatin, with the frequency of chromatin exchange correlated in a

Figure 8.43

A diagram of an instance of incomplete linkage in which the genes *A* and *B*—or their recessive alleles—are found on the same chromosome, and departure from randomness of distribution is obvious. See text for further details.

AABB × aabb — Parental cross

↓

AaBb × aabb — Testcross

Testcross progeny	Observed frequencies	Expected frequencies
AaBb	1,358	750
Aabb	149	750
aaBb	151	750
aabb	1,342	750
Totals	3,000	3,000

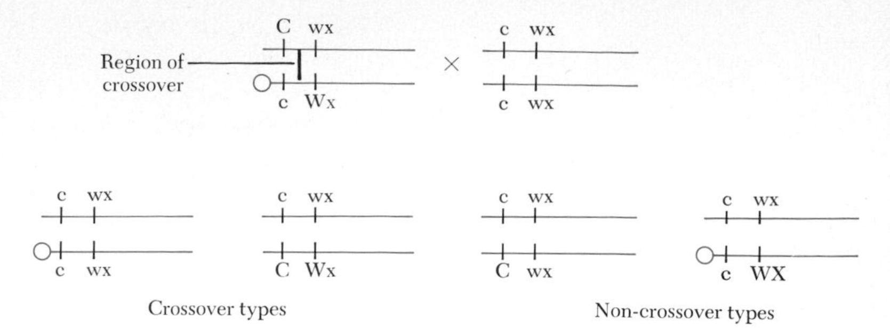

Figure 8.44

A modification of the Creighton-McClintock method for demonstrating that an exchange of genes, when both are on the same chromosome, is accompanied by an exchange of chromatin. The test plant on the left has both genes and chromosomes in heterozygous form, and an exchange of genes would put both recessive genes on the knobbed chromosome and the dominant genes on an unknobbed chromosome. The distance between *C* and the knob is too short to allow crossing over to occur at the same time that it takes place beween *C* and *Wx*.

quantitative way with the frequency of crossing over. The chiasmata visible in diplotene bivalents (Figures 8.19 and 8.20) are physical evidence that crossing over has taken place in the previous stages of synapsis.

It will be noticed in Figure 8.20 that the chiasmata are formed between chromatids of the two homologous chromosomes. Two pieces of evidence provide proof that crossing over (and chiasma formation) involves an actual exchange of chromatin between homologous chromosomes, and that it takes place after the chromosome has replicated itself and is longitudinally double. Chromosome 9 in maize has two forms: with and without a large terminal chromomere. This chromosome also carried the genes *C* or *c* (colored or uncolored seeds) and *Wx* or *wx* (starch or waxy endosperm). When the cross is made as indicated in Figure 8.44, with both genes and chromosomes in a heterozygous state, any testcross individual showing both recessive genes should show the chromomere on one homologue and those showing both dominant genes should be without the terminal chromomere. No exceptions to these expectations were found. Crossing over between these two genes was, therefore, always correlated with an exchange of chromatin. (Figure 8.45). Data from *Neurospora,* the pink bread mold, provided proof that crossing over takes place after, and not before, replication of the chromosomes (Figure 8.46).

The events taking place in meiosis—synapsis, chiasma formation, segregation of homologous centromeres and chromosomes at ana-

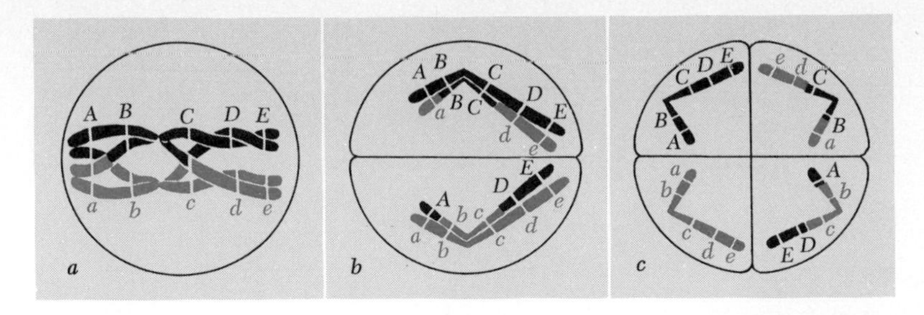

Figure 8.45

The genetic consequences of crossing over. (a) A bivalent, consisting of paternal (black) and a maternal (color) homologue, has formed, and crossing over has taken place between genes A and B, and C and D. (b) At anaphase the two chromatids in each segregating chromosome are no longer alike genetically. (c) The chromatids are now separated, and two of them have a different genetic composition, while the other two remain as before.

phase I and of sister centromeres and chromatids at anaphase II—ensure that the haploid gametes or spores resulting from meiosis will have a variable combination of genes. Since these cells will contribute directly or indirectly through fertilization to the next generation, the individuals of that generation must exhibit a comparable genetic variation. It is this inherited variability upon which natural selection acts to bring about the evolution of organisms. Sexual reproduction, with its complementary phenomena of meiosis and fertilization, is a means not only for the production of new individuals but of new individuals that vary among themselves. In this sense, meiosis differs greatly from mitosis, which, in its production of similarly endowed cells, is a conservative process of reproduction.

Earlier the question was raised as to why two divisions in meiosis were necessary to accomplish a reduction in chromosome number. The problem is actually a more complicated one. If a reduction in chromosome number was the only meiotic event of genetic importance, one division would suffice, but the segregation of genes and the phenomenon of crossing over are involved. Thus, if a particular gene were heterozygous, and a crossover occurred in the four-strand stage and between the gene and the centromere, a heterozygous gamete consisting of two chromatids would result (Figure 8.45). The involvement of such gametes in fertilization would make for additional complications, both chromosomally and genetically. In the first place, the first division of a zygote would have to take place without prior DNA replication or a tetraploid cell would result, and each generation would witness the doubling

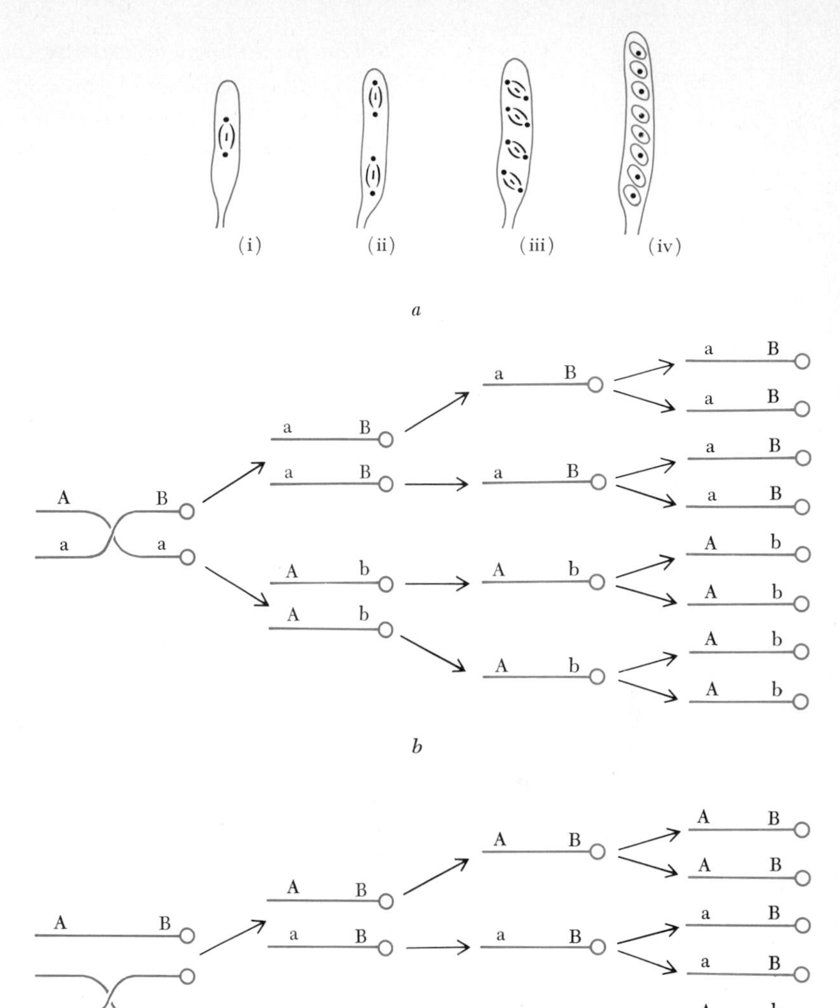

Figure 8.46

Meiotic divisions and segregation in *Neurospora*. (*a*) The two meiotic divisions followed by a mitotic division gives eight ascospores in each ascus. These can be isolated in serial order and their genotypes readily determined. (*b*) The sequence of events and the genotypes and serial order of the eight ascospores that would result if crossing over took place in the two-strand stage and before replication of the chromosomes; that is, if a crossover took place, all eight ascospores would show new combinations of genes from the original A B and a b combinations of the parental strains. (*c*) The sequence of events and the genotypes and serial order of the eight ascospores that would result if crossing over took place in the four-strand stage and after replication had occurred. Since a crossover between two genes would produce two crossover and two noncrossover chromatids, the map distance between two genes cannot exceed 50 map units when based on such a test. The circumstances in (*c*) are typical of such experimental testings, while that in (*b*) is not observed, leading to the conclusion that crossing over takes place in the four-strand stage.

of the chromosome number. Secondly, heterozygous gametes could lead to the formation of genetical chimeras, that is, individuals whose somatic cells have different genotypes, with these bringing on a variety of developmental problems. Meiosis, with its two divisions of the cell but only one replication of the chromosomes, avoids all of these difficulties.

The mechanism of crossing over The fact that crossing over involves the exchange of chromatin means that the helices of DNA somehow must be broken, with the broken ends rejoined in such a manner as to bring about recombination. The process of synapsis brings homologous chromosomes into close register with each other and makes crossing over possible, but from that point on the events are unclear. However, since all events taking place in the cell are chemical events mediated by enzymes, the most reasonable assumption is that crossing over is similarly determined. At least the requisite enzymes are known to exist. The events believed to take place are indicated in Figure 8.47, with only the two chromatids involved in crossing over being considered.

Enzymatically induced "nicks," caused possibly by an enzyme, endonuclease, in single strands of each double helix permit the double helices to open up through the breaking of hydrogen bonds, following which a rejoining takes place. The rejoining is exactly complementary, but this obviously leaves gaps in the double helices, as well as free, unpaired strands. The unpaired regions can be removed by an enzyme exonuclease, the gap can be filled by the action of a DNA polymerase, an enzyme that can add nucleotides in a complementary way to a growing strand, and, finally, a polynucleotide ligase can close the gap by uniting the growing strand to the old one. The actual number of new nucleotides need not be great, and the small amount of DNA replication detected in zygotene-pachytene of meiotic cells could be related to these phenomena.

There is very little direct experimental evidence to prove that the indicated enzymes are actually involved in crossing over, and there is no reason for assuming that synapsis sets up the enzymatic circumstances for crossing over. Rather, it would appear that nicks are frequently produced in double helices and in all kinds of cells, with the cells having a set of repair enzymes to keep the DNA sufficiently intact to perform properly. This is suggested by the observation that mitotic cells show reciprocal exchanges between the two chromatids of a single chromosome, indicating that breaks and rejoinings took place some time during the cell cycle (see Figure 1.15). Synapsis, leading to the formation of a synaptinemal complex, then may be viewed as a means of bringing homologous chromosomes close enough so that the nicks in their respective chromatids may interact with each other, thereby affecting crossing over by increasing the frequency of chromatid exchange and ensuring its reciprocal nature. But since synapsis is characteristic primarily of meiotic cells, it is in these cells that crossover events take place normally, even though it is possible that the number of nicks are the same in both mitotic and meiotic cells.

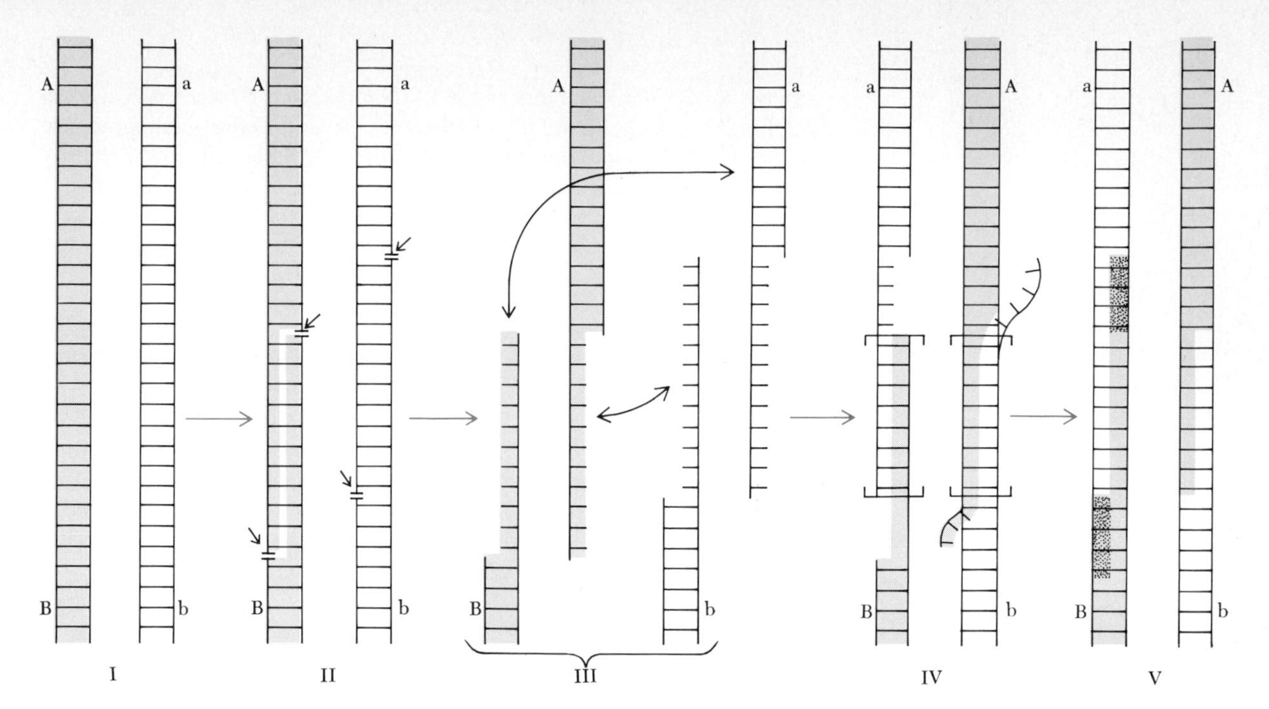

Figure 8.47

A mechanism of crossing over that is consistent with what is known of the process. (I) The two nonsister, but homologous, chromatids that will undergo crossing over between the two linked genes (the other two uninvolved chromatids are not included in this illustration). (II) The enzymatically induced "nicks" in the sugar-phosphate backbones of the polynucleotide strands of the two double helices. (III) The unraveling of the strands, which results from the breakage of hydrogen bonding and which precedes the rejoining to form new genetic combinations. (IV) The rejoined chromatids in a complementary fashion (those areas within the brackets), with gaps remaining to be filled and loose and excess polynucleotide strands remaining to be excised. (V) The finished result of crossing over, with the gaps filled (the darkest areas), the loose ends excised and the breaks in the sugar-phosphate backbone repaired. This is an enzymatically determined process of DNA breakdown and repair, which probably occurs in all cells (see Figure 1.15) but which, in meiotic cells and at the time of synapsis, leads to crossing over.

Bibliography

BAKER, W. K. 1965. *Genetic Analysis.* Houghton Mifflin Co., Boston. A paperback volume providing a sound treatment of cytogenetic information.

KOLLER, P. C. 1971. *Chromosomes and Genes: The Biological Basis of Heredity.* W. W. Norton & Co., Inc., New York. A brief, well-written, and introductory treatment of the interrelations of genes, chromosomes, and inheritance.

LEWIS, K. R., and B. JOHN. 1963. *Chromosome Marker*. J. and A. Churchill, Ltd., London. A difficult book for the undergraduate student, but containing extensive discussions of the role and behavior of chromosomes in inheritance.

MCLEISH, J., and B. SNOAD. 1958. *Looking at Chromosomes*. St. Martin's Press, New York. A small book describing mitosis and meiosis in the lily, and illustrated with a superb collection of photographs.

PETERS, J. A., ed. 1959. *Classic Papers in Genetics*. Prentice-Hall, Inc., Englewood Cliffs, N. J. A group of research papers upon which genetics and cytogenetics were founded; useful for gaining an historical perspective.

WILSON, G. B. 1966. *Cell Division and the Mitotic Cycle*. Reinhold Publishing Corp., New York. A small paperback volume dealing primarily with mitosis, but also including a good account of meiosis.

Note: The student might well consult any of the more recent textbooks of genetics, which generally carry good accounts of meiosis and the transfer of information by genes and chromosomes.

The cell in development

In the past few chapters, we have mentioned that organisms "develop" from a fertilized egg into a plant or animal of adult proportions. Each of us knows in a general way what is meant by development: it is a continuous and gradual process of change that takes time to be fully realized; it is generally accompanied by an increase in size and weight; it involves the appearance of new features and new functions; and it eventually slows down when mature dimensions are reached. Humans, for example, develop from the fertilized egg stage through embryonic and prenatal life, childhood, adolescence, sexual maturity, physical maturity, middle age, old age, and death. Development is, of course, one of the most prominent features in the early life of an organism, but the formation of new blood cells, gametes, and wound tissue, which may take place up to death at an advanced age, are also aspects of development. So, too, are those processes we associate with aging; for example, excess formation of collagen in the extracellular spaces and the calcification of joints. These, it would appear, are normal processes of development continuing beyond the point of a functional and developmental optimum. The terms we have used, however, are only broadly descriptive. They tell us very little about the mechanism of development as a biological phenomenon.

Development of a single-celled zygote into a multicellular organism, whether it be a human being or an oak tree, involves the processes of *growth* and *differentiation*. Growth can be simply defined as an increase in mass, and it results from assimilation of matter. It can involve an increase in cell size and an increase in cell number, the original cell taking from its environment the raw materials it needs and converting them into more substance and more cells like itself. Let us consider the human egg. It weighs about 1×10^{-6} g (grams), and the sperm, at fertilization, adds to it only another 5×10^{-9} g. At birth, however, a child will weigh around 7 pounds, or 3,200 g, which is an increase of about one billion times during the nine-month prenatal period. Another twentyfold increase in mass occurs between birth and the achievement of full size of an average adult.

Increase in mass, however, is not sufficient to account for the particular *form* of an organism. Development of form requires differential growth; that is, different rates of growth and different rates and patterns of cell reproduction are involved in the determination of form, resulting in the majesty of a giant redwood or the "fearful symmetry" of Blake's tiger. In other words, some parts of the body grow at a faster or slower rate than others, and in development some features come into existence earlier than others. Figure 9.1 shows how the growth rate in the human being alters the relative proportions of bodily parts to one another. The head and

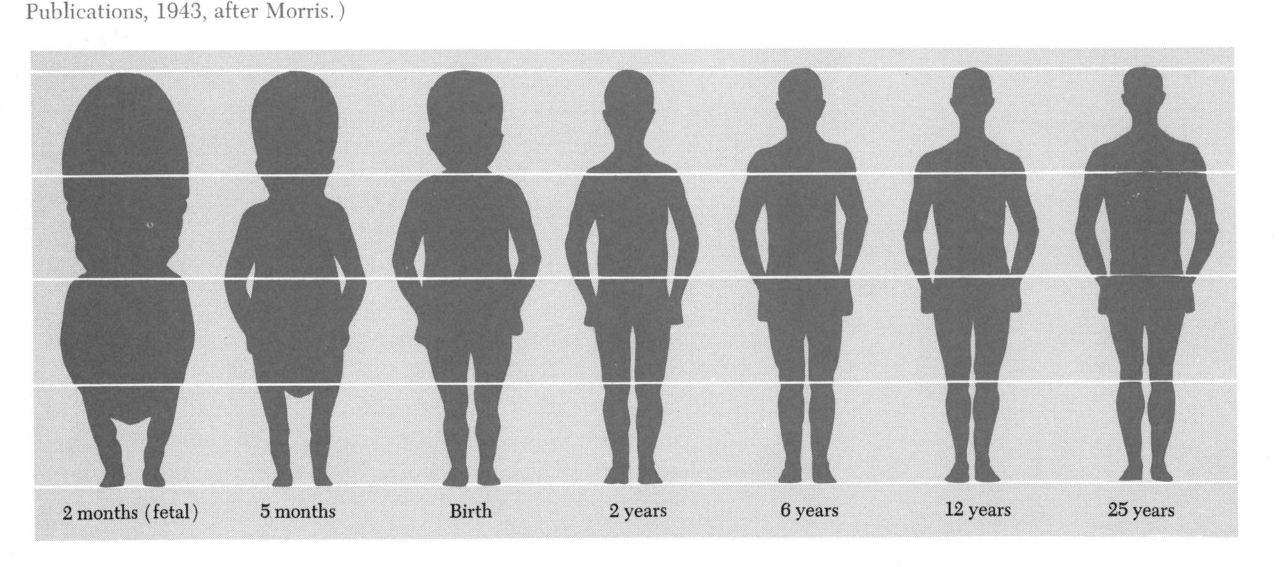

Figure 9.1
Changes in the form and proportion of the human body during fetal and postnatal life. (From H. B. Glass, *Genes and the Man.* New York: Columbia University Teachers College Bureau of Publications, 1943, after Morris.)

neck increase in size rapidly during the early period of gestation, the arms grow faster at an earlier stage than do the legs, whereas the trunk progresses at a more or less steady rate until maturity.

Developmental growth, therefore, is not just the enlargement and multiplication of cells; it is a complicated pattern, with different centers of growth being active at different times and with different rates of development. These centers are coordinated to produce an unfolding of form, and it is form—as well as function, of course—that distinguishes humans from other animals, one human being from another, and an orchid from a lily.

Development also involves differentiation, the acquisition (or loss) of specific structural and functional properties by different cells, such that these cells become specialized in different ways to carry out the various activities associated with living things. A generalized cell, therefore, is gradually transformed by a process of successive changes into a specialized one, and diversity is thereby introduced into a functioning organism (Figure 9.2). In the human, for example, growing cells are transformed into the myriad of different cells that makes up the human body (Figure 9.3): cells of the nervous, muscular, digestive, excretory, circulatory, and respiratory systems.

Differentiation, therefore, is a process of directed change. It is a

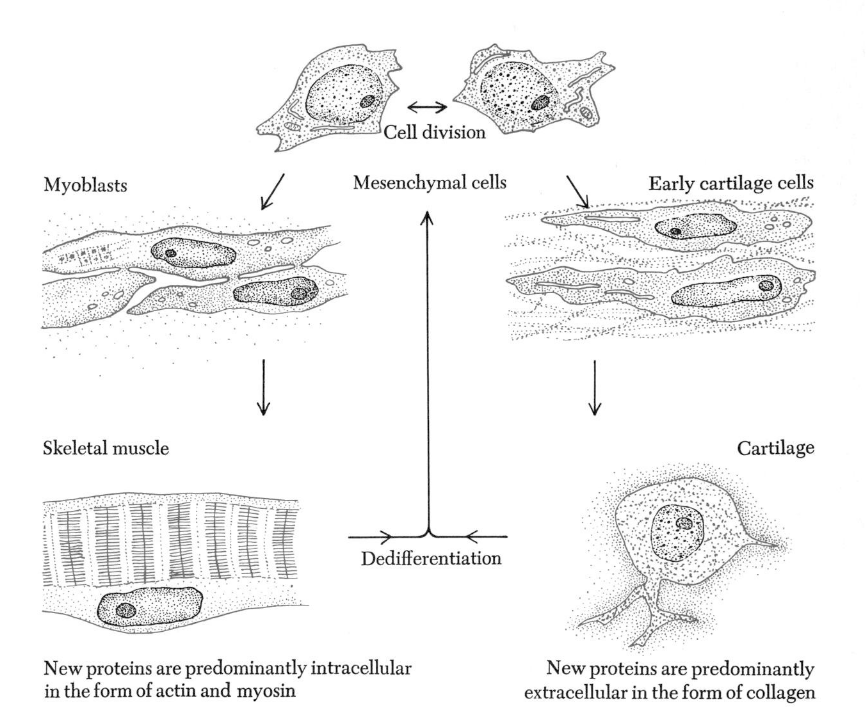

Figure 9.2
Differentiation of generalized mesoderm cells (mesenchyme) into two kinds of specialized cells, muscle and cartilage. Both these cells are similar in that they produce substantial amounts of protein, but in muscle cells the proteins (actin and myosin) are retained internally for contractile purposes, while in cartilage cells the protein is deposited as collagen outside the cell, where it plays a supportive role. The ultimate shape of the cells also changes as a result of differentiation. (After C. H. Waddington, *Principles of Development and Differentiation.* New York: The Macmillan Company, 1966.)

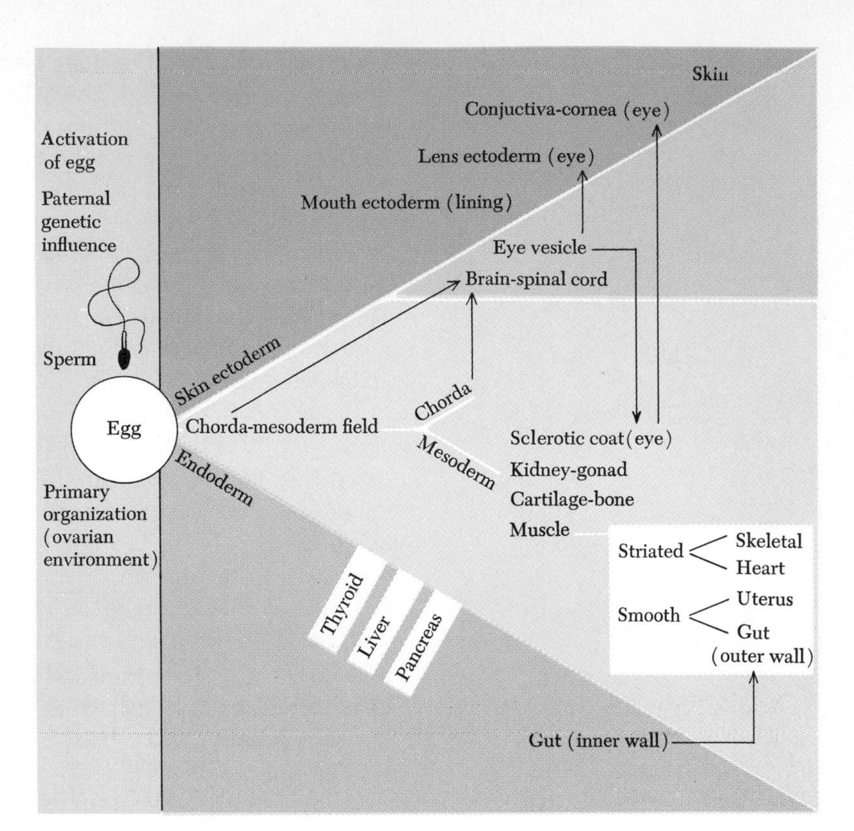

Figure 9.3
Diagrammatic representation of the pattern of progressive differentiation from unfertilized egg to mature tissues in a vertebrate. The three major tissue layers (ectoderm, mesoderm, and endoderm) originate early and progressively give rise to the cells of the major organs. Dashed lines indicate an influence of one tissue on another during the course of development. Note that the eye has a double origin from both ectoderm and mesoderm. (Courtesy of Dr. B. H. Willier.)

phenomenon that has no counterpart in the nonliving world, and any information we have about it has been derived from observations of living systems. This process is creative in the sense that life is creative, for out of the general features common to all cells arise structures and functions that are peculiar to specialized cells. Differentiation, therefore, is to development what mutation is to biological inheritance and what imagination is to human endeavor; it provides variety of form, function, and behavior without at any time destroying the unity of an organism as an individual.

Both growth and differentiation occur during the life cycle of unicellular organisms as well as in multicellular organisms. However, an additional factor is involved in development in multicellular types. For such development to proceed in an organized manner (which indeed it usually does), growth and differentiation of one cell must be coordinated with growth and differentiation of other cells. Otherwise chaos would ensue. Such coordination requires some sort of communication between different cells in the developing organism. Development, therefore, depends on *inte-*

grated growth and differentiation, and it is this integration that permits harmonious development and maintenance of unity of the whole organism.

The problems of development are many and complex, and we cannot deal with all its aspects. However, the cell is the basic building block of life, and since it is the cell that undergoes developmental changes, we must consider the cellular basis of development. Let us begin by examining the nature of the changes that cells undergo as they differentiate and become specialized.

What a cell is, and what it can do, are consequences of the chemical reactions that take place in that cell and that have taken place in the past. As we have seen in Chapter 2, the reactions that occur in a cell are determined by which enzymes are present and functionally active in the cell. Thus, we can think of differentiation of cells as being essentially a change in cellular proteins, in particular a change in the enzyme complement of the cell. However, we also have seen that because of the exact replication and segregation of DNA during cell reproduction, all of the cells in a multicellular organism contain the same type and amount of nuclear information. Since the nucleus is the control center of the cell, all cells should have the same potential and be capable of manufacturing all of the proteins of the body. The obvious fact, however, is that they do not do so. Red blood cells produce hemoglobin, nerve cells do not; certain cells of the pancreas produce insulin, others do not; the mesophyll cells of a plant leaf contain the enzymes involved in photosynthesis, root cells in general do not. Since particular sets of proteins are determined by particular sets of genes, differentiation must be a result of particular sets of genes functioning in different cells.

It can be demonstrated in several ways that differentiation is not a consequence of loss of genetic information. Regeneration of whole organisms from small parts of an original individual is commonly encountered in plants and in lower animals. For example, a stem cutting can produce functional roots, usually from tissue formed by reproduction of cells at the cut surface. Thus, the differentiated cells of the stem have not lost the information necessary for the functioning of a root cell, since this information is expressed in their descendants. A more dramatic demonstration that differentiated cells retain the information required for all of the activities of the whole organism comes from experiments in which segments of phloem parenchyma of a carrot are explanted and the cells induced to divide. The resulting callus tissue can then become organized in such a way that root and shoot meristems are formed and an intact, fertile carrot plant is produced (Figure

9.4). Such experiments tell us that once a differentiated cell is freed of the physical and chemical constraints imposed upon it in the intact organism, its full potential may be expressed.

Nuclear transplanation experiments carried out on certain amphibians also show that differentiation need not be a consequence of loss or addition of genetic information. Nuclei from intestinal epithelial cells of *Xenopus* tadpoles can be transferred by careful micromanipulation into eggs from which the original nucleus previously has been removed. These eggs can then give rise to tadpoles that ultimately develop into adult toads, showing that the nuclei of the epithelial cells retain the full potential of the nucleus of the zygote (Figure 9.5).

Heterokaryons formed by fusion of cells of different types also provide evidence for the reversible nature of differentiation. Chicken erythrocytes, unlike those of mammals, do not lose their nuclei when mature; however, the nuclei eventually become inactive and no longer synthesize any DNA or RNA. When such chick red blood cells are fused with Hela cells (a human cell line derived from a tumor), their nuclei respond by synthesizing DNA and RNA, and chick-specific proteins, which normally are not made by the mature, differentiated erythrocyte, are formed by the heterokaryon.

Differentiation, therefore, is not a result of changes in the genetic potential of a cell, but rather of differential expression of that potential. The above examples suggest that it is the environment

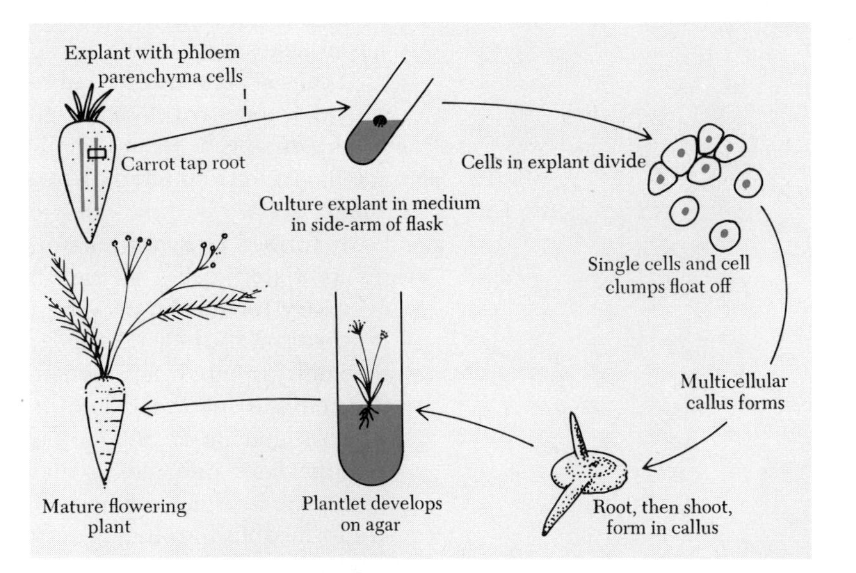

Figure 9.4
Diagram of development of whole carrot plant from cells derived from phloem parenchyma.

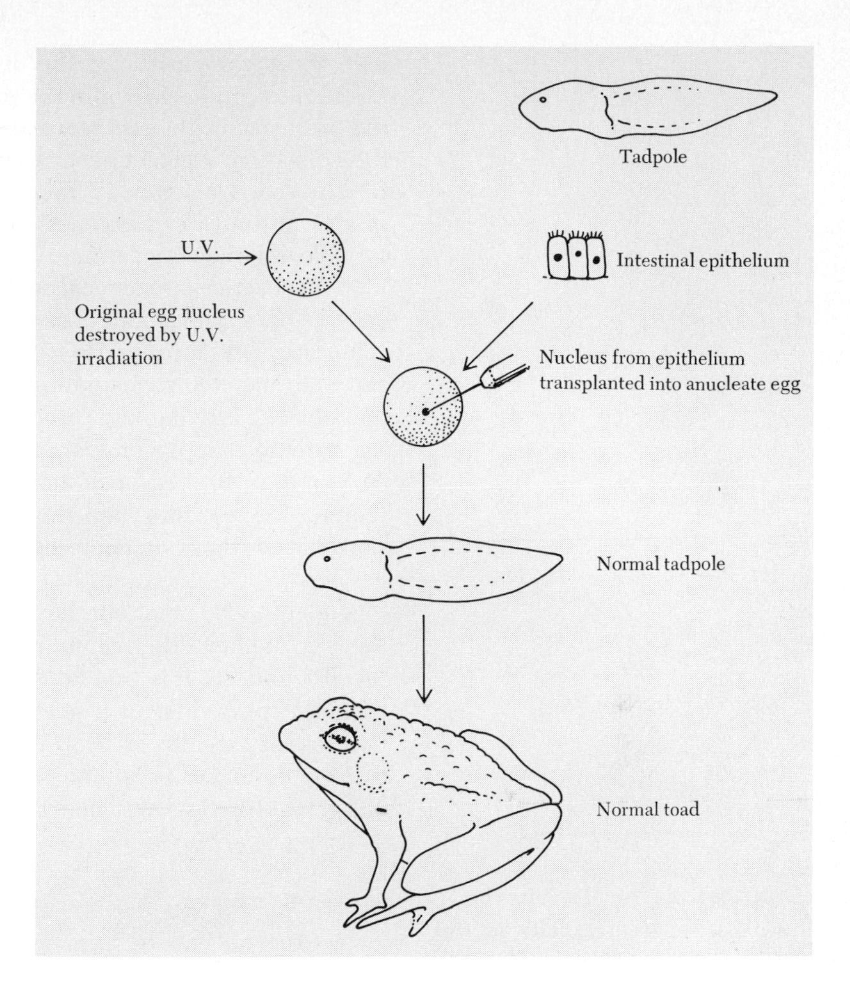

Figure 9.5
Diagram of nuclear transplant from tadpole cell into anucleate egg. The egg will develop normally into a mature toad, with the characteristics of the donor.

in which the cell, and more specifically the nucleus, finds itself that determines how the potential is expressed. Several other examples will make this clear.

1. A plant grown from seed in the dark will not produce chlorophyll. The plastids are present in an undeveloped form but are nonfunctioning in terms of photosynthesis. The addition of light to the environment will cause the plastids to develop and eventually to begin photosynthesizing. The genes responsible for the development of chloroplasts are expressed only in the presence of light. Light, therefore, as a controlling factor governs the differentiation of the plastids and leads to the selective and coordinated synthesis of a distinctive set of proteins.
2. The bacterium *E. coli* can be grown in a culture medium

with the sugar glucose as the only source of carbon and energy; *E. coli* also can be grown in the presence of the sugar galactose. In the presence of glucose, however, the enzyme system required for the utilization of galactose is absent; it appears only when galactose is provided. Galactose is, therefore, a specific inducer and is required to activate the genes responsible for the enzyme system capable of utilizing galactose.

3. By a series of chemical reactions, each mediated by a specific enzyme, *E. coli* manufactures its own histidine (an essential amino acid) at a rate appropriate to the needs of the cell. If histidine is added to the culture medium, the genes responsible for the formation of these enzymes are turned off, and they remain inactive until the external supply of histidine is exhausted and the necessary enzymes are formed once again. Externally supplied histidine is, therefore, a repressor, and the cell by means of a feedback repression controls the activity of some of its constituent genes.

We must also point out, however, that many of the changes cells undergo as they differentiate may be difficult, if not impossible, to reverse. Indeed, it is generally true that the more specialized a cell type, the more difficult it is to change the pathway of differentiation, either directly or by transplantation of the nucleus (Figure 9.6). Differentiation in multicellular organisms is a progressive affair requiring several generations of cells before the final and

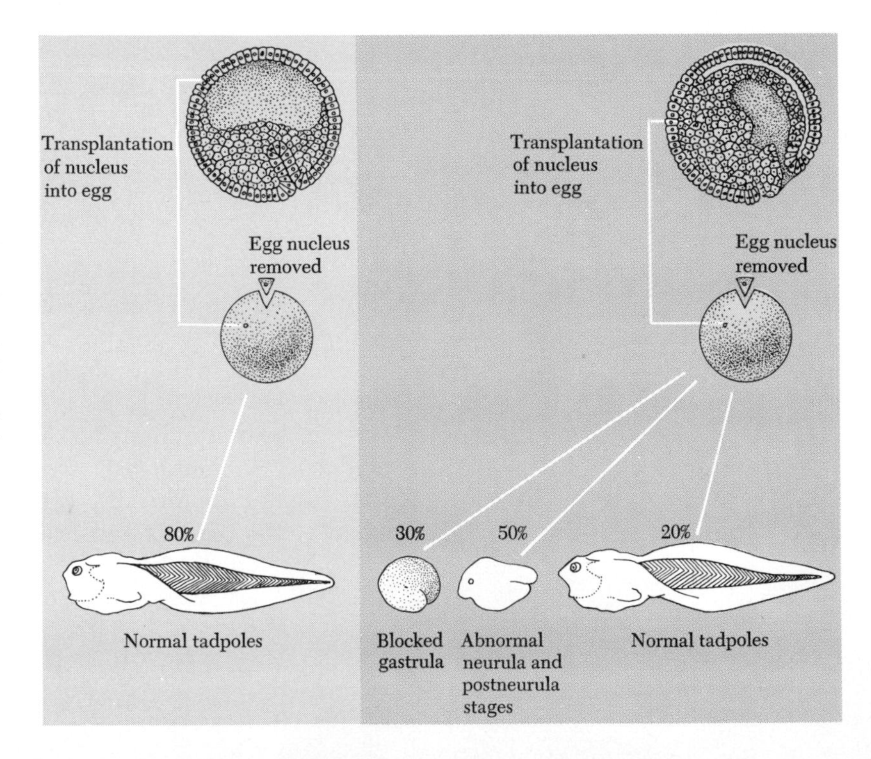

Figure 9.6

A comparison of the ability of two kinds of nuclei to induce development when transplanted into enucleated eggs. Only about 40 percent of the eggs develop regardless of the source of the nuclei, but in those derived from peripheral cells, 80 percent of the developing eggs form normal tadpoles (left), while only 20 percent are normal when the nuclei are obtained from cells in the interior of the gastrula. The nuclei are, therefore, varied in their potentiality as a result of differentiation. (After L. J. Barth, *Development: Selected Topics*. Reading, Mass.: Addison-Wesley Publishing Co., Inc., 1964.)

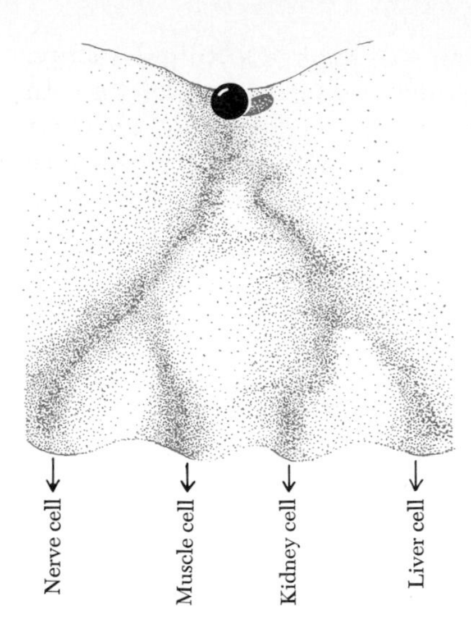

Figure 9.7
How an uncommitted cell (represented by a ball) may become committed by rolling down one of the channels of differentiation. (After C. H. Waddington, *The Strategy of the Genes*. New York: The Macmillan Company, 1957.)

stable differentiated state is achieved. At some point in the sequence, cells become committed to a course of action from which they cannot readily be diverted. C. H. Waddington, an English embryologist, has expressed this idea of commitment by his diagram of a developmental landscape (Figure 9.7). He visualizes a generalized cell as a ball rolling downhill toward its final destiny, a destiny that depends on which of the many valleys the ball rolls through. The farther the cell penetrates into the developmental landscape, the greater the loss of general properties, the greater the acquisition of special features, and the less likely is it able to return to an undifferentiated state.

Let us express this in more specific terms. The embryologist can approach the problem by cutting certain cells out of an embryo and transplanting them to other embryos (Figure 9.8). If a group of young, undifferentiated cells are transplanted to the future head region of another embryo, the transplanted cells become part of the head region; if they are transplanted to the back, they will become part of the back musculature; if to the posterior part, they become part of the tail. But if the embryologist transplants "committed" cells from an older embryo in the same way, instead of becoming an integral part of the region to which they are transplanted, they tend rather to retain their own identity and even to modify the surrounding cells. This is well illustrated by an experiment done in the chick embryo. If the leg bud, which has no resemblance to a mature leg in any way, is removed from a young chick embryo and is transplanted to the body cavity of another embryo, the cells in the bud live, continue to increase in number, and eventually form therein a very well-developed leg with bones and muscles (Figure 9.9). Yet, the bud at the time of transplantation had no obvious bone or muscle cells. In terms of Waddington's

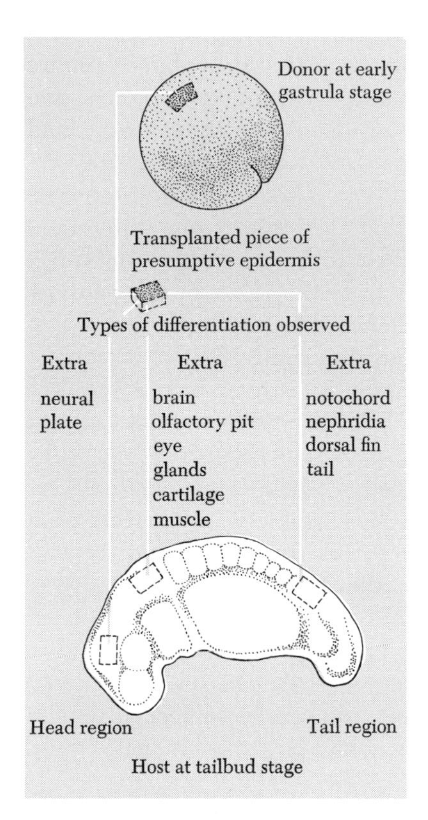

Figure 9.8
An experimental demonstration that cells that would ordinarily form epidermis can, if transplanted to other regions of an older embryo, be incorporated into those regions. Differentiation, therefore, occurs after the transplantation and not before. (After L. J. Barth, *Development: Selected Topics*. Reading, Mass.: Addison-Wesley Publishing Co., Inc., 1964.)

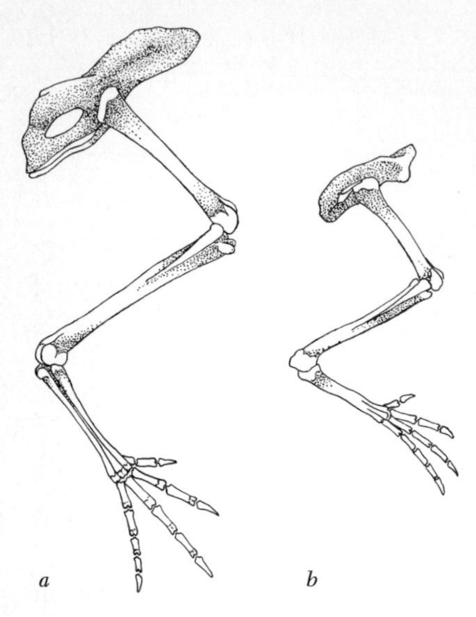

Figure 9.9

An example of a structure that develops after transplantation in a reasonably normal fashion. (*a*) Normal leg bones of a chick 18 days after incubation. (*b*) A slightly smaller but reasonably complete set of leg bones that developed after the hind limb bud (similar to the limb buds shown in Figure 10.12) was transplanted to the body cavity. At the time of transplantation, the limb bud showed no evidence of bone or muscle, but the cells had already been "committed" to leg formation, a process of differentiation that continued even though the limb bud had been removed to a foreign location. (From V. Hamburger and M. Waugh, *Physiological Zoology, XIII* (1940), 367–380.)

landscape, however, the cells had already entered a "valley" leading to leg formation, a valley down which they continued to roll and from which they could not escape. Indeed, the limb bud cells can be separated from each other and then allowed to reassociate. Following reassociation, the tissue is still able to develop into a normal limb.

Experiments carried out on ferns also show that tissues become partially autonomous early in development. If very young leaf primordia are removed from the apex of *Osmunda* and cultured on a synthetic medium, they develop into shoots and ultimately give rise to whole plants. Older leaf primordia, however, develop into leaves, as they would have had they not been explanted. At some point, therefore, the cells acquire the capacity to develop as a leaf independently of the rest of the plant.

From experiments such as these, we can conclude that the activity of the nucleus changes during differentiation; such changes are not necessarily permanent and may be reversed once the environmental milieu of the nucleus changes. Other changes, however, especially in multicellular organisms, result in a differentiated state that is stable and can persist over several generations of cells in a lineage. The pathway of development of a given cell, therefore, depends not only on its genetic constitution, which supplies its potential, but also on its past history and present physical and chemical environment, both of which determine those features of the potential that will be realized at any specific time.

When we say that differentiation is a consequence of differential gene expression, we are saying that it results from differential utilization of genetic information. As we have seen in Chapter 2, the pathway of information flow is from the DNA of the chromosome to the messenger RNA, and hence through the enzymes that are synthesized in the cytoplasm and that catalyze the reactions of the cell. The problem of differentiation, therefore, is how the flow of information is controlled in such a way that different enzyme complements can be established in different cells or in the same cell at different times during its lifetime. It now appears unlikely that there is a single solution to this problem, but rather that different control mechanisms operate in the cell and at several different levels. One such level in the information flow pathway in the cell at which control can be exercised is that of initial transcription of the DNA code into the base sequence of the RNA; the subsequent events leading to translation of that base sequence into the amino acid sequence of a protein also can be subject to regulation at several points in the process, while posttranslational modification of the protein itself can play a part in determining the phenotypic characteristics of the cell.

Differential transcription

Several studies suggest that selective expression of the information encoded in the genes of a cell can be achieved by activation or repression of transcription of these genes; in other words, by selective synthesis of the primary gene product, the RNA that contains the information to be carried to the cytoplasm. Let us begin by returning to the case we mentioned earlier—induction by galactose of the enzyme system required for the utilization of galactose by cells of the bacterium *E. coli.* Three different enzymes are involved, and they are coded for by three adjacent genes on the bacterial chromosome (Figure 9.10).

In cells grown in the absence of galactose, a *repressor* protein molecule binds to the bacterial DNA at a specific region at one end of the base sequence coding for the three proteins, thereby preventing initiation of RNA synthesis. In the presence of galactose, however, the repressor molecules bind to the galactose instead, and the genes are made available for transcription, the necessary mRNA now being synthesized. This *operon* system clearly shows how enzyme formation in prokaryotic cells can be regulated at the level of the genes themselves.

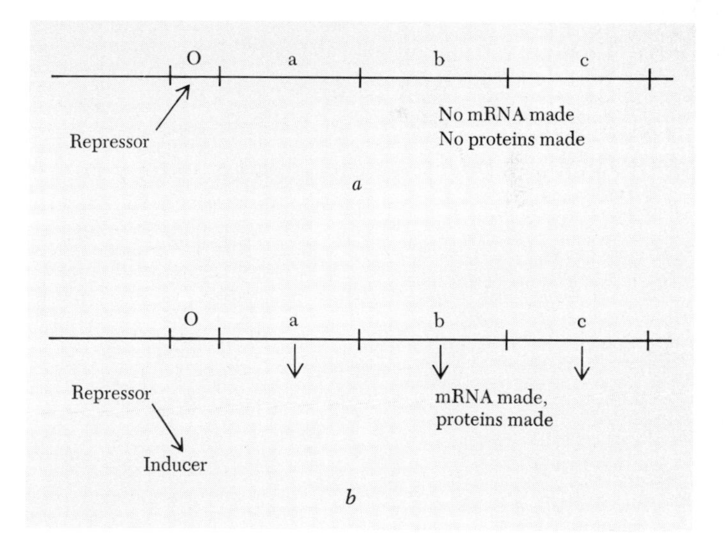

Figure 9.10
Diagram of operon model. (*a*) In the absence of the inducer—for example, galactose—the repressor protein binds to an operator region of the DNA, and prevents initiation of transcription of the three structural genes, *a*, *b*, and *c*. (*b*) In the presence of the inducer, the repressor protein binds to the inducer rather than to the operator region, and the structural genes can be transcribed as a unit and the proteins they code for can be synthesized.

While no such operon system has been demonstrated in eukaryotic cells, there is evidence that in such cells genes can be switched on and off. Comparisons between the types of RNA synthesized in different cells can be made, using the technique of competitive nucleic acid hybridization. The rationale behind this technique is that the base sequence of an RNA molecule is complementary to that base sequence of the DNA strand from which it was transcribed. The two strands of a DNA double helix can be separated by various methods, such as raising the temperature or the pH. The single-stranded DNA molecules are then immobilized on a filter or a column so that they do not come back together again and reanneal. If the bound DNA is incubated in the presence of radioactive RNA, any RNA molecules with base sequences that are complementary to base sequences in the DNA will anneal with those sequences, forming a double-stranded, hybrid DNA/RNA molecule. Any single-stranded RNA that has failed to find a complementary DNA sequence is then washed out, and the amount of radioactivity left can be used as a measure of how much hybridization has occurred. Results obtained with such techniques show that in any one cell type not all of the DNA is being used as a template for RNA synthesis at any one time, and that different types of RNA are synthesized in different cell types and at different stages of development (Figure 9.11).

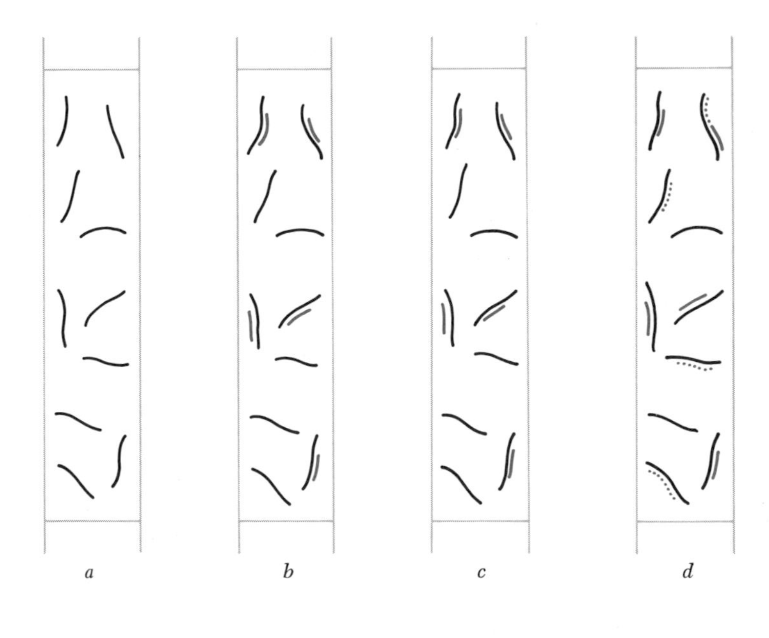

Figure 9.11
Diagram of DNA/RNA hybridization. RNA transcribed in liver cells is complementary to a certain amount of DNA. RNA transcribed in kidney cells includes sequences that are different from those in liver cells.

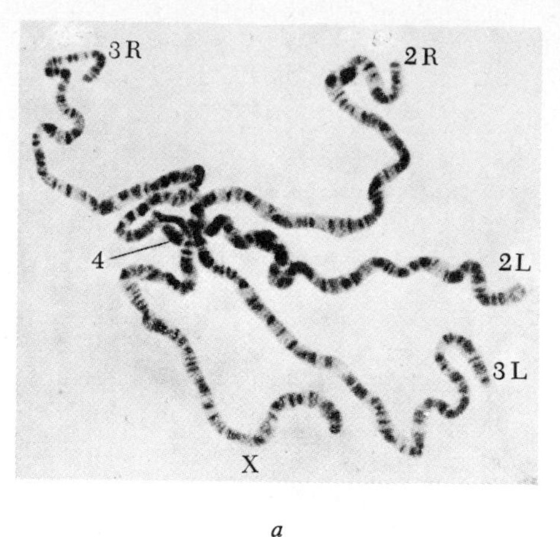

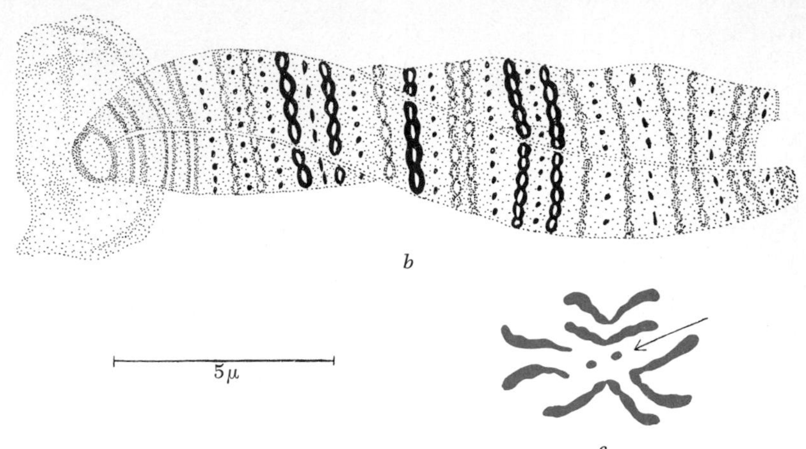

Figure 9.12
Salivary gland chromosomes of *Drosphilia melanogaster.* (*a*) A smear preparation from the salivary gland of a female, showing the X chromosome, the arms of the two autosomes (2L, 2R, 3L, and 3R), and the small chromosome 4. The diploid number of chromosomes is present, but the homologues are in intimate synapsis and are united by their centric heterochromatin into a chromocenter. (*b*) Enlarged drawing of chromosome 4, showing the banded structure; the diffuse chromocenter is at the left, and the two homologues are intimately paired. (*c*) Metaphase chromosomes from a ganglion cell, with an arrow pointing to the chromosomes 4 and with a scale to indicate differences in size between the two types of chromosomes. [(*a*) Courtesy of Dr. B. P. Kaufmann; (*b*, *c*) courtesy of C. B. Bridges, *Journal of Heredity,* 26 (1935), 60–64.]

Differential gene activity also can be observed directly at the chromosomal level in certain cells of the larvae of dipteran insects. Cells of the salivary glands in particular contain very large chromosomes formed by successive duplications of the chromosome material without intervening segregation. These giant, polytene chromosomes display particular banding patterns (Figure 9.12), which are characteristic of the species, and the arrangement of bands is believed to reflect the linear sequence of genes along the chromosome. Tissues other than salivary glands also contain giant chromosomes, and the arrangement and sequence of bands is the same in all tissues.

At some, but not all, of the bands the chromosome material is less condensed, and "puffs" are seen (Figure 9.13); these can be shown to contain RNA. If autoradiographs are prepared following administration of radioactive precursors of RNA, the puffed regions are labeled, indicating that the puffs are sites of RNA synthesis and represent regions of gene activity. While the banding pattern is constant within a species, the puff patterns change during development. Administration of the hormone ecdysone, which induces molting in insects, results in the formation of developmentally specific puffs occurring in a regular time sequence. The hormone acts, therefore, by initiating a sequential series of gene activations that lead to subsequent developmental changes; how this is achieved by the hormone remains a mystery. Furthermore, different tissues show different patterns of puffing, indicating that some puffs are tissue-specific. An example of such differential gene activity can be seen in the salivary glands of the larval stage of the midge, *Chironomus.* In a few specialized cells of the gland, but

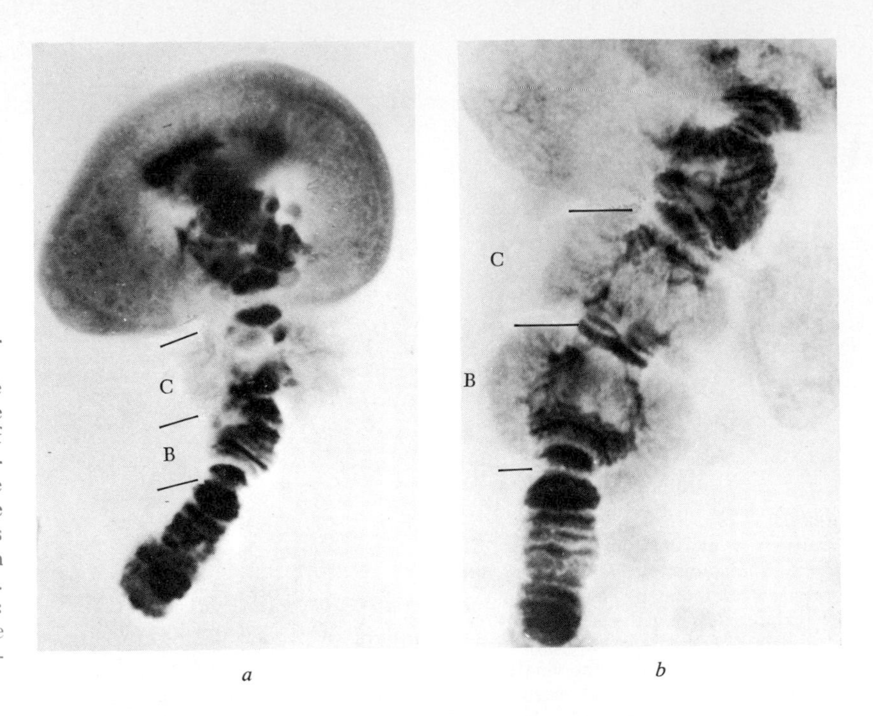

Figure 9.13
Differential puffing activity in chromosome IV of *Chironomus thummi.* (*a*) A puff at region C is present in late pre-pupal larvae. (*b*) In larvae treated with juvenile hormone a puff is induced at region B. The appearance of this puff coincides with the synthesis of RNA in this region. The upper portion of each chromosome is associated with the nucleolus. (From Laufer, H., and T. K. H. Holt, 1970. *Journal of Experimental Zoology* **173**: 341–351. Copyright, Wistar Institute Press, Philadelphia. With the permission of the authors and publisher.)

not in others, a certain type of protein granule is formed. A specific puff on a particular chromosome is apparent only in the cells in which the granules are present; this puff is not present on the chromosomes of the other salivary gland cells, nor on the chromosomes of a related species that does not form granules at all. Genetic tests have shown that the gene (or genes) for granule formation is (are) localized within the region of the chromosome from which the puffs arise. These giant chromosomes, therefore, allow us to see that different genes are active during different stages of development and in different tissues; furthermore, we can correlate the appearance of a particular protein in a cell type with the synthesis of RNA at a specific region of the chromosome, that is, with the activity of a particular gene.

Another type of chromosomal differentiation is shown in the behavior of the sex chromosomes of mammals. For example, humans possess an XX/XY sex-determining mechanism, with females being XX and males XY. Early in the prenatal life of the female embryo, one of the X chromosomes in all cells, except those destined for the germ line, becomes transformed into a heterochromatic, inactive body. This chromosome continues to replicate prior to each cell division but no longer synthesizes RNA. Direct confirmation that the genes on the heterochromatic X chromosome are not expressed comes from experiments in which cultured cells taken from females who are heterozygous for sex-linked genes are used to form clones. One such gene codes for the enzyme glucose-6-phosphate dehydrogenase (G-6-PDH). Two alternative allelic forms of this gene code for enzymes with slightly different properties, and heterozygous females produce both types of enzyme. However, if single cells from such a female are cloned, each clone produces only one of the two types of molecule. This, along with similar results for other genes of the X chromosome, shows not only that one of the two X chromosomes in a female cell is inactive, but also that once the inactivation has occurred, all of the descendents of that chromosome remain inactive. Clearly, in this case the capacity for repression of particular genes is stable and is passed on from cell to cell during development.

Regulation of gene transcription

Heterochromatin The behavior of the X chromosome in female mammals provides a clue as to a mechanism of gene inactivation at the chromosomal level. Chromatin can exist in two states: heterochromatin, which is tightly condensed, and euchromatin, which is more dispersed. Various lines of evidence suggest that genes in heterochromatic regions of the chromosomes are inactive; furthermore, heterochromatin can be physically separated from euchromatin and shown to be relatively inactive compared to the euchromatin in supporting RNA synthesis. It appears likely, therefore, that certain regions of the chromosome, and perhaps even specific genes, can undergo heterochromatization and thus be rendered inactive. Unfortunately, we do not yet know the molecular basis for such differential condensation of the genetic material.

Chromosomal proteins The major non-DNA component of the chromosome is the histone protein, which complexes with the DNA itself. If histones are removed from the DNA, the ability of the DNA to support RNA synthesis increases, suggesting that histone proteins may repress those regions of the DNA with which they are tightly complexed. Although histones can inhibit transcription of genes, it is unlikely that they can be specific in their action, since only a few histone types are present in cells, and these are remarkably uniform from tissue to tissue and even from species to species.

Other nuclear proteins, known as acidic proteins, are also associated with the chromosomes. These proteins are much more heterogeneous than the histones and show some tissue-specificity. Nucleohistone can be "reconstituted" by combining DNA and histones; the reconstituted DNA-histone complex is then capable of supporting RNA synthesis (although at a very low level) under appropriate conditions in the test tube. When nonhistone proteins that have been extracted from different tissues are added to DNA-histone complexes, new RNA molecules are synthesized, and these appear to be characteristic of the cell type from which the nonhistone proteins were obtained. For example, if chromatin is reconstituted in vitro from DNA, histones, and nonhistone proteins from cells that make hemoglobin, the mRNA that codes for the hemoglobin protein is produced. However, if the nonhistone proteins are derived from cells that normally do not form hemoglobin, no globin mRNA is synthesized. These nuclear proteins, therefore, may be involved in some way in the selection of which genes in the cell are to be transcribed.

Although we do not know the details of the changes that occur in the configuration of the chromatin or which are responsible for the specificity of its "template" activity, it seems likely that interactions between chromosomal proteins and DNA are involved, and that different regions of the DNA can be somehow "unmasked" and made available for transcription.

Nucleocytoplasmic interactions

Whatever the mechanisms might be that regulate transcription of genes at the chromosomal level, they must be influenced at least in part by the cytoplasm, the milieu in which the nucleus functions. An example of what appears to be cytoplasmic determination of nuclear activity is seen during development of the pollen grain in higher plants. Following meiosis, each haploid spore undergoes a mitotic division, forming a binucleate pollen grain. One of these nuclei becomes very condensed and elongate, and is completely inactive in RNA synthesis. The sister nucleus, on the other hand, maintains a normal appearance, synthesizes RNA, and supports all further development of the germinating pollen grain (Figure 9.14). Prior to the mitotic division, however, a high degree of polarity is established in the immature pollen grain, such that the two telophase nuclei end up in very different types of cytoplasm (Figure 9.15) and subsequently behave very differently. Similar cytoplasmic influences are thought to operate in development within embryos. Successive cleavage divisions of the egg result in the formation of a multicellular blastula; however, since the egg cytoplasm is complex

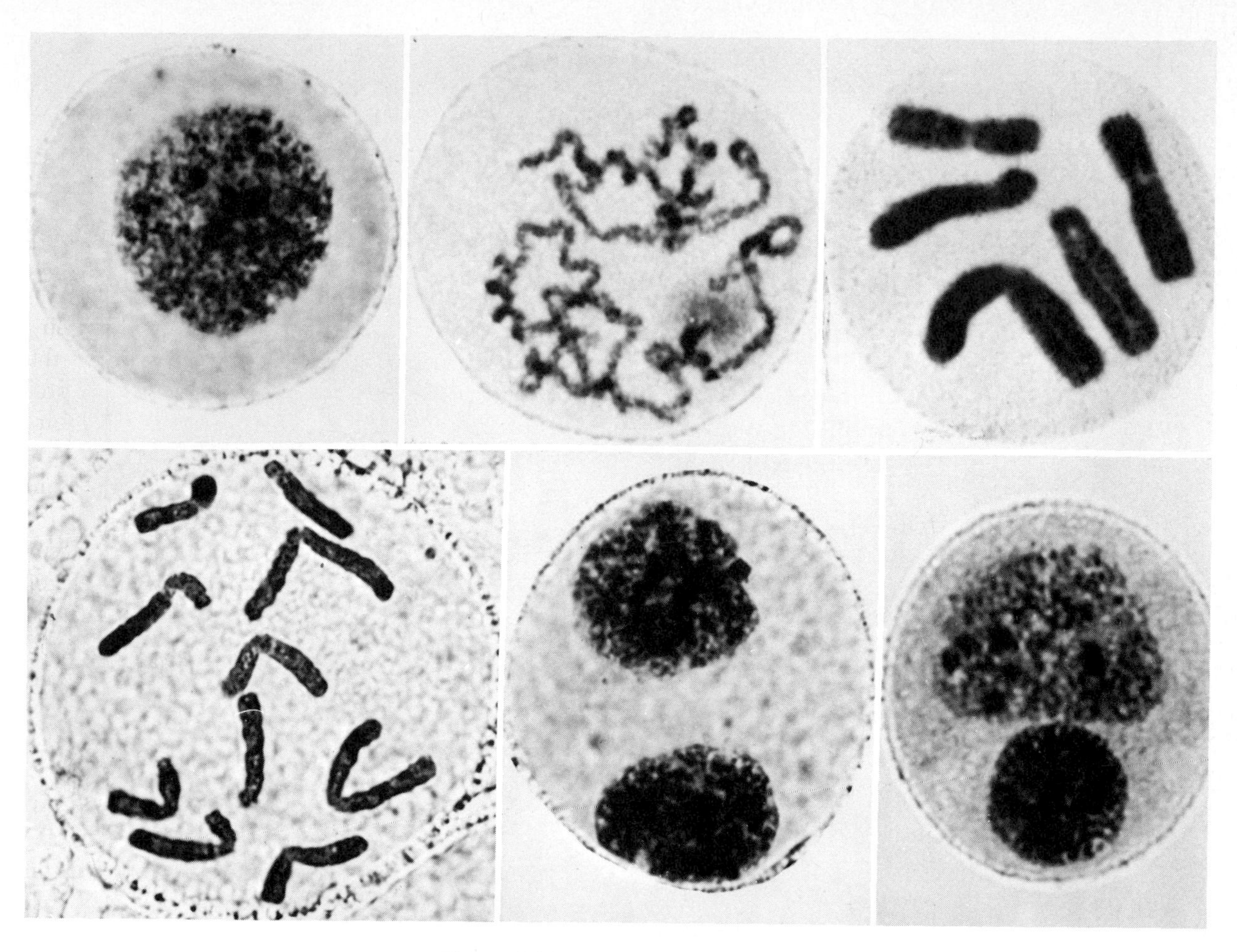

Figure 9.14
Mitosis in the haploid microspore of *Trillium erectum*. Following telophase the vegetative nucleus (top) and the generative nucleus (bottom) begin differentiation. (Courtesy of Dr. A. H. Sparrow.)

and heterogeneous, the genetically identical nuclei of the blastomeres find themselves in different cytoplasmic environments, and they may respond to these environments in different ways, leading to different pathways of gene activation and differentiation.

We have already discussed how nuclei from cells of the embryo of *Xenopus* can support normal development when transplanted into an anucleate egg. Let us examine this clear case of cytoplasmic control over nuclear activity in a little more detail. During the early cleavage stages of normally fertilized eggs, very little RNA of any kind is synthesized by the nuclei. (The RNA required

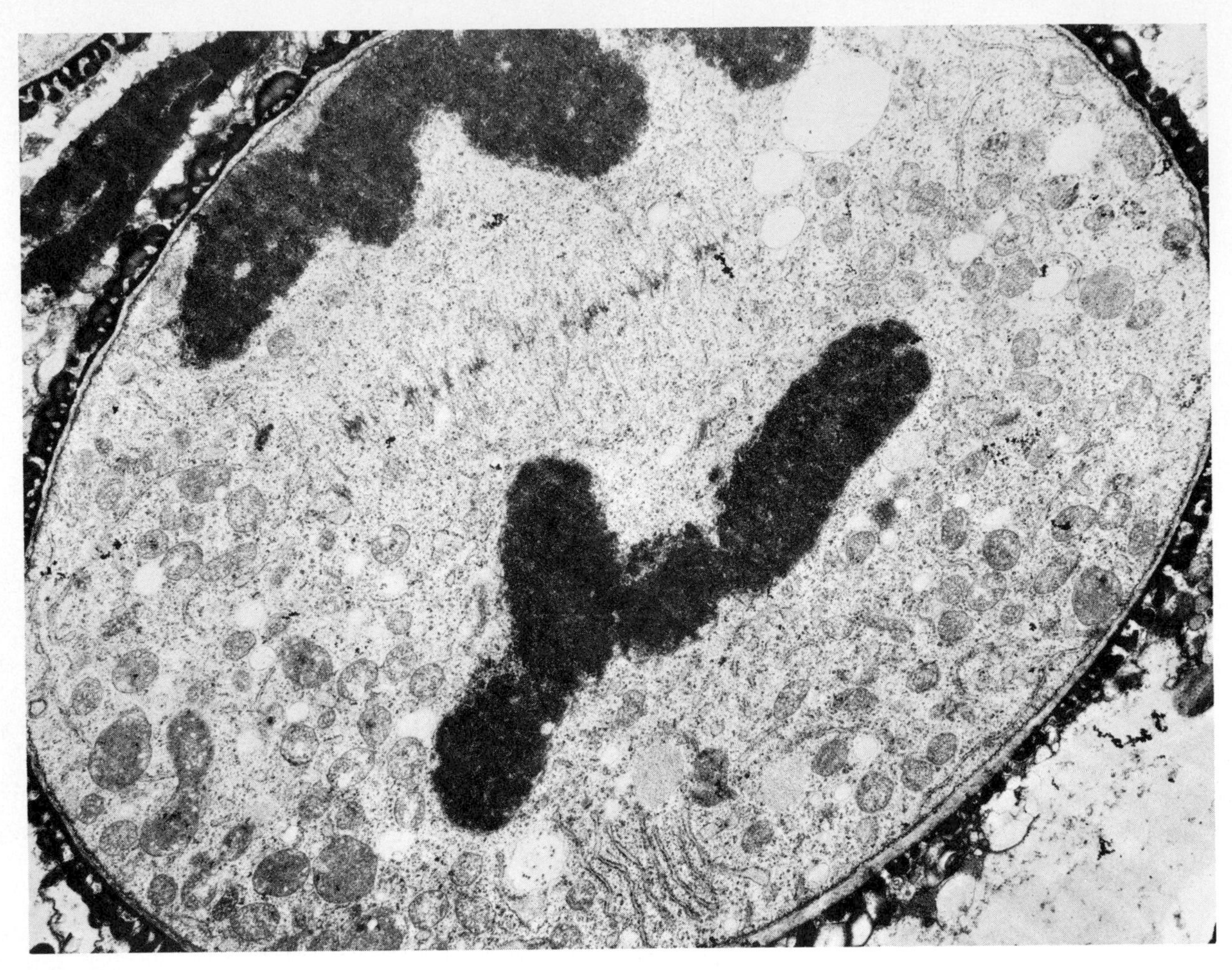

Figure 9.15

Electron micrograph of pollen grain (microspore) mitosis in *Tradescantia.* Note the unequal distribution of cytoplasmic components.

by these cells has already been synthesized during prophase of meiosis in the oocyte and is present in the egg prior to fertilization.) However, by the neurula stage of development of the embryo, all classes of RNA are again being synthesized in significant amounts. When nuclei from neurula cells of *Xenopus* are transplanted into *Xenopus* eggs, they undergo morphological changes, RNA synthesis stops, and the nuclei synthesize DNA and undergo cleavage divisions. Later in development, the patterns of RNA synthesis characteristic of normal embryogenesis are established again. Clearly, it is the cytoplasm that determines the patterns of transcription in nuclei during early development.

We discussed the direction of flow of genetic information as being from nucleus to cytoplasm, but information from the cytoplasm must somehow elicit responses from the nucleus. In the case of the nuclear transplants into amphibian eggs, it has been shown that cytoplasmic proteins do become localized in cleavage nuclei. Similar cytoplasm-to-nucleus migration of proteins has been demonstrated in *Amoeba*. By transplanting nuclei from cells that have been supplied with radioactive amino acids into other, unlabelled cells, it can be shown that some nuclear proteins move rapidly into the cytoplasm and become concentrated in the nucleus of the host cell. Although we do not yet know the nature of the proteins that migrate between nucleus and cytoplasm, such movement could be related to differential gene activity.

An example of a specific gene being activated following movement of molecules from cytoplasm to nucleus is seen in the response of chick oviduct cells to the steroid hormone progesterone. Progesterone treatment results in the synthesis, by the goblet cells of the oviduct, of the egg-white protein, avidin. The hormone can be shown to enter the cells, where it forms a complex with a cytoplasmic protein. The protein-hormone complex then enters the nucleus, reaches the chromosomes themselves, and synthesis of avidin-specific mRNA is initiated. Exactly how the complex from the cytoplasm interacts with the chromatin in the nucleus is not clear, but the interaction appears to involve the nonhistone proteins, which, as we have seen, may be involved in the specificity of gene transcription.

Differential translation

Expression of cellular information also can be modified by regulation of translation of the RNA that has been transcribed by the genes. We know that the RNA synthesized in the nucleus undergoes considerable modification before it is translated on the ribosomes. For example, much of the RNA made in the nucleus is degraded before it leaves the nucleus. This degradation appears to be related to some kind of "processing" of the RNA which is to be translated on the ribosomes and involves the breakdown of very long RNA molecules to shorter pieces. Following this breakdown, a stretch of RNA containing only the base adenine is attached in the nucleus to those RNA species destined to be translated. Now the mRNA with its attached "poly-A" sequence moves to the ribosomes, where its information is translated into the amino acid sequence of a protein. Although the significance of such types of posttranscriptional modifications is not clear, and although not all mRNA undergoes such modification, selective degradation or selective adenylation of RNA in the nucleus could regulate which messages reach the ribosomes and which are to be translated. An important question, therefore—and one for which we have no answer—is how the cell knows which gene transcripts to select for conversion to functional mRNA.

Differences in the stability of different mRNA species could also play a role in establishing which proteins are synthesized in a cell. For example, germination of seeds following long periods of dormancy is accompanied by the formation of polyribosomes and the initiation of protein synthesis. These events occur even in the absence of any RNA synthesis, indicating that the necessary mRNA already must have been present. Long-lived mRNA is also present in unfertilized eggs, although it is not used to support protein synthesis until after fertilization. Clearly, some mechanism must exist that prevents destruction of this RNA and/or prevents its utilization until a specific stage in development. Once again, however, we must admit ignorance of how the longevity of different mRNAs might be regulated, although the poly-A sequences may be involved.

Gene amplification

Regulation of the rate of synthesis of a specific gene product, ribosomal RNA, can be exercised in what appears to be a unique manner. During the extended prophase of immature oocytes of *Xenopus*, ribosomal RNA synthesis proceeds at a rapid rate. By the time the oocytes mature, rRNA synthesis can no longer be detected. The high rates of rRNA synthesis during this stage in development of the oocyte are achieved in part by the production during pachytene of large numbers of copies of the genes that code for rRNA. This differential replication of a specific part of the genome is followed by release of the DNA into a large number of nucleoli, each of which is then capable of making the rRNA. These amplified genes can be isolated from nucleoli and photographed (see Figure 2.27, Chapter 2). Although selective production of extrachromosomal ribosome genes is found in oocytes of several different species, we do not know how widespread is this method of increasing the number of DNA templates for specific RNAs during differentiation, or if genes coding for RNA other than rRNA are amplified in this way.

We have discussed some examples of how the flow of genetic information might be modified at different levels during the processing of this information by the cell, resulting in differential expression of genetic potential. Our knowledge of this important aspect of cellular behavior is still fragmentary, however, and we do not yet understand how the necessary control is exercised.

As we pointed out earlier, such differential expression represents the response of cells to external influences. Let us now consider a type of environmental stimulus to which cells respond by undergoing some form of developmental change, and how these signals are perceived.

Hormones

Coordinated development in multicellular organisms depends on communication between different cells and tissues. Hormones are chemicals involved in such communication, being produced in one part of the organism and exercising their effect on cells in a different part. Cells that respond to a hormone must possess receptors with which the hormone can interact, plus the ability to respond to the interaction in a way that results in specific biochemical changes in the cell. Some hormones may exercise their effects without entering the cell at all. They interact with specific binding sites at the cell surface, and subsequent changes occur inside the cell. For example, increased synthesis of the compound cyclic AMP results from activation by hormones of an enzyme located at the inner surface of the cell membrane. The cyclic AMP is then able to modify other proteins in the cell with a resulting change in cellular activities. Other hormones enter the cell and may even reach the chromosomes themselves. Such is the case with the hormone progesterone, which, as we have seen earlier, stimulates the production of a specific protein, avidin.

Our discussion of what are certainly only some aspects of differentiation should illustrate that many important questions remain to be answered. For example, what is the nature of the organization of chromatin? How can it be modified in such a way that some genes are "read" and others not? How are specific regions of chromatin recognized for transcription? How are the differentiated states that are acquired gradually and maintained over several cell generations inherited? A full understanding of the cellular basis of development requires a knowledge of organization and interaction at the molecular level that we do not yet possess. However, a beginning has been made in this exciting area of cell biology toward the goal of explaining differentiation in molecular terms, and, therefore, in understanding how the organism accomplishes the spatial and temporal division of labor necessary for it to carry out its various activities and to cope with its environment. And at a vital, practical level, our ability to control cancer, which results from the escape of cells from regulative control, rests on our comprehension of the functional cell and the processes of development as rigidly governed systems of chemical checks and balances.

Bibliography

BEERMAN, W., and U. CLEVER. 1964. Chromosome puffs. *Scientific American* V. **210**, No. 4, 50–58. Direct visualization of genes in action.

DAVIDSON, R. G., H. M. NITOWSKY, and B. CHILDS. 1963. Demonstration of two populations of cells in the human female heterozygous for glucose-6-phosphate dehydrogenase variants. *Proceedings of the National Academy of Sciences* V. **50**, 481–485. Direct evidence in support of the X-inactivation hypothesis.

GURDON, J. B. 1968. Transplanted nuclei and cell differentiation. *Scientific American* V. **219**, No. 6, 24–35. A clearly written account of nuclear-transplant experiments during early amphibian embryogenesis.

LYON, MARY F. 1961. Gene action in the X-chromosome of the mouse (*Mus musculus* L.) *Nature* V. **190**, 372–373. A brief presentation of the hypothesis of differential inactivation of mammalian X-chromosomes.

MARKERT, C. L., and H. URSPRUNG. 1971. *Developmental Genetics*. Prentice-Hall, Inc., Englewood Cliffs, N. J. An excellent introduction to basic aspects of development and differentiation.

STEIN, G. S., T. C. SPELSBERG, and L. J. KLEINSMITH. 1974. Nonhistone chromosomal proteins and gene regulation. *Science* V. **183**, 817–824. A review of evidence for the role of nonhistone proteins in regulation of gene activity.

STEIN, G. S., J. S. STEIN, and L. J. KLEINSMITH. 1975. Chromosomal proteins and gene regulation. *Scientific American* V. **232**, No. 2, 46–57. A clearly presented account of the experiments that suggest roles for nuclear proteins in switching genes on and off.

SUSSMAN, M., ed. 1971. *Developmental Biology—Its Cellular and Molecular Foundations*. Prentice-Hall, Inc., Englewood Cliffs, N. J. An extremely well-written introduction to, and statement of, current problems in development.

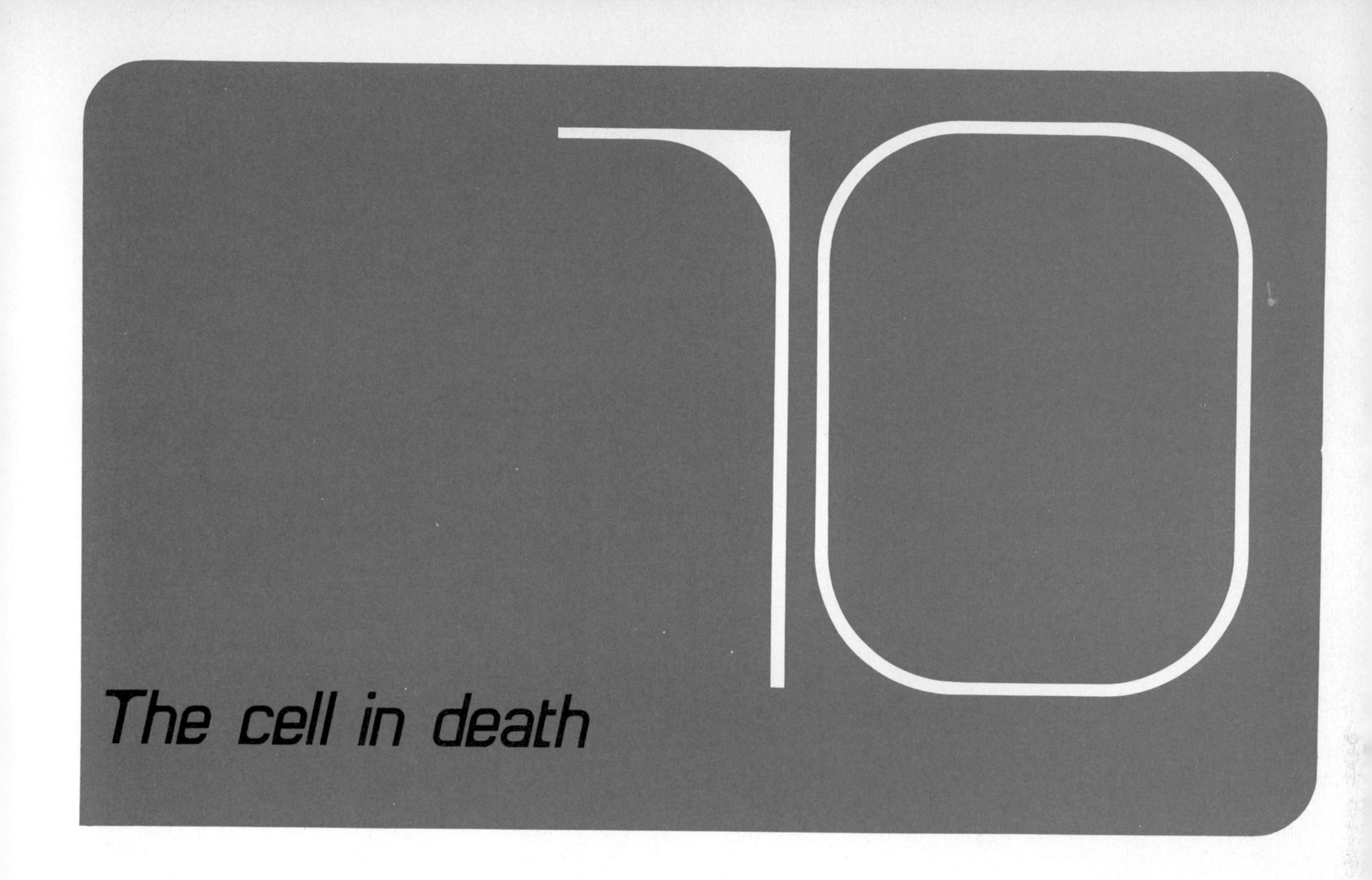

10 The cell in death

The old aphorism states that death and taxes are inevitable. Civilization, with its cities dependent upon and to a degree parasitic on an agrarian economy, made taxes inevitable. In a somewhat comparable manner, once life depended upon other life for its existence, death also became inevitable. Viruses invade and destroy bacteria; bacteria and fungi attack and destroy higher forms of life; animals are classified by their dietary habits as herbivores, carnivores, or omnivores; and even green plants, which utilize sunlight, CO_2, H_2O, and mineral salts in their processes of growth and development, are also dependent on humus, the dead remains of other organisms, as a soil conditioner. Life, therefore, recycles life through the medium of death, but some species and individuals are recycled more often than others.

Any individual, barring accidents, has a life-span that is characteristic of the species to which it belongs. In some instances, the life-span is rather exactly delimited—17 years for certain periodic locusts, 18 months for a mole, 15 months for a shrew, 7 weeks for a fruit fly (Figure 10.1). Usually, however, we think of the life-span

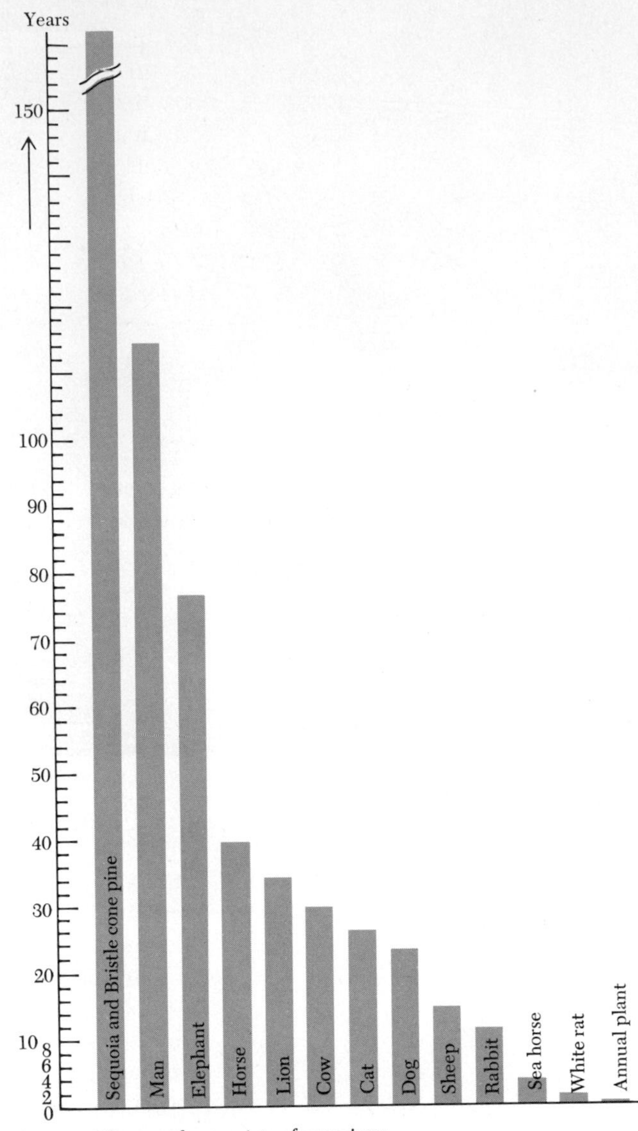

Figure 10.1
The maximum life-spans of a number of different organisms. Sequoias and bristle cone pines may live for several thousands of years, and their full life-span would extend well beyond the limit of this page.

as an average figure—a few days for certain insects, a few months for annual plants, three score and 10 years for humans, 250 to 300 years for an oak tree. The redwoods of our Pacific Coast and the bristle cone pines of California's White Mountains are probably the longest-lived organisms; some of these trees reach an age of several thousand years. But whether the time is long or short, death must inevitably intervene. If accidents or disease do not abbreviate the life-span, death comes about as a normal event of development. As one investigator has expressed it, a "genetic countdown" mechanism ticks off the remaining hours, days, or years of life, and the terminal developmental stage is lethal.

Cells, too, have a life-span they complete, after which they die And, like organisms, particular kinds of cells, even in the same organism, have characteristic long or short life-spans, the length of time being determined by the pattern of differentiation they have undergone. Yet it is entirely reasonable to consider some cells to be immortal. When a unicellular organism divides, the life of the single cell becomes part of the life of two new cells, and as long as the species lives, so in a sense does the original cell. The life of a number of such a species, then, stretches in an unbroken chain back to some original cell in the past. Among sexually reproducing organisms, only the cells of the germ line can lay claim to immortality, for they are the only cells that connect succeeding generations, contribute through fertilization, division, growth, and differentiation to descendent organisms, and thus keep the species alive. We are all familiar with death as it involves whole organisms, but among the cells of the body, death is a normal and necessary process because if its role is altered, the functioning of the organism will be affected to varying degrees. As a biological problem, and apart from the death of an organism, there are two broad categories of cellular death: (1) that resulting from the wear and tear of existence, which, during much of the life-span of an organism, is generally counterbalanced by an equivalent amount of cell replacement; and (2) that resulting from the normal processes of development and differentiation.

Cell replacement

It has been estimated that a human being has a new body every seven years, the time it takes for the old cells of the body to be replaced by new ones. Even if statistically accurate, this figure is very misleading since it does not provide a reasonable picture of what is taking place among the varied cells of the body at any particular stage of its existence. The brain, for example, reaches its mature size within a few years after birth, while the body as a whole undergoes considerable growth. By this time, most of the nerve cells have been laid down even if they have not yet achieved mature size and differentiation, and barring loss through accident or disease, they will continue to function throughout life at some level of performance. If a nerve cell is destroyed, it cannot be replaced by another. Once delineated as a neuroblast, and long before its attainment of final form, the potential nerve cell loses its capacity to multiply (Figure 10.2). No general nerve cell replacement center exists, as is true for blood cells, but there is some experimental evidence to suggest that enriched environments during early childhood can lead to either an increase in small, interconnecting neurons or in the number of connections between neurons in the brain. Recent studies also suggest that muscles are capable of limited replacement when losses occur. Dividing cells have been found among those forming smooth muscles, but as for striated skeletal or heart muscles, it is far more likely that injury to such tissues is repaired by connective cells forming scar tissue, which lacks contractile properties. That an organ remains constant in size is not, therefore, indicative of its rate of cell replacement; unless the actual turnover rate of cells is known, the constancy of size merely indicates that there is no net gain or loss of cells. The death of old cells can be equalled by the production of new ones.

Some biologists estimate that each day the human body loses 1 to 2 percent of its cells through death. Body weight, therefore, would double every 50 to 100 days if no cells died, and if cell division proceeded normally. If the body weight remains constant, therefore, dead cells must be replaced by new ones, by billions of

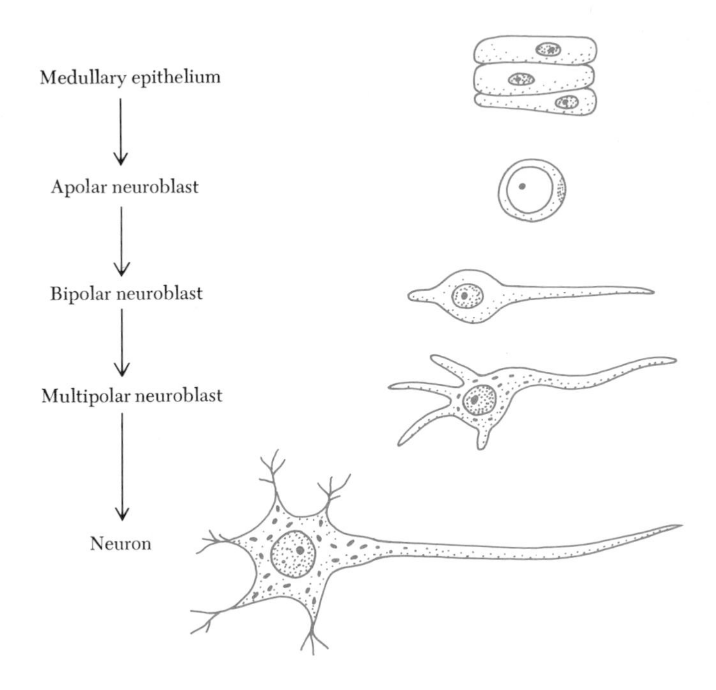

Figure 10.2
The human nerve cell, or neuron, which lasts generally as long as does the individual and which does not divide once delineated as a neuroblast, has its origin in a tissue known as the medullary epithelium. This epithelium produces all the varied cells of the nervous system, but the neuron acquires its polarity —with its axon and dendrites—as it is differentiated from a neuroblast to a neuron. The axon, if cut or damaged, can regenerate itself, but if the cell body with its nucleus is destroyed, it cannot regenerate. (After W. Bloom and D. W. Fawcett, 1962.)

cells every day. Since virtually none of these is produced in the muscles or nervous tissue, and since muscles and nervous tissue must be equal to about one-half or more of the body weight, there must be active centers of death and replacement elsewhere. These would include the protective layers and regions of the blood-forming, digestive, and reproductive systems. The other organs of the body have much slower replacement rates; a liver cell, for example, has been estimated to have an average life-span of about 18 months. Consequently, if we were to look at a slice of normal liver under a microscope, we would expect to find very few cells in division.

Mammals, unlike amphibia, cannot regenerate a lost limb, but the behavior of the liver reduced in size by surgery indicates that some aspects of regeneration remain. If two-thirds of the liver of a rat is removed, it regains full size within a few days as a result of the remainder of the cells undergoing waves of division. The mature liver cell is a differentiated structure, and one can ask whether it is the mature cell that divides without further change of structure of function, or do these cells revert to an embryonic-like status prior to division? The latter appears to be the case, as judged from the following experimental evidence. During the development of a fetus, cells destined to become part of the liver produce the kinds of RNAs appropriate to particular stages of fetal development, which are different from those of mature and differentiated liver cells. Different genes are active, therefore, at different stages of development. Regenerating liver cells apparently revert to a fetal-like stage as reflected by the kinds of RNAs they form, and then gradually assume a mature status when regeneration is complete. Cell replacement, therefore, involves much more than simply the formation of similar cells.

The outer surface of the human body is also an area of constant cell division and cell replacement. This is the epidermis, which also includes the lining of all openings of the body externally, the cornea of the eye, and, in modified form, the structures that give rise to nails and hair. The cells of these structures are constantly being lost through death: the skin sloughs off, and the growing nails and hair are composed of dead cells. Cells of the oral epithelium are easily scraped off for examination under the microscope. The process of replacement, then, must be a relatively rapid one. The underlying cells are constantly dividing and are pushed outward toward the skin surface, while the outermost cells, particularly of the skin, become cornified (hardened) as they die (Figure 10.3). It takes approximately 12 to 14 days for a cell in the skin of the forearm to move from the dividing innermost layer of the epithe-

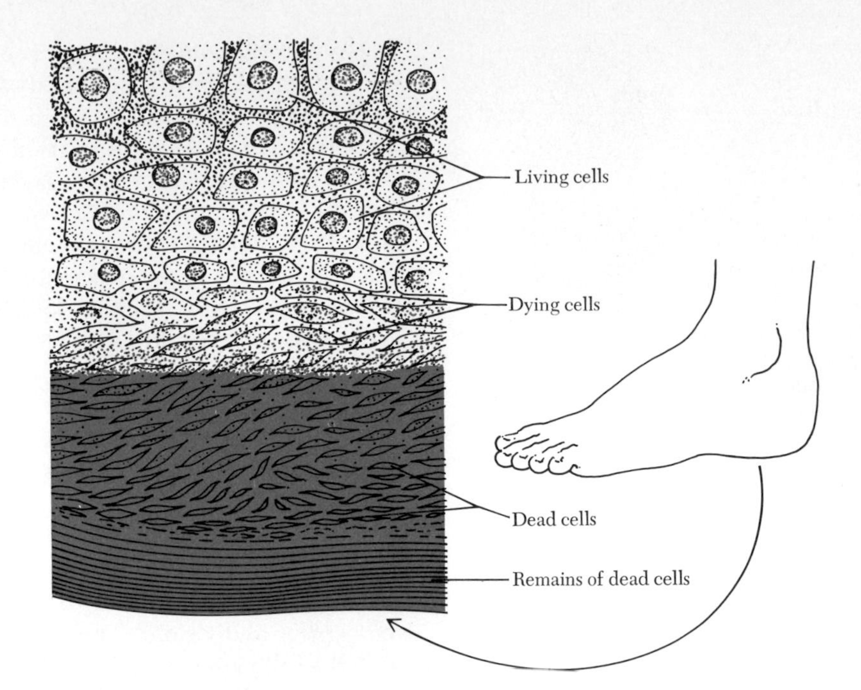

Figure 10.3
A section through the skin of the human foot would show that cells are continually dying at the surface but are continually being replaced by cells beneath them.

lium to the outermost layer of the skin. Calluses on the hands or feet are thickened areas of dead cells, and a needle can be pushed through these areas without causing pain or drawing blood.

The cornea of the eye is a special type of skin in which the rate of cell death and replacement is high. In fact, the cornea is an excellent type of tissue to examine for active cell division. Since it is only a few cell layers thick, it can be stripped off (the salamander and rat are good animals to use for this purpose), fixed, stained, and mounted intact on a microscope slide. The dying cells can be seen at the outer surface, while the cells underneath are in active division.

Cell replacement is extraordinarily high in the blood of the mammalian body, a tissue of many varied kinds of cells. The blood, however, is a derived tissue with its cells formed elsewhere and then freed into the circulatory system. *Myeloid tissue,* or bone marrow, is the source of the red blood cells, as well as the several types of white cells, or leucocytes, of a granular nature. All apparently arise from a single kind of stem cell, the hemocytoblast, but follow different routes of differentiation before being released into the bloodstream (Figure 10.4). The red blood cells carry oxygen from the lungs to the tissues of the body, and carbon dioxide in the reverse direction; the leucocytes play important roles

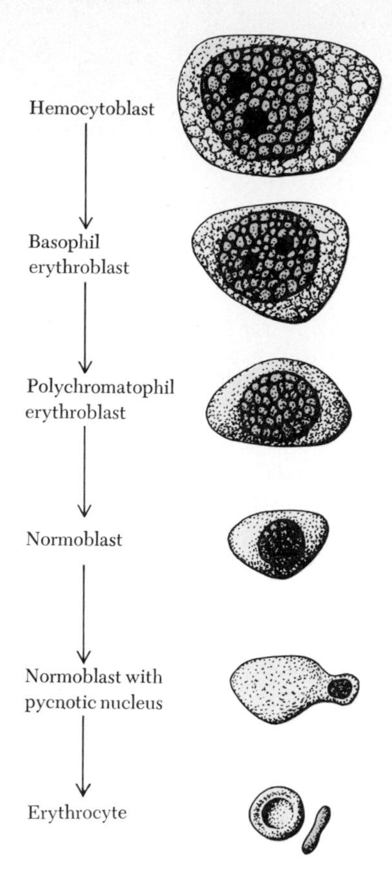

Figure 10.4
The circulating red blood cell with no nucleus originates from a bone marrow cell, the hemocytoblast, which progressively becomes smaller in size as it passes through several stages as an erythroblast, then to a mormoblast, and finally to an enucleated erythrocyte. Mitosis can occur during any stage up to the normoblast; at the latter stage, the nucleus becomes pycnotic (condensed) and is eventually extruded from the cell to produce the circulating erythrocyte. The erythrocyte, lacking a nucleus, has a limited life-span; it is depicted in both face and side views. The hemocytoblast, through other avenues of differentiation, gives rise to several kinds of leucocytes, each with its own characteristic structure and behavior, and its own limited life-span.

in disease and infection, cleaning up damaged tissues and invading bacteria.

Lymphatic tissue, located in the bone marrow, lymph vessels, thymus, and spleen, produces the several sizes of agranular leucocytes, which include the lymphocytes and plasma cells. These serve an important role in immunologic reactions. Taken together, these cells and the *plasma* constitute the blood, which has an average ratio of one white cell to every 400 to 500 red cells. When the leucocyte-forming areas overproduce, leukemia, a form of blood cancer, results. In myeloid leukemia, the blood contains excessive numbers of granular leucocytes due to overactivity in the bone marrow and spleen, while overactivity in lymphatic tissue, particularly in the spleen, leads to lymphatic leukemia. The blood-forming areas must, therefore, maintain a balance between the several kinds of cells, but this is complicated by the fact that there is a very high rate of loss of cells from the wear and tear of circulation and the cleaning up of cellular debris.

Very little is known about the life-spans of the kinds of white cells, but since it is likely that each type of cell dies off at a relatively constant rate, we need consider here only one—the red blood cells. Their life-span is about 120 days. They lack a nucleus and virtually all other cytoplasmic organelles, having lost these as they passed into the bloodstream. The wear and tear on the plasma membrane as a result of passage through the vessels cannot be repaired on an indefinite basis, and the cells grow fragile. As they reach the end of their life-span, they are engulfed by specific white cells, the macrophages, which break down the hemoglobin and isolate the iron for re-utilization in subsequent hemoglobin production. Certain types of illness may shorten their life-span. In a patient with pernicious anemia, the life-span is reduced to about 85 days; with sickle-cell anemia, to 42 days (Figure 10.5). The rate of replacement cannot keep up with the loss of cells; the red cell count falls below normal and results in an anemic state. The cause of the shortened life-span of these cells is not known but is possibly related to the fragility of the cells, and hence to the character of the plasma membrane.

The digestive system also has a very high cell death rate. It has been estimated that the cells lining the intestine of the rat are replaced every 38 hours, while the surface epithelium of the stomach is replaced every 3 days. In humans, a total replacement of the epithelium of the small intestine occurs every 7 to 8 days. The cells at the tips of the villi are constantly being sloughed off and are replaced by the division of cells lower down, which then move up toward the tip in the process of replacement.

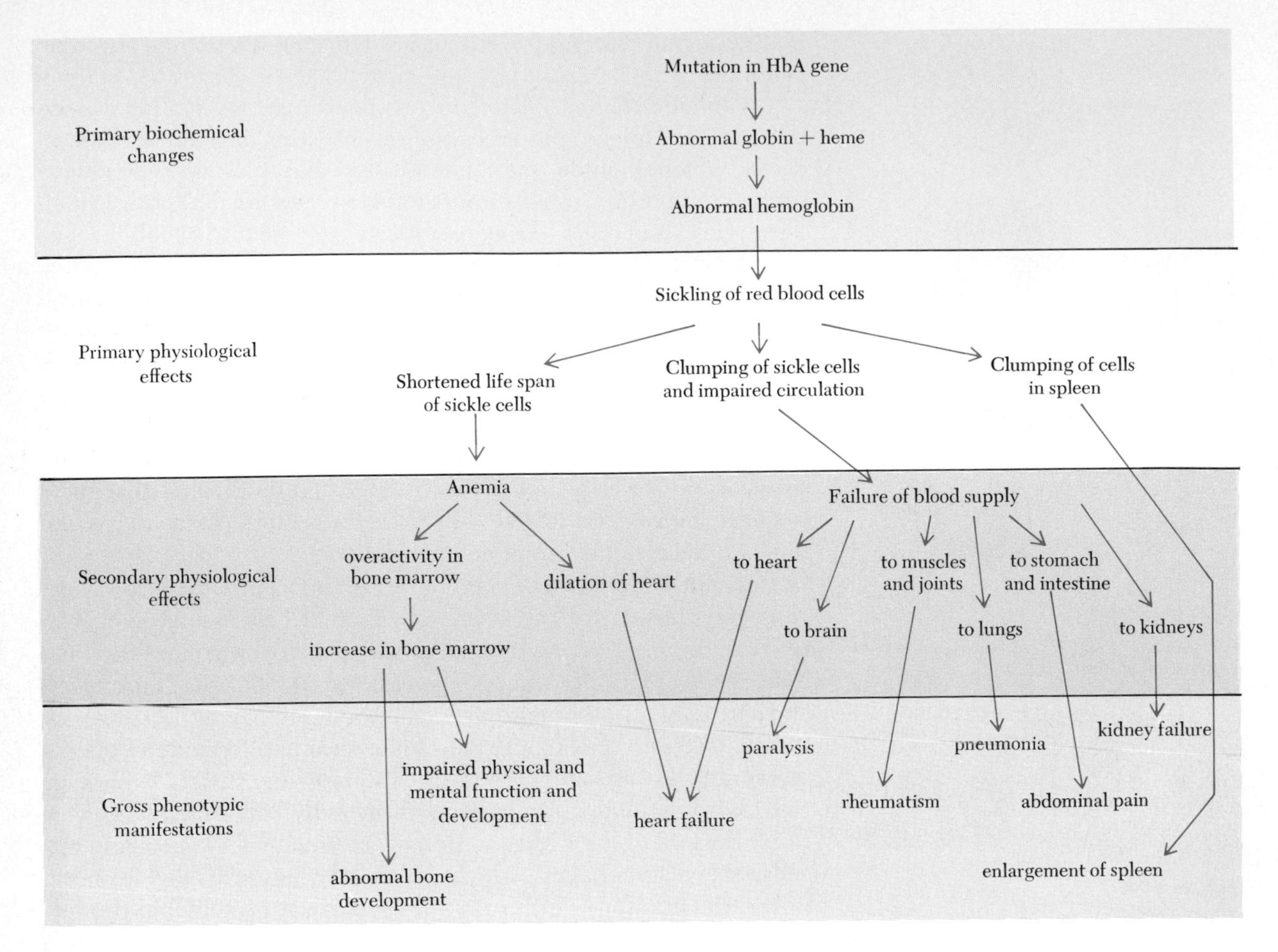

Figure 10.5

A single defect in the gene responsible for the formation of hemoglobin A, when in a homozygous state, leads to a wide variety of manifestations in virtually every organ of the body. (After J. V. Neel, and W. J. Schull, *Human Heredity*. University of Chicago Press, Chicago, 1954.)

If we consider the human body as a whole, it consists of about 60,000 billion cells. Roughly, 250×10^9 of these are red blood cells. If their life-span is about 120 days long, 2 billion of them must be produced anew each day. Another 70 billion are estimated to be lost and replaced each day in the intestine. These numbers merely emphasize the fact that the daily wear and tear of existence is a constant interplay between the loss and gain of cells, and that if one aspect of the process is somehow out of step—for example, the overproduction of white cells—the body itself falls into a state of disrepair.

In the plant kingdom, we find that the lower plants—algae and fungi, in particular—have a rather low loss of cells through death. In the higher plants, however, the rate is enormous. In herbaceous plants, all the cells aboveground are lost every season. But consider a large tree. The annual loss of cells in the leaves, flowers, and fruits (only the seeds remain alive) is high enough, but when you add to that number all the cells going to form dead wood and bark, it is readily apparent that the loss of cells through death in animals is small by comparison. Yet a high rate of cell death in the seed plants is as much a normal pattern of existence as is the continuation of living cells.

The difference between higher plants and animals, regarding longevity and cell replacement, is even more profound than it appears at first glance. A plant such as a tree preserves its "youthfulness" and vigor in two ways: (1) by maintaining a thin sheath of meristematic cells (apical, lateral, and root meristems) over the entire exterior of the plant, and thus continuing, at least during the growing season, the processes of cell division and differentiation indefinitely; and (2) by paralleling these processes with the equally continuous death of differentiated cells, either by discarding these cells at the end of the growing season, as in the case of leaves, flowers and fruits, or by constantly converting them into dead conductive and supportive tissue, as in the case of wood. A mammal, on the other hand, achieves its longevity by an exactly opposite process, that is, by preserving the majority of its differentiated cells in a living, functioning state. As was pointed out earlier, however, a cell that does not divide is destined to die; the result is that the life-span of a mammal, or indeed any multicellular animal, is short compared to that of the longest-lived trees.

Cell death as a developmental process

When we think of normal development, we think of an increase in the number of cells, their subsequent differentiation into specialized cells, and the grouping of these cells into organs and organ systems. This process is dynamic and creative, so it may seem incongruous to characterize cell death as a vital and necessary aspect of development. Cell death, however, plays two very significant roles in development. The first, *metamorphosis*, has long been known; the second, the role of cell death in the shaping of organs and body contours, has only recently been appreciated and investigated as a phase of development.

Metamorphosis involves a change in shape or form (the transformation of a larval form of an organism into an adult) and a change in organs when one mode of life is exchanged for another. It is, therefore, the removal of one phenotype and replacement of it by another, and in its intricacies and because the organism must continue to function even while undergoing change, it is a program of cellular gains and losses, beautifully timed and executed. Two well-known examples are the metamorphosis of a tadpole into a frog and a caterpillar into a *pupa,* then into a butterfly or a moth.

During the time of metamorphosis, a tadpole is transformed into a frog without any great change in size, and in the common American leopard frog the process takes about a year. The tadpole that emerges from the egg, and the large tadpole about to metamorphose, have the same general shape. In its conversion into a frog, it grows legs and loses its tail, which is devoured by wandering cells, or *phagocytes,* that are carried by the bloodstream to the tail region where they gradually consume the muscles, nerves, skin, and other tissues. The skin shrinks, and eventually the tail is reduced to a mere stump. In addition, the tissues in the respiratory, digestive, and excretory systems are extensively reorganized. There is, in fact, virtually no tissue or organ system that is unaffected, and in the process the gilled, tailed, ammonia-excreting, vegetarian tadpole is metamorphosed into a lung-breathing, tailless, four-footed, urea-excreting carnivorous frog (Figure 10.6). The ancient and ancestral tadpole becomes the more evolutionarily modern frog. The process can be speeded up or slowed down experimentally, for in the frog, metamorphosis is, at least in part, under the control of an iodine-containing hormone, thyroxine, from the thyroid gland. More thyroid hormone accelerates the process, less reduces the rate of change and may even

Figure 10.6
Extensive tissue alterations take place as the metamorphosis of the tadpole into the adult frog progresses: the destruction of old tissues occur as new tissues come into being. Gills disappear and are replaced by lungs, and the circulatory system has to be altered accordingly. The intestinal tract undergoes change as the vegetarian diet of the tadpole shifts over to the more or less carnivorous diet of the frog. As the mode of locomotion changes, the tail of the tadpole is resorbed by cellular scavengers as legs make their appearance. Comparable changes occur in the excretory system. Cell death, therefore, is as much a part of the process of development as is the appearance of new organs and altered physiological phenomena.

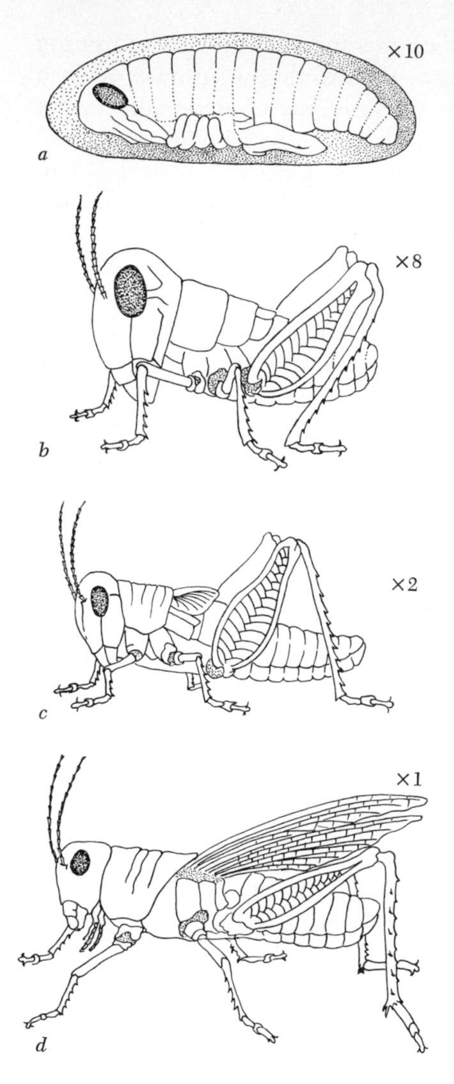

Figure 10.7

Incomplete metamorphosis as it occurs in the grasshopper. A gradual transformation with an increase in size takes place rather than a complete change of form as in the larva-pupa-adult transformation of flies, moths, and butterflies. No great tissue alterations occur, and the organs appropriate to a given stage, such as wings and mature sexual organs, make their appearance progressively.

prolong larval life and shape indefinitely. An interesting aspect of such cellular changes in the frog is that the same thyroxine that governs cell death in some tissues is responsible for the growth and differentiation of other tissues contributing to adult form and function. The capacity of a cell to degenerate or to grow actively in response to the hormone must therefore, be a function of the state of the cell.

The character of metamorphosis in insects varies quite widely, and cell death is not always a major aspect of change. In the simplest type of metamorphosis, the cells of a particular immature, or nymphal, tissue are retained and enlarged to form the corresponding tissue in the adult, and only minor differences in growth and differentiation are needed to bring this about. In these cases of *incomplete metamorphosis,* the form of the insect is only slightly altered—in particular, the completed growth of its reproductive system and of fully developed and functional wings and flight muscles—as the juvenile form matures to adult proportions. Good examples of this type of metamorphosis are found in locusts, grasshoppers, and cockroaches (Figure 10.7).

In *complete metamorphosis,* the larval and adult forms are totally different from each other. The larva, or caterpillar, is converted into a pupa, usually encased within a cocoon. The larval skin hardens and shrinks into the outer skin, or puparium, of the pupa, and the larval tissues are almost completely destroyed. The adult develops during pupation, making use of the cellular debris of the larva, and adult tissues arise from *imaginal buds* that form in the larva and that escape cell death. These buds can be regarded as zones of persistent embryonic tissue in which the potentiality for growth and differentiation is suppressed during most of the larval life. It is realized only when the juvenile hormone (which controls larval growth and prevents metamorphosis) lessens its normal activity and ecdysone (the hormone concerned with molting and metamorphosis) becomes fully effective. The action of ecdysone in insects is similar to that of thyroxine in amphibians in that its target cells behave variously. When a molt occurs under the influence of ecdysone, the old cuticle is freed and discarded because the immediate epidermal cells die and are reabsorbed, while the underlying epidermal cells are stimulated both to divide and to produce a new cuticle. The programming of a cell determines its response to a stimulus.

This particular point is beautifully illustrated by what takes place in the developing silk moth. As the juvenile hormone lessens in amount and influence, and ecdysone exerts its effect, diapause is broken and adult development is initiated. As this occurs, the

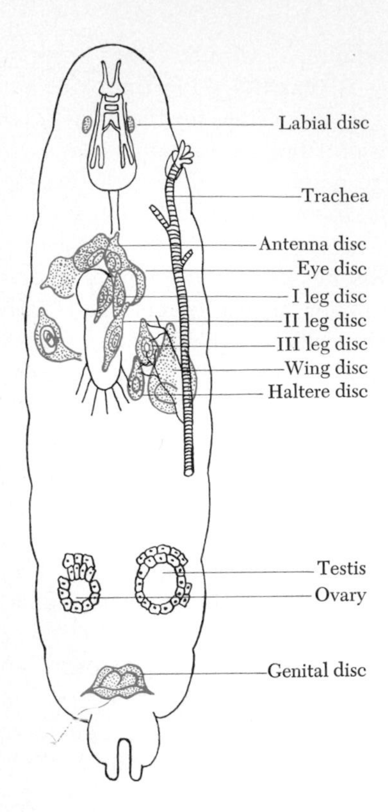

Figure 10.8
Imaginal buds in a larva of *Drosophila* just before pupation. During metamorphosis most of the larval structures except the nervous system will undergo destruction, while the adult tissues will arise from the imaginal buds, some of which are indicated. The buds form during larval life but do not undergo differentiation until the influence of the larval hormones wanes. Since there is no intake of nutrients during metamorphosis, the new tissues of the adult are constructed from the digested debris of larval cells. (With permission of D. Bodenstein, from M. Demerec, *Biology of Drosophila*. New York: John Wiley & Sons, Inc., 1950.)

larval intersegmental muscles, which are of no use to the flying adult, begin a process of disintegration that continues over a three-week period. The muscles, however, are functionally useful until the moth emerges from its pupal case and begins to unfold and spread its wings. The intersegmental muscles aid in developing a bodily compression, which forces haemolymph into the wings, causing them to attain their normal size and shape. Once this is accomplished, the destruction of the muscles hastens to completion. However, the very same endocrine situation that causes the intersegmental muscles to undergo cellular death also promotes the development of the whole assembly of flight muscles, these becoming fully functional when the intersegmental muscles have fulfilled their final role in spreading the wings. The death of cells, therefore, like their production for organ formation, is a programmed affair, and the target cells respond differentially to the hormonal stimulus.

Figure 10.8 shows the location of certain imaginal buds of the larva of the fruit fly, *Drosophila*. Much of the brain and nervous system will survive cell destruction, but the larval intestine, blood system, muscles, and skin will be totally destroyed and then replaced by appropriate adult tissues and organs.

The last type of cell death we shall consider is that involved in the shaping of organs. Form can be achieved by relative rates of cell growth or number. As organs develop during morphogenesis, for instance, excess cells are often a hindrance, and those transient cells that were of use to the embryo or larva, but not to the adult, must be removed (the tail and gills of the tadpole are cases in point, as is the embryonic tail of humans that is formed, but resorbed, before birth). Failure to get rid of unwanted cells often leads to embryonic tumors of various sorts. Or when an organism forms secretory ducts, cells often die instead of pulling apart to provide for the central hole, or lumen. Many organs form by the infolding of tissues that then fuse along their edges—for example, the eye and part of the nervous system—and the seams where fusion takes place are removed by cell death. Within the mouth, lips are freed from the gums by cell death. Fingers and toes are separated from one another in the same way; if cell death and cell resorption fail to take place, and the separation is incomplete, a webbed condition, or *syndactyly*, results (Figure 10.9). The pedigree indicated in Figure 10.10 points to syndactyly as a dominant trait. The difference between the webbed foot of a duck and that of a chicken is also largely a matter of the retention or death of cells between the digits.

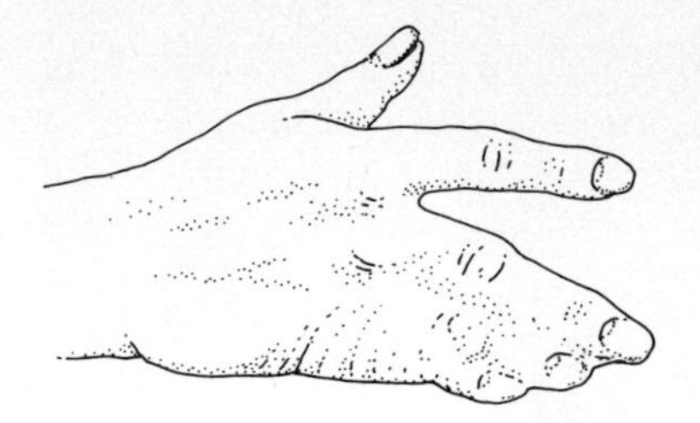

Figure 10.9
Syndactyly, as it occurs in humans. As Figure 10.10 indicates, it behaves as a simple autosomal dominant. The expression of the mutant condition is variable: the ring and little finger may be involved, sometimes three fingers are united, sometimes the expression is much less than that indicated above. The variability relates, at least in part, to the degree to which cell destruction of the tissues between the digits occurs. How the particular pattern of cell destruction is governed is not known.

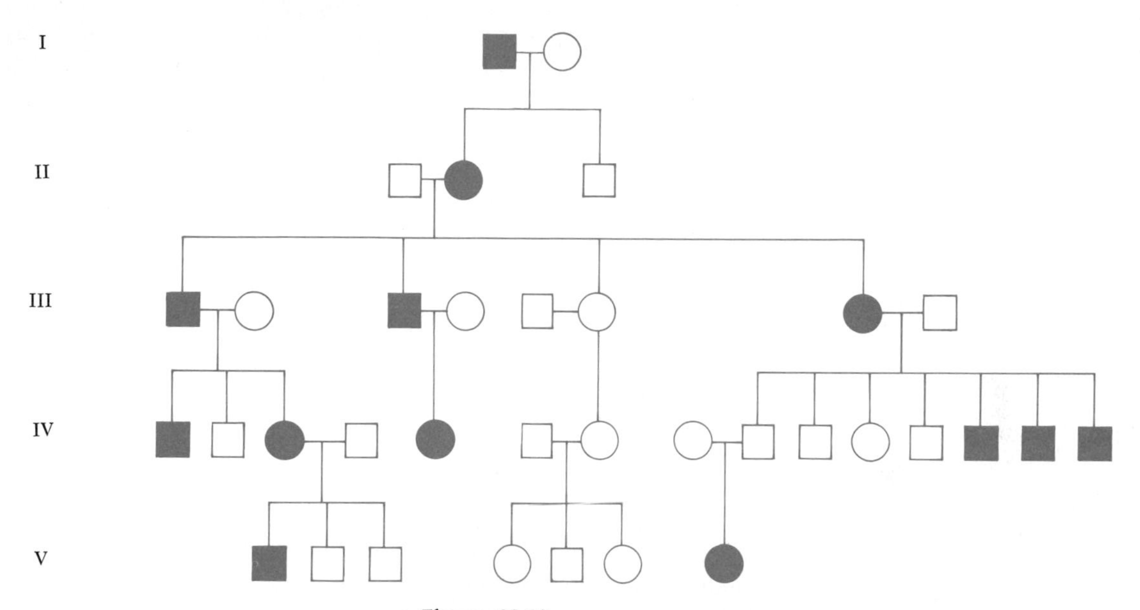

Figure 10.10
A selected portion of a pedigree of a family in which the appearance of syndactyly can be traced with a fair degree of accuracy. Squares represent females, circles males, and filled-in symbols those individuals exhibiting the trait. The phenomenon behaves as a simple autosomal dominant since, with one exception, any individual showing the trait also had a parent who had the trait. The fact that it appears in both sexes indicates that it is not a mutation found on one of the sex chromosomes (that is, the X or the Y). The exception, the son in the V generation at the right has normal parents, but from other known information it seems most likely that the son is a heterozygous individual in which the trait has not manifest itself.

One of the most arresting instances of cell death as a morphogenetic phenomenon is the one that frees the elbow of the wing of the chick from the body wall and gives the wing its characteristic shape as development takes place. Figures 10.11 and 10.12 illustrate the region in question. These cells die as a line of cellular

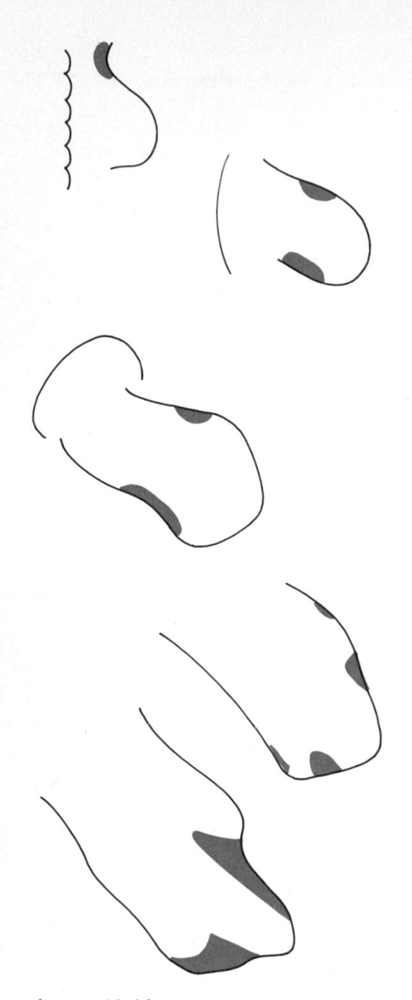

Figure 10.11
Diagrammatic representation of the stages in the development of the wing bud in the embryo of a chick, showing the areas (shaded) where massive cell death and necrosis occurs, a process that gives shape to the developing organ. The numbers accompanying each figure are recognizable stages that occur during development. If cell death and cell removal failed to occur, the wing would not be freed from the body wall, and it would also lack its characteristic shape.

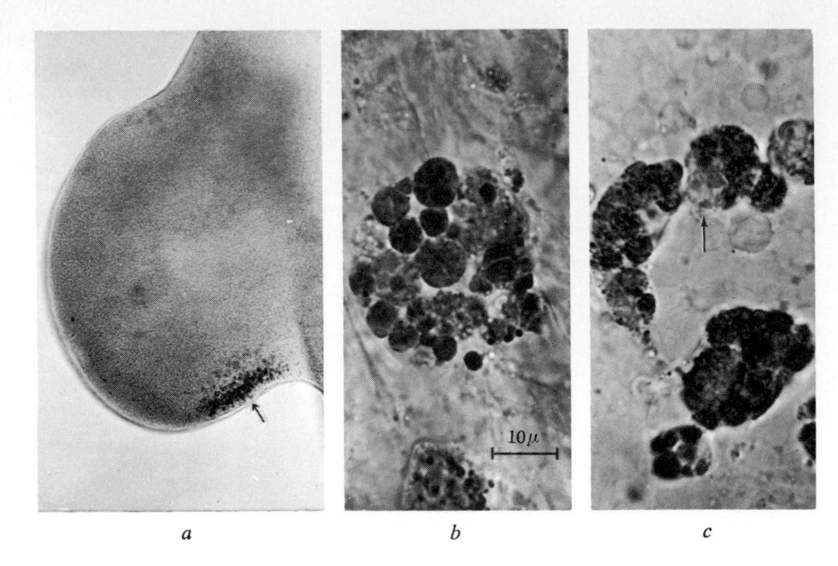

Figure 10.12
(*a*) A wing bud showing the area of cell death and necrosis that normally takes place as the wing undergoes development and a change in shape. (*b*) and (*c*) A higher magnification of the necrotic areas showing the dead or dying cells. (Courtesy of Dr. John Saunders.)

destruction moves from the body area along the front and back of the wing toward its tip, separating in the process the region from the elbow to shoulder from the body wall, shaping and separating the digits, and thus leading to a sculpturing of the form of the wing. What is most fascinating in this process, however, is that when these cells at a particular stage of development are removed and transplanted to another part of the embryo, the cells *die on schedule* (at approximately four days of age) as if they were still in their original site. Had they been transplanted somewhat earlier they would not have died, but rather would have become part of the region to which they were transplanted. Their time of death already had been precisely determined by some unknown change that had taken place within them, and once embarked on their course of destruction, they could not escape. Such programmed cell death, is, of course, a specific instance of differentiation that includes elimination.

We may well ask why cells die, since this seems an extravagant waste of materials that took substance and energy to form. Dying cells are not hard to recognize. The nuclei become compact and dense, the lysosomes and the destructive enzymes in them increase, and macrophages (scavenger cells) move in to dispose of the dying cells. In developing systems, where cell death contributes to the emergence of form, the causes are many. The death of cells in the wing bud, in addition to being time-dependent, is dependent, at least in part, on underlying tissues. If placed alone in tissue culture, they grow indefinitely, but if the normal underlying tissues are placed adjacent to them, they soon die. Some diffusible substance would appear to be the causative agent of death. In the instances of metamorphosis we have mentioned, hormones are involved; a tadpole remains a tadpole if thyroxine is decreased, and a caterpillar remains a caterpillar if juvenile hormone is plentiful and ecdysone is absent. In other instances, a genetic influence on cell death is evident, and this kind of death can be as sharply definitive as that caused by hormones. In *Drosophila,* female sterility results when certain genes cause an early death of the nurse cells that normally surround the developing eggs; in the mouse, a variety of genes cause selective destruction of parts of the tail; in humans, the absence of cell death, resulting in syndactyly, is an inherited abnormality. Without positive evidence, we can only assume that the nucleus controls cell death as well as the normal living activities of the cell. As pointed out in Chapter 2, development is an extraordinarily precise phenomenon, consisting of many coordinated events and processes. Even though our knowledge of the causes of cell death is meager, we must view such death as part of the normal processes and as important as growth and differentiation.

We know even less of cell death in organisms that die naturally of old age. Old cells look different morphologically from young cells. In humans, aging cells accumulate a pigment that contains proteins and enzymes, but the relation of the pigment to aging is not known. About the only thing we can say at the present time is that to stay young, a cell must divide. If it differentiates, it writes its own death sentence.

Bibliography

Glücksmann, A. 1951. Cell deaths in normal vertebrate ontogeny. *Biological Reviews* **26**, 59–86.

Hayflick, L. 1975. Cell biology of aging. *BioScience* **25**, 629–637.

Hoffman, J. G. 1957. *The Life and Death of Cells.* Hanover House, Garden City, N. Y.

Lockshin, R. and C. M. Williams. 1964. Programmed cell death. II. Endocrine potentiation of the breakdown of intersegmental muscles of silk moths. *Journal of Insect Physiology* **10**, 643–649.

Saunders, J. W., Jr. 1966. Death in embryonic systems. *Science* 154, 604–612.

Saunders, J. W., Jr., and J. F. Fallon. 1966. Cell death in morphogenesis, in *Major Problems in Developmental Biology,* ed. M. Locke. Academic Press, New York.

Saunders, J. W., Jr., M. F. Grassland, and L. C. Saunders. 1962. Cellular death in morphogenesis of the avian wing. *Developmental Biology* **5**, 157–158.

Whitten, J. M. 1969. Cell death during early morphogenesis: parallels between insect limb and vertebrate limb development. *Science* **163,** 1456–1457.

Cells through time

The point has been stressed that the cell is the most elementary unit of organization through which life is expressed. Stated somewhat differently, it can be said that life manifests itself when matter achieves a special state of organization, one that evolved from a noncellular condition to a degree of complexity represented by cells. Admittedly, we do not know with any degree of certainty just how life began, or how cells were first formed, but there is a good deal of evidence to suggest that a variety of chemical and physical processes, taking place on earth or its surrounding atmosphere, led to the appearance of simple organic (carbon-containing) compounds. These compounds interacted with each other to give more and more complex chemical groupings and structures until that special state of organization made its appearance, and life as we know it expressed itself.

Life, therefore, and hence cells, have a history, one that the fossil record suggests began about 3.5 billion years ago. The cells we can examine today, even the simplest of which are highly complex and intricately organized, are products, consequently, of a long process of time and change. We can dismantle a cell, and by piecemeal examination learn a great deal about its structures and how these are related to functions, but it is quite a different thing to carry out the opposite task—to visualize how simple components came into being and ultimately organized themselves into a life-producing state of arrangement. Much is known, and a good deal of experimental evidence is available, but two major gaps (as well as many minor ones) exist in our knowledge of the evolution of cells: the origin of the prokaryotic cell from noncellular materials, which is the more elusive problem; and the origin of the eukaryotic cell from prokaryotic ancestors, the subject of much speculation.

Prebiotic conditions

The solar system and its planets, including the earth, were formed about 4.5×10^9 years ago from a diffuse dust cloud condensed through gravitational force. The early atmosphere of the earth was, in all likelihood, strongly reducing, with no molecular oxygen (O_2) present. Nitrogen was probably present as N_2 with ammonia (NH_3) in small amounts. Carbon could have been present as methane (CH_4), carbon monoxide (CO), carbon dioxide (CO_2), or a mixture of all three. Water seems to have covered a portion of the earth within a short time, derived possibly from volcanic activity. These molecules would react only very slowly with each other if left undisturbed, but energy in the form of electrical discharges (lightning) and ultraviolet light from the sun could activate these simple molecules and cause them to form more complex ones in the atmosphere or in water. Such molecules as those indicated in Table 11.1, which have been identified as existing in outer space, were almost certainly important intermediates to the larger macromolecules. Among these intermediates, hydrogen cyanide, formaldehyde, more complex aldehydes, and the acetylenes are thought to have been particularly important as precursor molecules, especially when dissolved in water and concentrated by the evaporation of lakes or tidepools, or by being adsorbed onto clays, which serve as activating sites.

This seems like an inauspicious group of molecules out of which to derive the proteins, nucleic acids, and carbohydrates so commonly associated with life, but during the 1920s the Russian Oparin and the Englishman Haldane put forth the proposition that life arose spontaneously from these meager beginnings. This remained no more than an interesting proposal until the 1950s, when

Table 11.1

A partial list * of molecules found in interstellar space that could have played a role as prebiotic precursors of amino acids, sugars, and nucleic acids.

Molecule	*Formula*	*Molecule*	*Formula*
Cyanogen radical	$CN-$	Carbon monosulfide	CS
Hydroxyl radical	$OH-$	Formamide	$HC(NH_2)O$
Ammonia	NH_3	Silicon monoxide	SiO
Water	H_2O	Carbonyl sulfide	OCS
Formaldehyde	$H_2C{=}O$	Acetonitrile	$CH_3C{\equiv}N$
Carbon monoxide	CO	Isocyanic acid	$HN{=}C{=}O$
Hydrogen	H_2	Methyl acetylene	$CH_3C{\equiv}CH$
Hydrogen cyanide	HCN	Acetaldehyde	CH_3CHO
Cyanoacetylene	$HC{=}C{-}CN$	Thioformaldehyde	$H_2C{=}S$
Methyl alcohol	CH_3OH	Hydrogen sulfide	H_2S
Formic acid	$HCOOH$	Methylene imine	$H_2C{=}NH$

* From Herbig, G. H. 1970. *American Scientist* 62, 200.

a number of investigators demonstrated that an electrical discharge passed through an enclosed mixture of ammonia, methane, hydrogen, and water vapor led to the formation of a number of amino acids, plus other molecules of biological interest. Ultraviolet light and very high temperatures, possibly from volcanoes or hot springs, also serve as sources of energy for bringing about similar molecular charges. With carbon dioxide, carbon monoxide, and nitrogen added to the above mixture, nearly all of the amino acids known from living organisms, some that were unknown in biological systems, and some simple proteins were also found. Seventeen of the 20 amino acids that occur in proteins have been obtained from experimental situations simulating prebiotic conditions, sometimes with yields of up to five percent. Sugars can be similarly formed from formaldehyde (H_2CO), and hydrogen cyanide (HCN) and related compounds are sources of the bases entering into the nucleic acids. Adenine, in fact, is formed in high yields, and perhaps it is not surprising that adenine would become a molecular species in centrally important structures such as DNA, RNA, and ATP.

The fact that amino acids can be formed under prebiotic conditions suggests that they should be found where life does not now exist but where the conditions are appropriate. As Table 11.2 indicates, amino acids are present in meteorites and in samples from the moon, in roughly the same kinds and proportions as in comparable simulated laboratory experiments. This suggests, of course, that the laboratory experiments are on the right track, and

Table 11.2
Proteinaceous amino acids from a variety of abiotic sources, compared with animal sources.*

Terrestrial lava	*Moon samples*	*Murchison meteorite*	*Laboratory samples from CH_4, N_2, H_2O, and traces of NH_3*	*By sparking: from CH_4, NH_3, H_2O, and H_2*	*Animal sources*
Glycine	Glycine	Glycine	Alanine	Glycine	Aspartic acid
Alanine	Alanine	Alanine	Glycine	Alanine	Glutamic acid
Glutamic acid	Glutamic acid	Glutamic acid	Aspartic acid	Glutamic acid	Glycine
Aspartic acid	Aspartic acid	Aspartic acid	Valine	Aspartic acid	Serine
Serine	Serine	Valine	Leucine		Alanine
Threonine	Threonine	Proline	Glutamic acid		Leucine
Isoleucine			Serine		Isoleucine
Leucine			Isoleucine		Proline
Valine			Proline		Valine
			Threonine		Lysine
					Threonine
					Arginine
					Cystine
					Phenylalanine
					Tyrosine
					Histidine
					Tryptophan
					Methionine

* Laval, lunar and animal data from Fox, S. W., 1975 (see bibliography); other data from Miller, S. L. and L. E. Orgel, 1974. A large number of amino acids of nonbiological significance, and other organic molecules, were also present in the laboratory samples. The meteorite and laboratory-derived amino acids are listed in order of prevalence of formation.

are not far removed from the terrestrial conditions just preceding the beginning of life. This also suggests that we exist in an orderly universe: the chemistry of 4.5×10^9 years ago on earth, on the moon today, on a recent meteorite from the asteroid belt, or in a planned laboratory experiment are strictly comparable to each other. The same physical laws operate independent of time and place.

With 1×10^9 years intervening between the formation of the earth and the first positive appearance of life, much could have happened. There were no organisms to use up these compounds as they formed, and with evaporation occurring in some areas, the molecules could have reached relatively high concentrations. It has often been argued that life began in the dilute molecular soups of a warm tidepool, but it now appears far more likely that it occurred in a cold, concentrated mixture in which the degradation of newly formed molecules was taking place at a slower rate than their spontaneous formation. Under these conditions, larger and larger molecules such as simple proteins and nucleic acids could have had a more favorable opportunity for coming into being, particularly in the presence of activating clays.

It is a far cry, however, from molecules in solutions to organized cells. One can, indeed, ask if there is a predisposition, due to chemical affinities or structures of these molecules, to assemble into cell-like forms. Sidney Fox has argued that this is so to a degree; that is, there is stereochemical information in these prebiotic protein molecules that caused them to assemble from amino acids in preferred sequences without the guidance of nucleic acid codes. He has demonstrated in the laboratory that these molecules possess the ability to aggregate upon contact with water, producing "proteinoid" droplets, which possess a membranelike outer surface and a fluid interior, and which can enlarge and divide. In addition, these proteinoids exhibit enzymatic and active transport activities of a primitive but qualitatively distinct character, just as do artificially prepared membranes composed of proteins and phospholipids. These discoveries would seem to point to an important step in our understanding of the evolution of the cell from inanimate beginnings, but unless the sequence of amino acids in these prebiotic proteins was preferentially ordered because of stereochemical relations, the proteins and the proteinoids would be merely statistical examples of what was possible, and the structural or enzymatic properties would be unpredictable, uncoordinated, and not likely to manipulate a steady-state flow of energy through the system. However, laboratory studies have shown that although any amino acid in solution can spontaneously link up with any other one with a loss of water (Figure 11.1), the process does exhibit a degree of specificity. In the presence of activated clay, which acts

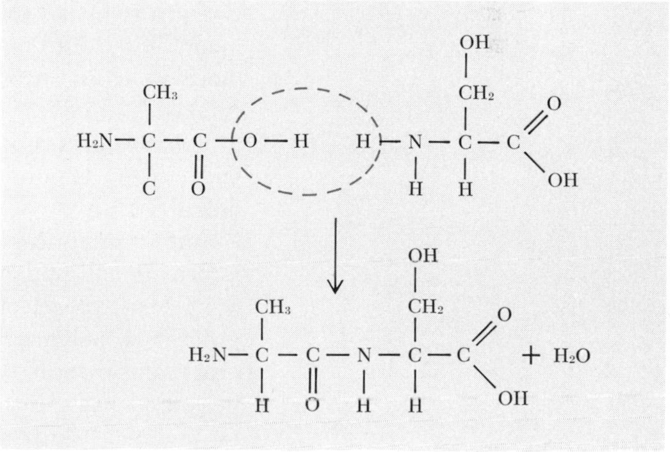

Figure 11.1
Linkage of the amino acids alanine and serine to form the dipeptide alanylserine with loss of water.

as a catalyst, proteins of 20 to 40 amino acids long are possible. The intriguing fact is that the sequencing of amino acids in these spontaneously formed proteins is statistically similar to those found in today's organisms. The significance of this fact is that it is the sequence of amino acids in a protein that determines both its three-dimensional shape through folding and its active site if it functions as an enzyme. The same property also permits the proteins to assemble into higher orders of structure. This can be readily demonstrated with artificial membranes or with the tobacco mosaic virus. This virus can be decomposed into protein and RNA molecules. If separated from each other, the proteins alone will reassemble spontaneously to form rodlike particles. If the RNA is added to the solution, the intact virus will be reconstituted and will be infective. A comparable situation occurs with ribosomes: if decomposed into their proteins and rRNA, spontaneous reassembly will occur under the proper conditions. There is, therefore, not only a selectivity in noncatalyzed sequencing of amino acids into primitive proteins, but also a selectivity in the spontaneous union of an amino acid with a nucleotide. It is this coupling that is central to the beginning of a genetic code, that is, the coupling of an amino acid to a specific tRNA.

The moon, meteorite, and laboratory data all point to the fact that given time, which was abundantly available, a primitive kind of life, represented by something like the Fox proteinoids, could have arisen many times and in many locales on earth, the only impetus being the ability of molecules to react selectively with each other. The subsequent development of primitive catalytic systems could enhance the survival of these proteinoids by improving the flow of energy. Such a catalytic function may have been based on the ability of metallic ions to act as catalyst. For example, the capacity of the ferric ion, F^{+++}, to catalyze the decomposition of H_2O_2 to H_2O and O_2 is increased many fold when the ferric ion is incorporated into an organic molecule such as a heme, and even more so when incorporated into a protein to form the enzyme catalase. Improved catalysis would favor survival of those proteinoids possessing it, and they could interact more successfully with their environment than their statistically less favored neighbors.

But a word of caution. A proteinoid without nucleic acids has no informational guidance system to ensure exact genetic continuity; the degree of selectivity of one amino acid for another in the process of polymerization would not be sufficient to specify an exact unit-by-unit sequencing to form proteins in the prebiotic period. It has to be assumed that this was a trial-and-error period of unguided "experimentation," with errors predominating and leading to short-lived proteinoids incapable of going beyond initial formation. Such errors could be tolerated, and even among the "successful" proteinoids there would be a high degree of variation. If these, with varying success, could draw upon the free molecules in the environment and, with even limited catalytic ability, make use of these molecules for energy and additional structure, a start toward a self-perpetuating system would have been under way. The interaction of proteinoids of varying capabilities with environments of varying concentrations of free molecules would set the stage for the operation of natural selection. A successful system, which then acquired some degree of exactness in its reproductive and replicative capabilities, would come to predominate over others as a consequence of natural selection. In fact, a reasonable definition of the beginning of life would be that point in time when natural selection became operative (Figure 11.2).

It is, in fact, difficult to imagine things happening the other way around, that is, a similar protocell being based initially on nucleo-

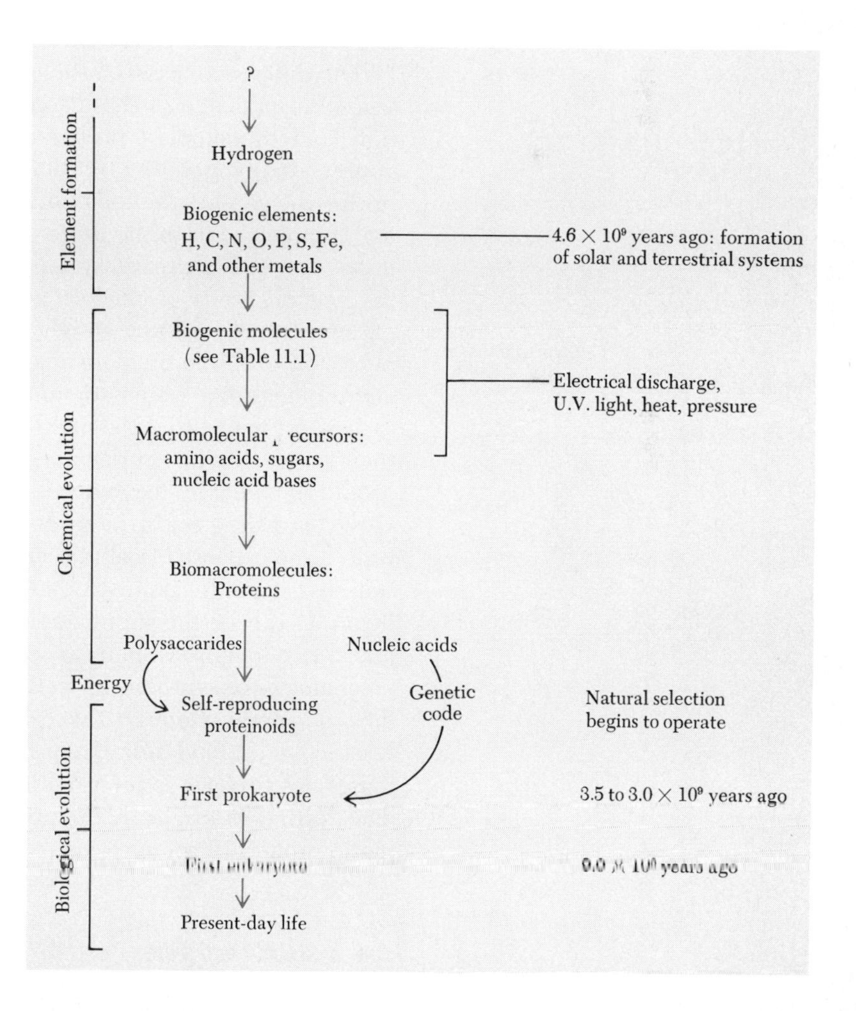

Figure 11.2
Time sequence of evolution from hydrogen to present-day life.

tides and nucleic acids rather than on proteins. Being informational to a large extent and lacking the ability to make use of free molecules in the environment, the nucleic acids have severe limitations as starting points for life systems. It is inconceivable to think of a coding system coming into being before the end products of coding are known, and these, of course, are proteins. In addition, the role of the nucleic acids as we know them today is far longer in range, that is, for genetic continuity and stability, whereas that of the proteins is the moment-to-moment metabolism of the cell. The fact that the two systems are so closely integrated today does not mean that they always have been so, but the universality of the genetic code argues for its high selective value when once formed and, very probably, its appearance in the very early biotic past. In some way, an interaction between nucleic acids and the proteinoid systems must have been established if an informational capability was to come into existence.

Primitive cells

The difference between the proteinoids of Fox, assuming for the sake of argument that these constitute a real step in cellular evolution, and the simplest prokaryote is enormous. This can be best appreciated by pointing out that among the mechanisms needed to bridge the gap are the following: a system for the precise sequencing of amino acids in the proteins for structural and enzymatic purposes; an equally precise system for the formation and assembly of the several kinds of nucleic acids; a genetic code that sets the pattern for all cellular activity through organized structure and behavior; and one or more means for extracting energy from the environment in a systematic and sustained manner.

It is presumed that the first organisms, when once formed, drew their sustenance for growth and reproduction from the prebiotically produced organic molecules in their immediate neighborhood. How long they could have depended upon such a supply, with little or no expense of cellular energy for synthesis, is open to question, but it can be assumed that as the demand for these molecules began to exceed the supply, and survival became precarious, more and more emphasis had to be placed on the development of mechanisms of synthesis as well as mechanisms to ensure a reliable flow of energy through the cell. Such metabolic transactions, at least as understood today, require two features: enzymes and other molecules that act as intermediaries in energy-exchange processes. The fortuitous origin of prebiotically formed proteins with enzymatic properties could serve as a start for promoting and maintaining crucial chemical reactions; any circumstances, intra- or extracellular, that would favor their formation and continuance would be evolutionarily advantageous. As to energy-exchange systems, it is no coincidence that adenine is not only a part of the informational system of all cells—that is, in DNA and the several RNAs—but is also a part of some of the most basic energy-related molecules: the mono-, di-, and tri-prosphonucleotides (AMP, ADP, and ATP), cyclic AMP, nicotinamide adenine dinucleotide (NAD), nicotinamide adenine dinucleotide phosphate (NADP), and flavin adenine dinucleotide (FAD). Adenine is very readily formed under prebiotic conditions if hydrogen cyanide is present, so its central role in energy exchanges must have been established very early in the history of life. ATP is, of course, the universal source of energy for so many basic reactions; cyclic AMP is involved in a host of cellular events; and the other adenine-containing molecules are intimately related to the activity of enzymes.

It is difficult to determine which of the energy-exchange systems are primitive; indeed, it is equally difficult to determine which, among present-day prokaryotic species, are most primitive and consequently closest in function to the earliest formed cells. Practically every known biochemical pathway exists in one or another species of the prokaryotes. But in view of the reducing character of the precellular atmosphere, it is probable that the first organisms that would have been recognized as cellular were prokaryotic anaerobes, with their catalytic systems based upon primitive iron- and sulfur-containing enzymes such as hydrogenase and ferredoxin (Figure 11.3). Of the three principal types of

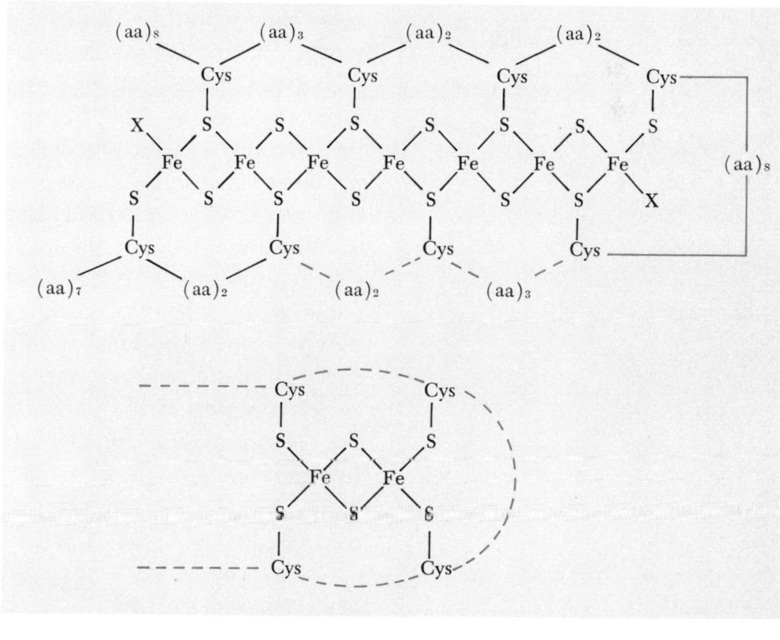

Figure 11.3
The structure of bacterial ferredoxin (above) and the proposed structure of hydrogenase and green plant ferredoxin (below). The former is a small molecule of about 5,000 molecular weight, built around 4 to 7 iron atoms held in place by the sulfhydryl groups of cysteine. The latter is a larger molecule, about 13,000 molecular weight, but with only 2 iron atoms linked together by 2 sulfur atoms and 4 cysteines; the amino acids of this iron-sulfur protein extend off to the left. (Above: after H. T. Yost, *Cellular Physiology*. Englewood Cliffs, N.J.: Prentice-Hall, Inc., 1972; below: after M. Calvin, *American Scientist* 63: 169–177, 1975.)

energy-yielding metabolism—fermentation, respiration, and photosynthesis—fermentation is the simplest, as well as the most inefficient, and it is widespread among prokaryotes. The fermentation of glucose—presumably a prevalent prebiotic molecule—to lactic acid or to ethyl alcohol and carbon dioxide is not an efficient source of energy as compared to the breakdown of glucose to carbon dioxide and water through oxidative phosphorylation. In fermentation, only two net moles of ATP are realized per mole of glucose, whereas 36 moles of ATP are obtained via the oxidative route. These facts, among others, would argue for the primitiveness of fermentation.

As prebiotically formed molecules in the environment were being depleted, other sources of energy and means of energy exchange had to be developed. Anaerobic photosynthesis, trapping solar energy for conversion into chemical energy, was probably an early development. This would require concentrations of chlorophyll and related light-trapping pigments; enzymes such as hydrogenase and ferredoxin, which could function anaerobically; and either internal membranes formed by extensions of the cell membrane, on which the photosynthetic apparatus was layered, or chromatophores, the photosynthetic particles of some bacteria. No oxygen would be involved, organic compounds other than carbon dioxide would serve as carbon sources, water would not be oxidized, and the only biosynthetic product would be the ATP needed for cellular fuel.

The photosynthetic process of green plants, yielding ATP, glucose, and molecular oxygen, would eventually evolve, but at a much later time. As O_2 became prevalent in the atmosphere, the accumulation of prebiotically synthesized compounds could not occur, since they would be oxidized as soon as formed, and since O_2 is a poison, means would have had to be evolved to handle it. The increasing concentration of oxygen in the atmosphere would force some species to shift from fermentation to respiration—many prokaryotes are facultative anaerobes and can function in the presence or absence of oxygen—and a more efficient energy-yielding system would gradually become established.

With the advent of photosynthesis and the production of photosynthetic products, the relations with the environment had to become more selective if only for the purpose of retaining these newly formed products within the cell. It can be presumed that the cell membrane underwent change with time, becoming a selectively permeable barrier between the inner cell contents and the environment, and eventually acquiring the structure and function appropriate to an active transport system. What remains to be explained, however, is how the cell, in order to attain a high degree of stability and to ensure the perpetuation of similar offspring, acquired the informational system necessary to bring this about.

Origin of the code

In considering the origin and evolution of the informational system of cells, we need first to be reminded that life makes use only of L-amino acids (left-handed) and D-sugars (right-handed) (Figure 11.4). This well may be an accident of history, for double-stranded nucleic acids, capable of replicating, can contain either D-ribose or L-ribose but not both. Similarly, spontaneous groupings of amino acids into short stretches of protein usually contain only D- or L- forms but not both. In addition, certain D-amino acids do not fit well into certain secondary forms of proteins such as helices, but there is no overwhelming and compelling argument as to why L-amino acids became the exclusive choice of living forms. It may be simply that L-forms competed better and won out.

There is very general agreement that the beginning of an informational system starts with the tRNAs. It is this molecule that is central in the coding process, for it couples in an exact manner with a particular amino acid, and brings that amino acid into position and in an activated state so that it can react with other amino acids to form proteins. The anticode, therefore, would seem to have preceded the code in evolution.

Figure 11.4
Right-(D-) and left-(L-) handed amino acids and sugars. The amino acids are mirror images of each other, but if rotated 180° cannot be superimposed upon each other.

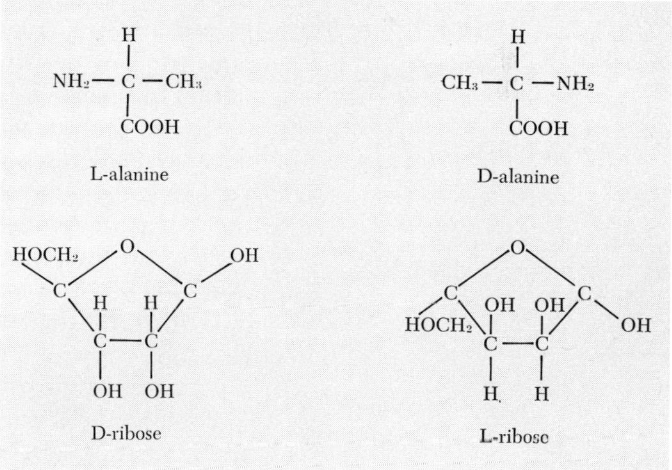

H
NH_2— C —CH_3
COOH
L-alanine

H
CH_3— C —NH_2
COOH
D-alanine

$HOCH_2$ O OH
C C
H H
C—C
OH OH
D-ribose

O
C C
OH OH
$HOCH_2$ OH
C—C
H H
L-ribose

Was the code established initially on the basis of a trinucleotide, or triplet, code sequence? There is no compelling argument suggesting that the code evolved from a singlet to a doublet to a triplet state. There is, on the other hand, evidence to suggest that it might have started initially as a doublet code, and then at a later stage became triplet in nature. As Miller and Orgel point out, a glance at the code as we now know it will indicate why this might be so (see Table 2.4). If we let *xy* stand for any pair of the four nucleotides, certain facts emerge:

1. xyU and xyC *always* code for the same amino acid.
2. xyA and xyG *usually* code for the same amino acid, but when they do not, one or both of the codes are message terminators (UAA, UAG, and UGA), or an initiator (AUG).
3. In eight cases, xyU, xyC, xyA, and xyG all code for the same amino acid.
4. Except for leucine, serine, and arginine, which are coded for by six triplets each, all codes for the same amino acid start with the same pair of bases. It seems most unlikely that the informational content of a code, so strongly dependent on the first two letters of the triplet, could have evolved in stepwise fashion from a singlet code, but it does seem probable that the third letter of the code, responsible for a lesser degree of selectivity, might have been a later addition.

Other pieces of information are extractable from Table 2.4. When U is the second letter, all of the amino acids are hydrophobic, that is, insoluble in water; and when a purine, A or G, is the second letter, the amino acids coded for are always charged instead of being neutral. The codes, consequently, are not a haphazard array of cryptic devices; rather, the regularities argue for some kind of precise chemical specificities, the nature of which resides in unknown fashion in the structure of the tRNAs. It is at this point that knowledge of the early evolution of the informational system ceases. The origin of the code is not known, nor is it yet known how the chromosome, the mRNAs, the rRNAs, and the ribosomes came into existence and were fitted into the scheme as parts of the transcriptional and translational machinery of the cell.

Origin of the eukaryotic cell

Evolution tells us that all living things share a common ancestry, and if what has just been stated about primitive cells is reasonably correct, that ancestral form was a prokaryotic cell. In this state, the prokaryote underwent a good deal of change to produce a variety of species differing in their energy-yielding metabolic processes. The first were probably anaerobic, deriving energy from chemical substances formed earlier by abiotic processes, and, as these chemicals in the environment became depleted, perfecting fermentative pathways that provided sources of carbon and nitrogen as well as energy. Some forms were anaerobic photosynthesizers, capable of forming chlorophyll for light absorption, deriving their hydrogen to reduce CO_2 from hydrogen sulfide and gaseous hydrogen rather than water, and producing ATP as a by-product, but no oxygen and no carbohydrates.

A further evolution of the photosynthetic mechanism produced a system in which water was split to yield the hydrogen that entered into the carbohydrates being formed; oxygen was released into the atmosphere as a waste product. Oxygen, however, is a metabolic poison to an obligate anaerobic species, with the result that as the O_2 in the environment increased in concentration, the anaerobic bacteria retreated into oxygen-free locations, while aerobic forms evolved and made their appearance. These were able to adapt to the presence of molecular oxygen and, in fact, to change their energy-yielding processes from a fermentative to a respiratory form. These changes must have taken millions and millions of years, and among the aerobes that evolved were the blue-green algae, the first photosynthesizing aerobes.

If we think of the first primitive cells arising about 3.5×10^9 years ago, blue-green algal fossils (stromatolites) from the Precambrian are at least 2.5×10^9 years old, and possibly even older. Fossils judged to be of eukaryotic origin are no older than 0.9×10^9 years ago. We must search, therefore, for eukaryotic origins from among the prokaryotes that preceded them in time. All eukaryotes are aerobic, so it is from among this group of prokaryotes that the search must be focused. Several hypotheses account for the origin of the eukaryotic cell, but before examining these speculations, let us consider again how these two types of cells differ from each other.

1. Prokaryotes are unicellular or filamentous forms, and the cells are small, not exceeding 10 μm in diameter. Eukaryotes have large cells generally, only a few kinds being under 10 μm in diameter. They include not only unicellular and filamentous forms, but also all of the varied two- and three-dimensional forms of the plant and animal world. The latter result from the division of cells in several planes, with the cells adhering to each other after division.

2. All eukaryotes are aerobic, although some function as facultative anaerobes; the prokaryotes are enormously varied meta-

bolically. The prokaryotes, nevertheless, are structurally simple; the eukaryotes are extensively partitioned into double-membraned organelles—nuclei, mitochondria, and plastids—and single-membraned cytoplasmic systems—endoplasmic reticulum, Golgi apparatus, vacuoles, lysosomes, and so on. The variety of eukaryotic membranes permits the separation of various metabolic pathways, while the membrane-associated pathways in the prokaryotes must be confined to the single cell membrane or to the membranes concerned with photosynthesis.

3. Both prokaryotic and eukaryotic cells have motile systems based on cilia or flagella, but they differ chemically and structurally. Those of the flagellated bacteria are made of the protein bacillin and are constructed as single strands with no discernable internal structure. Eukaryotic cilia and flagella are based on a 9 + 2 microtubular system, connected to a 9 + 0 basal body (Figure 11.5), with tubulin the basic protein component. No microtubules are found in any prokaryote, whereas they play significant roles in

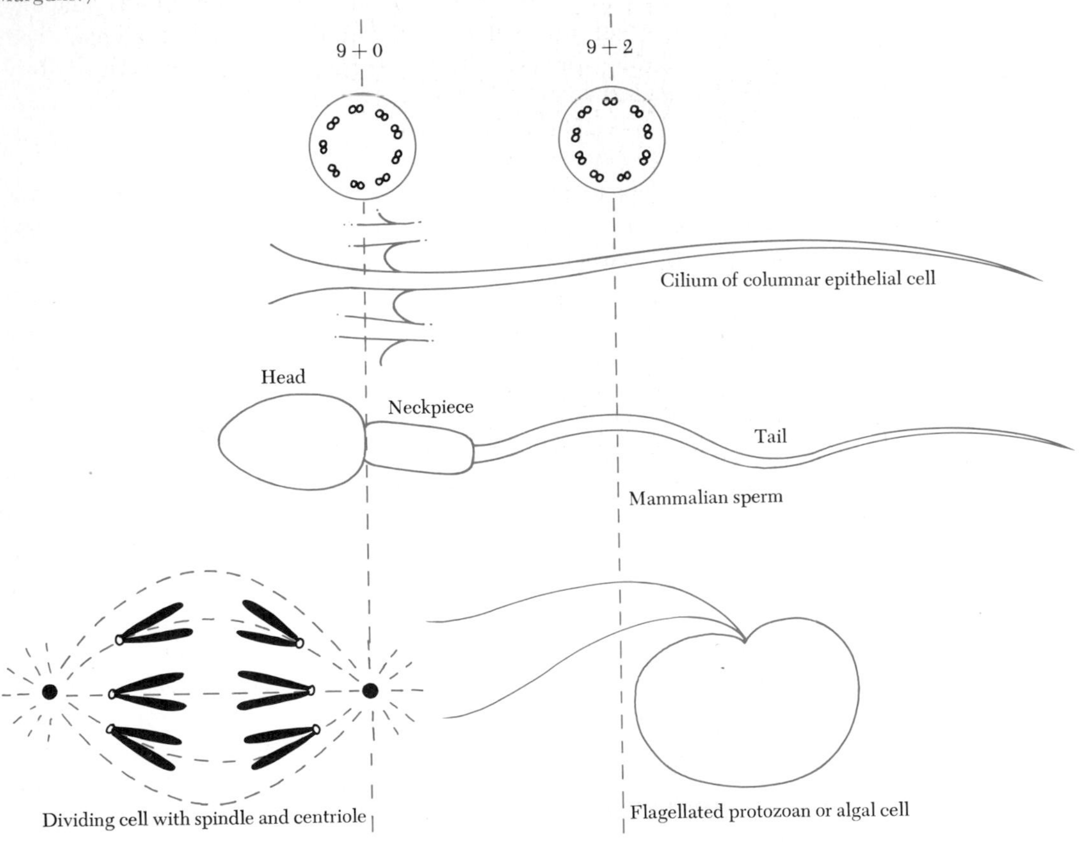

Figure 11.5
Arrangement of microtubular elements found in eukaryotic organelles concerned with movement. Left: the 9 + 0 arrangement characteristic of centrioles, basal bodies at the proximal ends of flagella and cilia, and of sperm just below the attachment of the middle piece to the head. Right: the 9 + 2 arrangement of the cross-sections of the main body of flagella, cilia, and the tails of sperm. (After Margulis.)

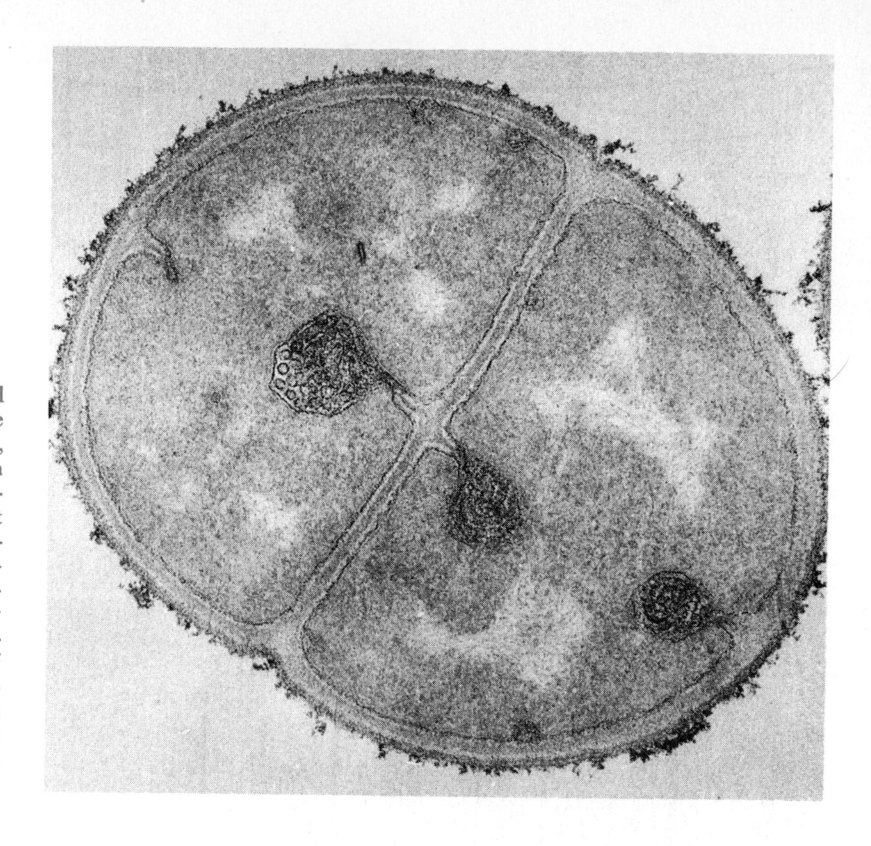

Figure 11.6
An electron micrograph of a bacterial cell in the process of dividing. The rounded structures are mesosomes, formed by a folding in of the plasma membrane. The function of the mesosome is not known for certain, but the different functions suggested include: a localized center of respiratory activity; a means for wall formation in dividing cells; and a means for the separation of nuclei after replication has taken place. In support of the latter suggestion is the observation, made in some bacteria, that the bacterial chromosome is attached to the mesosome. (Courtesy of Dr. Stanley C. Holt.)

movement, wall building, and spindle construction in the eukaryotes.

4. Multiplication of cells is by binary fission in the prokaryotes (Figure 11.6). The naked DNA of the nuclear area is believed to be attached to a mesosome, a structure in the cell membrane that seems to divide and pull apart as the DNA is replicating. The actual details, however, remain uncertain. Cell division in the eukaryotes is by mitosis with the spindle being composed of microtubules, and there is no similarity between the two kinds of divisional processes even though the end results are the same.

5. Inheritance in eukaryotes is biparental where sexual reproduction is concerned. Prokaryotes can, at least in some bacteria, transfer genetic information on a piecemeal basis from one cell to another, but not on a reciprocal basis; it is always a one-way transfer. All prokaryotes exist only as haploids, whereas eukaryotes show an alternation of haploid and diploid states, with fertilization and meiosis making sexual reproduction a workable system. Also, no prokaryote is known to possess more than a single naked chromosome per genome, whereas the eukaryotes have acquired a

multichromosomal state with the DNA associated very generally with histones to form microscopically conspicuous nucleoprotein structures.

It is obvious, therefore, that the eukaryotic cell architecturally is far more complex than its prokaryotic counterpart. In attempting to understand the origin of the eukaryotic cell, the question, consequently, is whether the prokaryotic ancestral cell, by the gradual development and differentiation of internal organelles, evolved into a eukaryotic status, or whether the structural complexity was acquired by a different evolutionary route. Three of the hypotheses that have been advanced are illustrated in Figures 11.7, 11.9, and 11.11.

The cell symbiosis theory, most recently advanced and elaborated upon by Margulis, holds that the eukaryotic cell is a composite structure, made up of several kinds of cells living in symbiotic relations with each other and within a common cell membrane (Figure 11.7). The concept is not new, but it can be supported today with greater conviction because of a more extensive knowledge of comparative biochemistry and ultrastructure.

Symbiosis can be defined as the cooperative existence of two or more organisms for mutual benefit. Symbiosis exists today at many levels of biological organization, but it is those of a cellular, and particularly of an intracellular, nature that are of interest here. The green alga *Chlorella* can be incorporated into the cytoplasm of the slipper animalicule, *Paramecium bursaria,* where it continues to photosynthesize and to provide nutrients to the host cell, even in a subminimal medium. The blue-green alga *Cyanocyta* has been similarly found in symbiotic relations with the protozoan *Cyanophora,* while certain snails have the ability to extract chloroplasts from plant cells and incorporate them intracellularly, where they continue to function photosynthetically. In these instances, the relationships seem not to be mandatory for the host, but may well be for the invading plant.

The cell symbiosis theory holds, therefore, that the particulate organelles of the eukaryotic cell had independent origins as prokaryotic cells. Thus, the plastids of the green plants were derived from a symbiotic algal cell capable of aerobic photosynthesis; this kind of cell was also ancestral to our present-day blue-green algae. The mitochondria of the eukaryotic cell were similarly derived, but in this instance from an aerobic bacterium. The host cell is presumed to be an anaerobic, nonphotosynthesizing cell, which, because of the necessity of adapting to atmospheric oxygen, acquired a symbiont capable of managing oxygen through respiratory pathways.

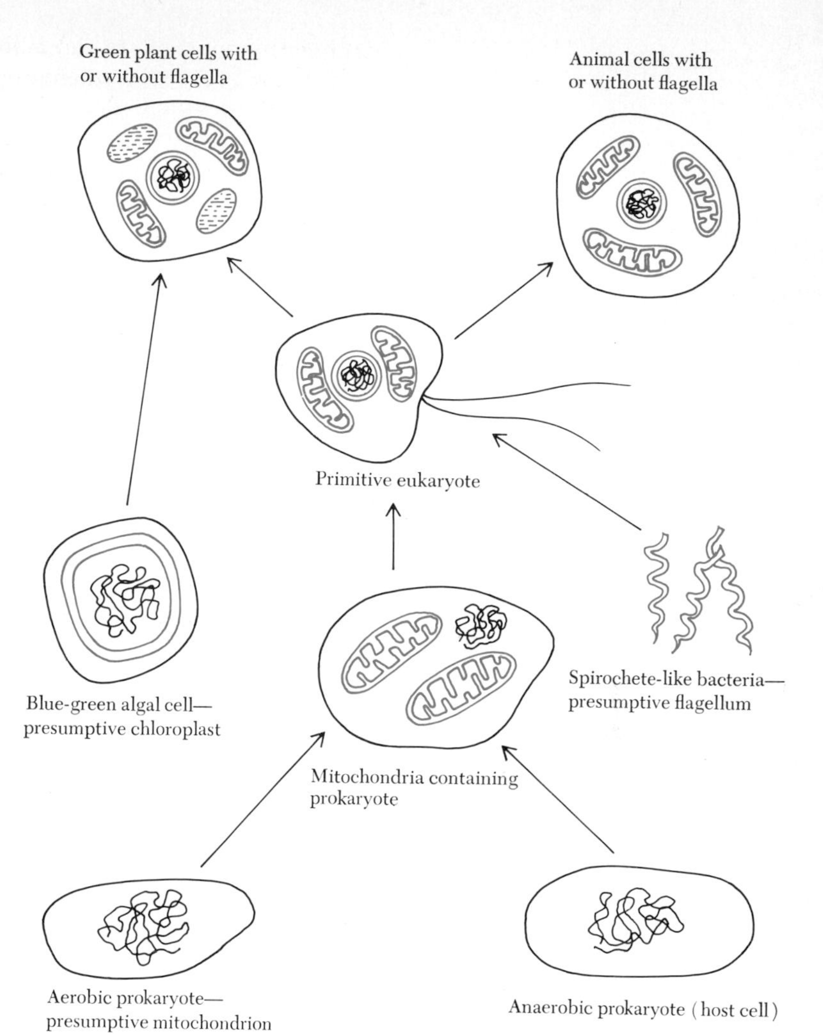

Figure 11.7

The origin of the several kinds of eukaryotic cells according to the symbiosis theory of Margulis. The host cell is presumed to be an anaerobic, prokaryotic bacterium. Aerobic bacteria are assumed to be the precursors of the eukaryotic mitochondrion, and a blue-green algal cell the forerunner of the chloroplast. The origin of the double-membraned nucleus is obscure, but it is probably the DNA of the host cell that became enclosed. Once the nucleus is formed, the endoplasmic reticulum and the Golgi complex could be derived from the outer nuclear membrane, with the lysosomes, vacuoles, and so on derived from the Golgi complex (see Figures 11.8 and 11.10).

How is a symbiont, acquired as described above, to be distinguished from an organelle that evolved intracellularly and progressively over long periods of time? Are there criteria that point to the symbiotic acquisition rather than to the evolutionary developments of these organelles? The arguments are compelling in favor of the symbiotic origin.

First, a newly acquired symbiont of cellular nature would possess its own hereditary system, although time might well have brought about alterations in hereditary function and dependence. Mitochondria and plastids fulfill these criteria; they possess DNA, which is usually circular in form, mRNAs that are complementary to the DNA, rRNAs, tRNAs, and the enzymes necessary for these to function. The tRNAs are different in certain nucleotide sequences from those in the cytoplasm of the host cell, the rRNAs of both mitochondria and chloroplasts are very similar in sequence to prokaryotic rRNA, and the ribosomes are smaller, being much more like those of bacteria and blue-green algae than the larger ones of either animal or higher plant species. In addition, protein synthesis is inhibited at the ribosomal sites by the antibiotic chloramphenicol, but is insensitive to cycloheximide. A similar situation occurs in bacteria and blue-green algae, whereas protein synthesis in the cytoplasm of the eukaryotic cell is inhibited by cycloheximide and unaffected by chloramphenicol. There are striking parallels, therefore, in the structure and physiology of eukaryotic organelles and of prokaryotic cells.

The hereditary system of present-day organelles is, however, insufficient in complexity and diversity of function to fulfill all of the needs of the organelle. This well might be expected, since the continued evolution of the symbiotic relation might lead to a loss or reduction of independent synthetic and informational capabilities, particularly if these were duplicated in the host cell. For example, in yeast, parts of two mitochondrial enzymes, ATP-ase and cytochrome oxidase, are synthesized on the mitochondrial ribosomes, other parts on cytoplasmic ribosomes. In green plants, the chloroplast enzyme, ribulose-1, 5-diphosphate carboxylase, consists of two subunits, one unit with a molecular weight of 14,000, the other about 50,000. Synthesis of the smaller subunit occurs on cytoplasmic ribosomes and is preferentially inhibited by cycloheximide, that of the larger on plastid ribosomes, and is preferentially inhibited by chloramphenicol. It would be expected, of course, that the ribosomes of an invading symbiont and those of the host cell would differ in their protein structure; a long evolutionary period, prior to the establishment of a symbiotic relation, would almost ensure such a difference. This is strikingly borne out in the flagellated, photosynthesizing unicell, *Chlamydomonas*. The chloroplast ribosomes contain a total of 48 different proteins in the two subunits; the larger cytoplasmic ribosomes contain a total of 65 proteins. Based on electrophoretic mobility, no more than four proteins from the two types of ribosomes show any degree of similarity, and even these may show differences when examined by more discriminatory methods.

A second argument in favor of symbiotic origin concerns the fact that the organelles exhibit a pattern of division very similar to that of prokaryotes but quite dissimilar to that of eukaryotes. Mitochondria, chloroplasts, and prokaryotes increase their numbers by binary fission, eukaryotes by mitosis with its structured spindle apparatus. Whether mitochondria and chloroplasts have their DNA attached to their inner membranes is not entirely clear, but individual plastids and mitochondria can possess a number of copies of its genomic DNA, so that division of an organelle and its hereditary content need not be done with the same precision as the eukaryotic genome. It is known that when cells contain only a single plastid, as in *Spirogyra,* or a single mitochondrion, as in *Microsterias,* division of the cell is accompanied by division of the organelle, but the great majority of eukaryotic animals and plants have sufficient numbers of these two kinds of organelles to ensure that each daughter cell receives a share at division.

The presence of DNA in organelles indicates that they are capable of mutation and that, with patterns of variation possible, patterns of inheritance of organellar traits should be discernible that depend upon modes of transmission from one generation to the next. A number of species of green plants exhibit a kind of heredity known as maternal inheritance; the trait in question is transmitted only through the maternal side. In barley, for example, a large number of chlorophyll or chloroplast mutants have been identified. The great majority of these are inherited in a Mendelian fashion, and the mutant genes are, therefore, believed to be in the nuclear genome. Some chlorophyll mutants, however, exhibit maternal inheritance, and these have been traced to mutant chloroplasts. Since chloroplasts are only rarely transmitted through the male pollen grain, the inheritance must of necessity be maternal through the embryo sac. Similarly, uniparental inheritance of mutant mitochondria in yeast has been identified with mutations occurring in mitochondrial DNA.

The theory of the symbiotic origin of the mitochondria and plastids rests, therefore, on a substantial body of circumstantial evidence. On less firm ground is the concept, also a part of the symbiosis theory, that flagella, cilia, centrioles, and the mitotic spindle also owe their origins to symbiotic organisms that attached themselves initially to the outside of the host cell, later to become incorporated into the internal organization of the cell. According to Margulis, these were motile, spirochetelike bacteria, which, upon external attachment, provided the host cell with mobility, an obvious adaptive mechanism if the food supply were in any way depleted in the neighborhood. A number of observations support this view. In the first place, the kind of symbiosis envisioned exists today. The protozoan, *Myxotricha paradoxa,* a parasite in the intestinal tract of termites, has a limited mobility of its own, but thousands of attached and motile bacteria, embedded in the thick outer pellicle, provide motion through symbiosis. In addition, other bacteria, internally located or externally attached, are also symbiotic, but their precise roles remain unknown at present.

Such attachments have been proposed as a first step toward the acquisition of eukaryotic motility and, eventually, replacement of fission with mitotic cell division. At the point of attachment of a flagellum or a cilium to a eukaryotic cell, there is a structure, a basal body, comparable to that of a centriole—an arrangement of microtubules in a 9 + 0 pattern (Figure 11.5). Since it is known that centrioles can function as basal bodies, it is presumed that they are homologous structures. In addition, both centrioles and basal bodies function as microtubule-orienting centers, the basal bodies being concerned with the orientation of microtubules along the length of the flagella and cilia, the centrioles adjacent to the outer edge of the nucleus with the formation of spindle fibers that extend from pole to pole. At a somewhat later time in evolution, the centromeres of the chromosomes also became involved in microtubule-orienting processes, interacting with the poles to form half-spindle fibers stretching from the chromosome at metaphase to one or the other of the two poles. Microtubules are, therefore, involved with movement, whether this is external to the cell by flagellar or ciliary activity, or internal and expressed as anaphase movement of chromosomes.

The above hypothesis is open to question on a number of grounds. Prokaryotes lack the 9 + 0 or 9 + 2 structure in their flagella or cilia; their ciliary protein, bacillin, is different from the tubulin of the eukaryotes; they possess no centrioles or basal bodies; centrioles seem able to appear de novo; and no DNA, and hence no independent inheritance system, has been shown to be positively associated with either centrioles or basal bodies.

No satisfactory solution has been advanced to account for the origin of the eukaryotic nucleus, and no physiological traits reminiscent of prokaryotic ancestry have been discovered that would provide appropriate clues.

It is remotely possible that the nucleus, like the mitochondria and plastids in being enclosed by a double membrane, is also the evolutionary remnant of another invading symbiont, which, in the course of time, lost most of its cytoplasm. If so, the host cell must have evolved in the opposite direction, retaining its cytoplasm but losing its hereditary features. Another possibility is that the prokaryotic genome, plus the mesosome that attaches it to the cell membrane, became enveloped by an extension of the cell membrane. Such an origin, however, does not account for the origin of the double membrane, even though it is obvious that many eukaryotic chromosomes are attached to the inner nuclear membrane for a good part of the cell cycle.

Given the double nature of the nuclear membrane, however, it is not too difficult to envision the origin of the endoplasmic reticulum

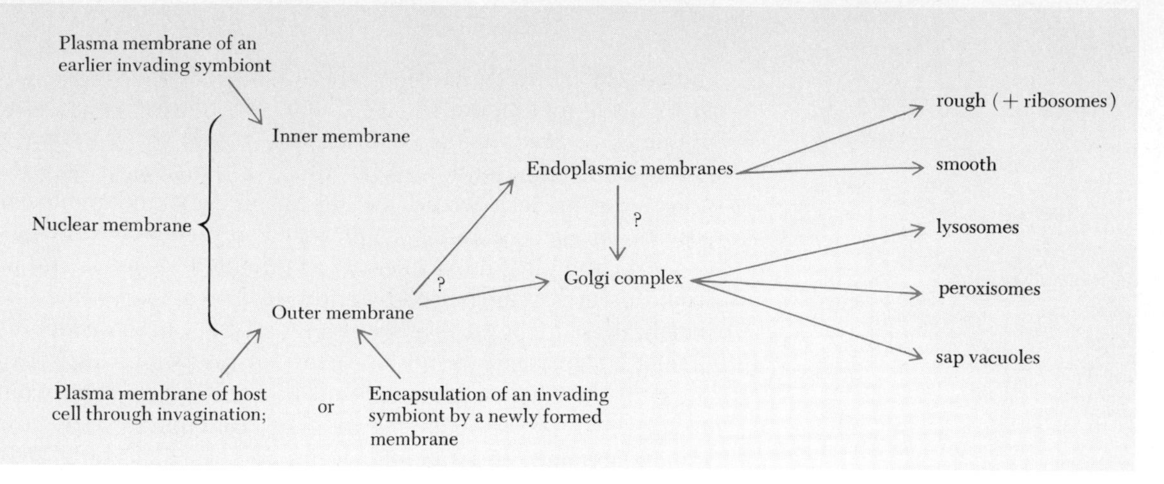

Figure 11.8

Possible origins and subsequent derivations of the inner membranes of the eukaryotic cell.

(Figure 11.8) and, by further elaboration, the Golgi apparatus and the several kinds of lysosomes and lysosomic-like vesicles.

Figure 11.9 illustrates a second hypothesis to account for the origin of the eukaryotic cell. This is an evolutionary model in which the several organelles are produced initially by invaginations of the cell membrane. The ancestral cell is assumed to be an aerobic

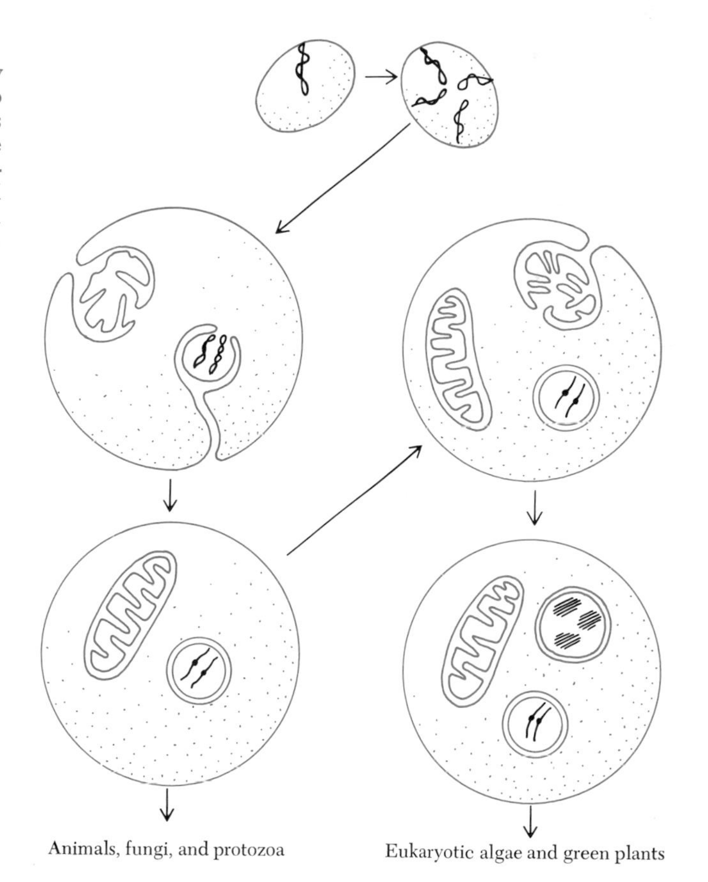

Figure 11.9

The origin of the eukaryotic cell by invagination of the cell membrane to form double-membraned organelles and nucleus. It is assumed that the invaginations did not occur simultaneously and that the nuclear and mitochondrial invaginations occurred first and led in the direction of the animal, fungal, and protozoan groups, while the photosynthetic invaginations occurred later and led in the direction of the eukaryotic algal and higher plant groups (see Figure 11.10). The ancestral form is assumed to be an aerobic bacterium that had acquired duplicate genomes prior to invagination, a feature seen in many present-day bacterial cells. M = mitochondria; P = plastids; N = nucleus. (After Uzzell and Spolsky, 1974.)

and multigenomic prokaryote, with each separate genome attached to the cell membrane. All of the invaginated bodies would be presumed to be similar initially and possibly possessing photosynthesizing capabilities. With time, a division of labor would occur: the nucleus would lose its respiratory and photosynthetic aspects, these becoming emphasized respectively in the mitochondria and plastids; surface area in the mitochondria and plastids would be greatly increased by additional invaginations of the inner membranes to form, respectively, cristae and thykaloids; the organellar genomes would lose many of the genic functions duplicated in the nuclear genome; and the nuclear genome would become much more complex as the functional and structural complexity and diversity of species increased. Figure 11.10 provides a family tree of ancestry, one based on the invagination

Figure 11.10
An interpretation of evolutionary trends according to symbiosis hypothesis of Margulis (left) and the invagination hypothesis of Uzzell and Spolsky (right).

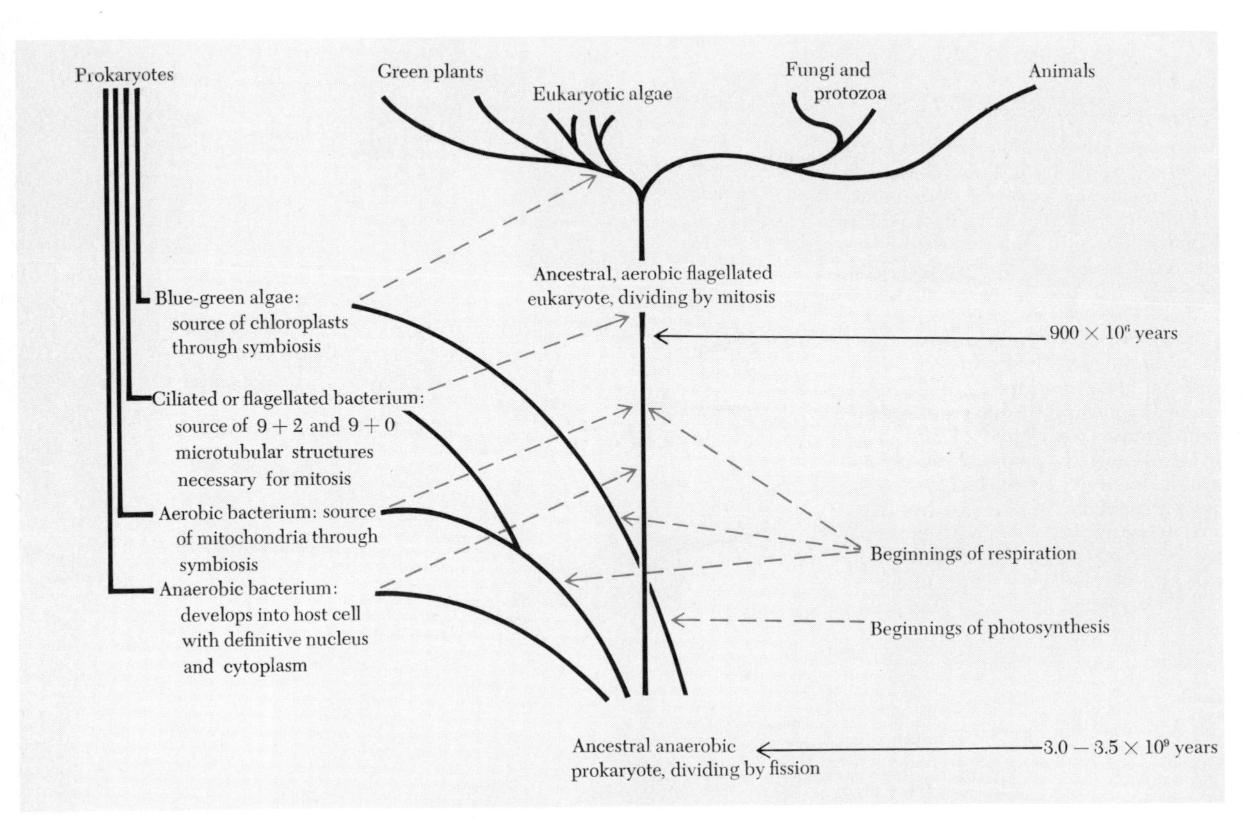

hypothesis and the other on the symbiosis hypothesis, and with acquisitions or losses of function indicated along the evolutionary route.

The invagination hypothesis, therefore, proposes a unicellular rather than a multicellular origin for the eukaryotic cell. The features that the organelles share with the prokaryotes—single, small, circular, and naked genomes; small ribosomes; chloramphenicol sensitivity; cycloheximide resistance; and so on—are regarded as retained primitive characteristics rather than ones acquired through symbiosis. A more positive feature of the invagination hypothesis is that it provides a plausible explanation for the origin of the double membranes of nucleus, mitochondria, and plastids. In addition, where intracellular membranes do exist in the prokaryotes, as in photosynthesizing bacteria, these are formed by invagination of the cell membrane. The larger size of the eukaryotic cell could have been an adjustment to maintain a proper genomic/cytoplasmic ratio.

A third hypothesis accounting for the origin of eukaryotic cells

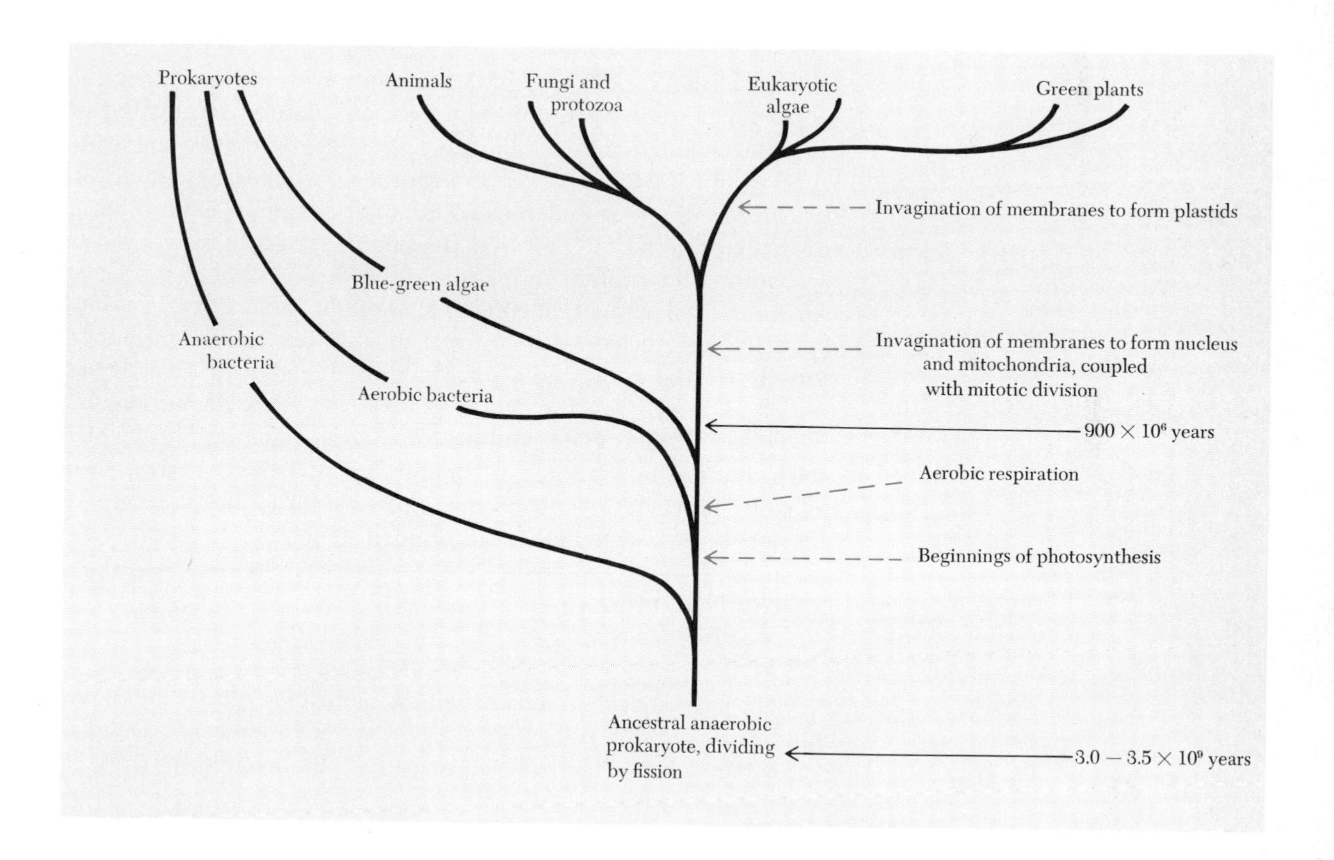

Figure 11.11
Possible origin of the eukaryotic cell through a system of cluster-cloning, beginning with a single genome within a prokaryotic cell. It is assumed that genes having special functions would become clustered within the genome, that is, those having to do with mitochondrial functions would become clustered in one part of the genome to form the D-E segment, those having to do with plastid functions becoming clustered in another part of the genome to form the F-G segment, leaving the A-B-C segment to govern the remaining functions of the cell. The genome would then become fragmented, with each special segment remaining intact to become enclosed within a double membrane, this leading to the formation of the nucleus, mitochondrion and plastid of the newly created eukaryotic cell. Further cloning of the mitochondrial and plastid systems would give multiple copies within a single cell. The microtubular system, in all of its manifestations, could be presumed to arise at a later time through mutational events. It is also assumed that the original prokaryotic cell giving rise to the above system was an aerobic, photosynthetic organism. In the diagram, the A-B-C system is nuclear, the D-E system mitochondrial, and the F-G system related to the plastids. (After Bogorad, 1975.)

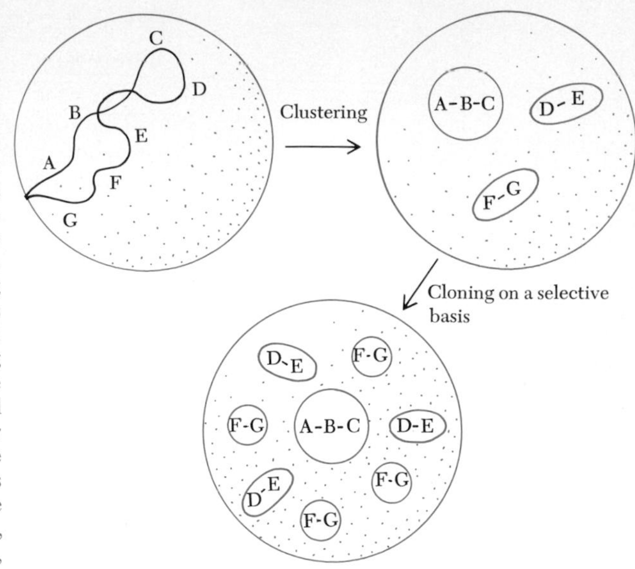

suggests that they arose by *cluster-cloning* of the elements in the genome (Figure 11.11). A basic assumption would be either that an ancestral prokaryote of multigenomic nature segregated its separate genomes into independent vesicles, or that a single genome would break up in parts, with these becoming isolated into vesicles of different functions. The multigenomic assumption is made plausible by the fact that the organellar and nuclear coding mechanisms for protein formation are basically similar, with control of organellar function coming eventually to be shared by both organelle and nucleus. It is, in fact, the shared nuclear-organellar control of respiratory and photosynthetic activities that is advanced as the reason for this kind of intracellular evolution, but since gene transfer, gains, and losses can occur among prokaryotes, and in unidirectional ways, it should also be a possible avenue of exchange between invading symbionts as well as between cloned and membraned segments of an original genome.

Origin of mitosis

Nuclear membranes, microtubules, centrioles, and mitosis are uniquely eukaryotic phenomena, and they are somehow related to each other in the process of cell division. The behavior of present-day eukaryotes during division makes this relation obvious, but questions of origin and of evolutionary integration of parts remain unanswered. It is clear, however, that the enclosure of the genome within a nuclear membrane, and the freeing of the nucleus in the cytoplasm, required the parallel development of a mechanism that would distribute the replicated chromosome(s) in a regular quantitative and qualitative manner.

A good many variations of the mitotic theme exist among the lesser known algal groups, the dinoflagellates and the ciliated protists. As these become better known cytologically, it well may be that the origins of mitosis can be ascertained with greater certainty than we now possess, but there is general agreement that a "closed" mitosis is primitive. This is a process in which the nuclear membranes never break down during the cell cycle, the chromosomes are continuously attached either to the nuclear membrane or to a spindlelike structure outside of the membrane, and separation of the replicated chromosomes occurs when the nucleus elongates and pulls apart as a result of elongation of the spindle. Such a spindle consists of microtubular fibers extending from pole to pole (centriole to centriole?) and following the outer contours of the nucleus. Later modifications led to complete or partial breakdown of the nuclear membrane in prophase and the origin of half spindle fibers extending from the poles to the centromeres. The breakdown of the nuclear membrane before metaphase and the absence of centrioles in many species indicate that the aggregation and behavior of microtubules alone can bring about the required separation of chromatids or chromosomes, but whether the microtubules are primitive or more recently derived structures is an actively debated subject.

The evolutionary significance of mitosis is undisputed, however. The precision with which chromosomes are segregated at anaphase guarantees a steady succession of genetically similar cells, a necessary condition upon which multicellularity is dependent. Meiosis is also a variant of mitosis made possible by diploidy. It is the chromosomal process basic to an alternation of generations in the plant kingdom, to sexual reproduction and the formation of eggs and sperm in all of the eukaryotes, to the whole gamut of patterns of biparental inheritance, Mendelian and otherwise, and to the recombination of genes that promoted diversity among individuals and made rapid evolutionary change possible. There can be no doubt that the introduction of mitosis and meiosis in eukaryotic organisms was an evolutionary innovation of far-reaching consequences.

Changes in eukaryotic cells

Eukaryotic cells originated less than a billion years ago, and the appearance of a vast array of different forms of life in the early Cambrian period would suggest that this newly arisen cell possessed enormous potential for evolutionary change from the time of its inception. Perhaps this is not surprising for two reasons. In the first instance, each eukaryotic cell had at least two genomes, nuclear and mitochondrial, either or both of which could undergo heritable change through mutation, while plant cells had an additional genome in the plastids that could do likewise. The genomes of plastids and mitochondria are small in size and gene content, being comparable in these respects to bacterial genomes, but they are present in multiple copies, for example, the chloroplast of *Euglena,* a single-celled green alga, has many copies of its genome, thus increasing mutational possibilities to the same extent. The nuclear genome of eukaryotes is, on average, many times greater in size than that of the prokaryotes, but this well may be a more recently derived feature. If it is assumed that the eukaryotic cell of a billion years ago contained nuclear, mitochondrial, and plastid genomes, all of a similar size and genetically self-sufficient—a not unreasonable assumption on either the symbiotic or invaginational hypothesis of eukaryotic origin—one can then inquire as to the changes that have led to present-day cells and their behavior. All of the genomes have retained the same basic mechanisms of transcription and translation, and with the triplet code universally applicable, so one can assume this to be a system relatively intolerant of major mutational change. In other respects, the genomes of mitochondria and plastids have probably undergone less change than has that of the nucleus. Neither organelle has a self-sufficient genome, which suggests that over long periods a loss of genetic function has occurred, the loss being compensated for by nuclear genes that were either present from the beginning, obtained by transfer from the organelles, or developed anew by mutational process. In any event, the mitochondria and plastids lost whatever independence they once might have had. Mitochondria, on the other hand, have retained an aerobic metabolic system that is basically similar throughout the eukaryotic world. The enzymes involved may be of mitochondrial or of nuclear origin, but the mitochondria remain the basic converters of chemical energy into useable form wherever they are found.

The several kinds of algae—red, brown, golden, yellow-green, and green—as well as all of the variety of higher green plants would suggest that the chloroplasts, the site of color, have undergone considerable changes. This is so only in part. The process of photosynthesis has undergone little alteration in essential biochemical steps, and chlorophyll *a* and phycobilin are the basic photosynthetic pigments in all forms of algal species. However, each kind of alga has additional secondary pigments, giving each form its characteristic color and, because the secondary pigments are also light absorbing, shifting the peak of photosynthetic efficiency to different wavelengths. The chloroplasts are also structurally characteristic. In all except the green algae (*Chlorophyta*), where the thykaloids are stacked in grana as in higher plants, the thykaloids run from one end of the plastid to the other (Figure 11.12). In the red algae (*Rhodophyta*), the thykaloids are single; in

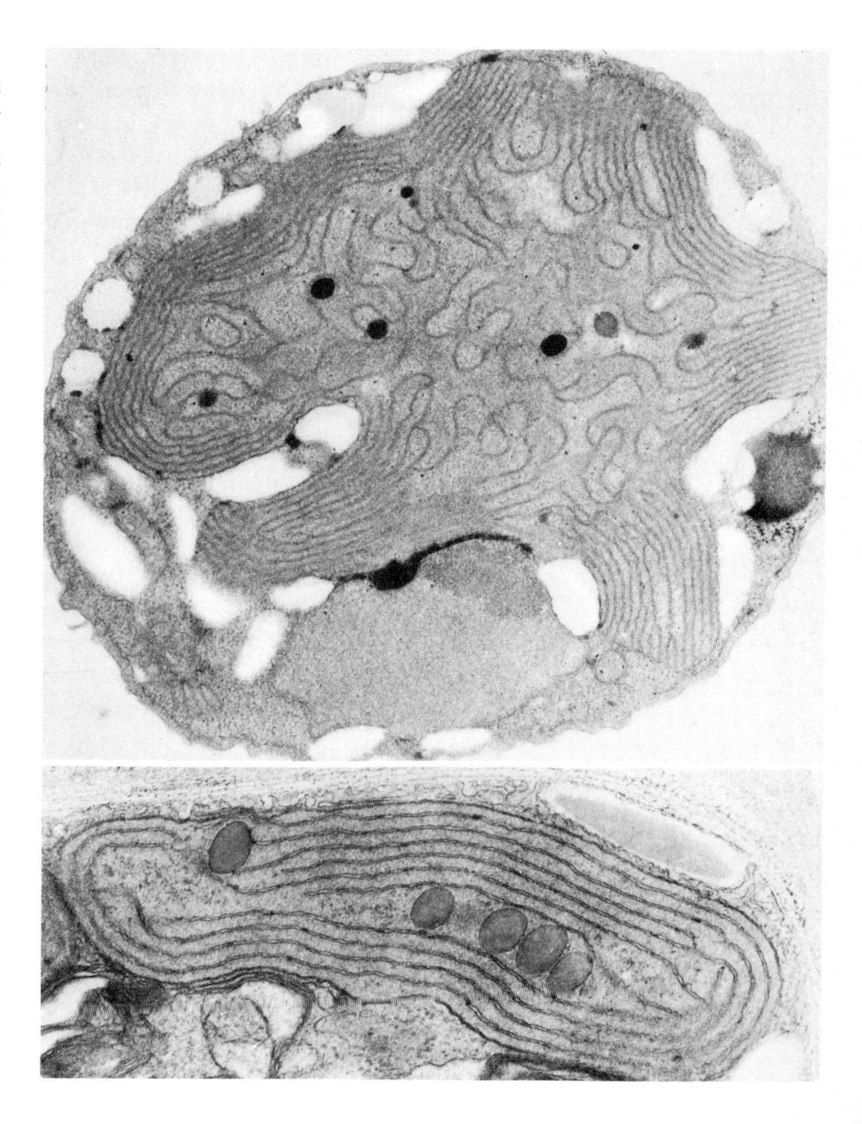

Figure 11.12

Chloroplast structure in two species of red algae in which the photosynthetic membranes exist singly. Top: *Porphyridium*, a single-celled species containing but a single chloroplast; bottom: a chloroplast from *Rhodophysema*, a multicellular species with each cell containing more than one chloroplast. (Courtesy of W. McDaniel.)

the Cryptophyta, in pairs; in the *Chrysophyta* (golden), in groups of three; and in the *Phaeophyta* (brown), in groups of three or four. It is quite probable on the basis of secondary pigment, thykaloid character, and lack of flagellated cells that the red algae are most primitive and most similar to the blue-greens, the supposed source of symbiotic plastids.

Just as the eukaryotic cell has its major biochemical systems compartmentalized into organelles and other membrane systems, so has it introduced comparable changes in its hereditary system. One of the most striking differences between prokaryotes and eukaryotes is the shift from single, naked DNA molecules to multiple, histone-clad chromosomes. These chromosomes vary in number, size, and genetic content, and are equipped with a centromere, which can interact with the spindle and the poles in the process of division.

There is little discernible evolutionary trend that relates to chromosome number. No eukaryotic organism is known that has a haploid number of one. A member of the plant family Compositae has a haploid number of two (*Haplopappas*, a tiny annual species), while an ancient fern, *Ophioglossum*, has a diploid number of over 1,000. Tables 11.3 and 11.4 list a number of common species of organisms and their haploid number, and Table 11.5 lists the chromosome numbers in the order Primates, to which humans belong. The most primitive members of the groups have the highest number, and the most specialized the lowest. Humans are about in the middle of the Primate groups.

Plants and animals differ to the extent that *polyploidy* has played an evolutionary role; this is when the basic genome is represented more than twice in somatic cells. For example, the grass family, Graminae, has a basic haploid chromosome number of 7, but more than 75 percent of all species exhibit some degree of polyploidy, a rather rare circumstance among animals. Among higher plants as a whole, about 35 percent are polyploid, but polyploidy is not evenly distributed throughout the plant families.

Chromosome size differs as widely as does chromosome number among the eukaryotes. Among the animals, the Orthoptera (grasshoppers, mantids, and their allies) and the Amphibia possess the largest chromosomes, while among the plants the monocots have generally larger ones than the dicots, with the lily and grass families particularly noteworthy.

The amount of DNA per nucleus is, as one would expect, correlated with chromosome number and size (Table 11.6), and these relations are generally but not absolutely correlated with evolutionary advancement. Evolution has led to greater and greater

Table 11.3

Haploid chromosome numbers in some common eukaryotic species of plants (see Moore, R. J., 1973. *Chromosome Numbers 1967–1971*, Oosthoek's *Uitgeversmaalschappij*, B. V. Utrecht; see also Table 11.6).

Species	Haploid number
Fungi	
Aspergillus nidulans (black bread mold)	8
Neurospora crassa (pink bread mold)	7
Saccharomyces cerevisiae (yeast)	15
Saccharomyces pombe (yeast)	2
Penicillium notatum	5
Fomes annosus (bracket fungus)	7
Ustilago maydis (smut fungus)	2
Algae	
Amphidinium carteri (dinoflagellate)	25
Spirogyra weberi (green alga)	2
Oedogonium cardiacum (green alga)	19
Triploceras verticillatum (desmid)	ca. 15
Chlamydomonas reinhardi (protozoa)	15
Chara braunii (green alga)	14
Bryophyta	
Marchantia polymorpha (liverwort)	9
Sphagnum sp. (moss)	19
Anthocerus husnotii (liverwort)	6
Funaria hygrometrica (moss)	14, 28
Psilopsida	
Psilotum nudum	104
Tmesipteris sp.	208
Lycopsida	
Lycopodium lucidulum (club moss)	132
Selaginella kraussiana (club moss)	10
Sphenopsida	
Equisetum arvense (horsetail)	108
Pteropsida	
Onoclea sensibilis (sensitive fern)	37
Ophioglossum petiolatum (ancient fern)	ca. 520
Osmunda regalis (royal fern)	22
Pteris aquilina (bracken)	58
Gymnospermae	
Juniperus virginiana (red cedar)	11
Larix laricina (larch)	12
Pinus sp.	12
Tsuga canadensis (hemlock)	12
Angiospermae	
Allium cepa (onion)	8
Brassica oleracea (cabbage)	9
Raphanus sativa (radish)	9
Zea mays (maize or Indian corn)	10
Lilium canadensis (common lily)	12
Lycopersicum esculentum (tomato)	12
Triticum diccocum (macaroni wheat)	14
Triticum aestivale (bread wheat)	21

Table 11.4

Haploid chromosome numbers of some common species of animals, with the Protozoa included for convenience (see Table 11.5 for chromosome numbers among the Primates).

Species	Haploid number
Protozoa	
Amoeba proteus	ca. 250
Barbulanympha ufalula (flagellate)	26
Trichonympha sp. (flagellates)	24
Invertebrata	
Strongylocentratus purpuratus (sea urchin)	18
Drosophila virilis (fruit fly)	6
Nemobius fasciatus (ground cricket)	8
Astacus trowbridgei (crayfish)	188
Tityus bahiensis (scorpion)	2
Phigalia pedaria (geometrid moth)	112
Papilio sp. (swallowtails)	30
Pisces	
Torpedo ocellata (shark)	ca. 20
Lepidosiren paradoxa (S. African lungfish)	19
Petromyzon marinus (lamprey eel)	84
Salmo satar (Atlantic salmon)	30
Salmo trutta (brown trout)	40
Amphibia	
Ambystoma tigrinum (salamander)	14
Amphiuma means (Congo eel)	12
Triturus cristata (newt)	12
Bufo bufo (common toad)	11
Hyla aborea (tree frog)	12
Rana catesbiana (bullfrog)	13
Reptilia	
Alligator mississippiensis	16
Chelonia sp. (turtles)	28
Natrix sp. (water snakes)	20
Aves	
Columbia livia (pigeon)	40
Sirinus canarius (canary)	ca. 40
Numida meleagris (guinea hen)	33
Mammalia	
Macropus rufus (kangaroo)	10
Trichosurus vulpecula (oppossum)	10
Bos taurus (cattle)	30
Canis familiaris (dog)	39
Equus caballos (horse)	32
Felis catus (cat)	19
Ovis aries (sheep)	27
Rattus rattus (rat)	21
Sus scrofa (pig)	20

Table 11.5
Haploid chromosome numbers of some species of Primates, including humans.

Prosimii		*Anthropoidae*	
Tupaiidae (tree shrews)		Cebidae (New World monkeys)	
Tupaia glis	30	Lagothrix ubericola (wooley monkey)	31
T. chinensis	31	Ateles sp. (spider monkeys)	17
T. minor	33	Alouatta seniculus (red howler)	22
T. montana	34	A. caraya (black howler)	26
T. palowensis	26	Cebus capucinus (organ-grinder monkey)	27
Lemuridae (lemurs)		Saimiri sciureus (squirrel monkey)	22
Lemur macaco	22	Callimico jacchus (marmoset)	23
L. catta	28	Cercopithecidae (Old World monkeys)	
L. mongoz	30	Macaca mulatta (Rhesus macaque)	21
L. variegatus	23	Papio sphinx (mandrill)	21
Lorisidae (loris and bushbaby)		P. hamadryas (baboon)	21
Nycticebus coucang (slow loris)	25	Presbytis entellus (common langur)	22
Loris tardigradus (slender loris)	31	Colobus polykomos (colobus monkey)	22
Galago senegalensis (lesser bushbaby)	19	Nasulis larvatus (proboscis monkey)	24
G. cassicaudatus (thick-tailed bushbaby)	31	Cercopithecus aethiops (green vervet monkey)	30
Tarsioidae (tarsiers)		C. mitis (guenon)	36
Tarsius bancanus	40	Hominoidae (great apes)	
T. syrichta	40	Hylobates lar (gibbon)	22
		Symphalangus syndactylus (siamang)	25
		Pongo pigmaeus (orangutan)	24
		Gorilla gorilla	24
		Pan troglodytes (chimpanzee)	24
		Hominidae	
		Homo sapiens (humans)	23

structural, functional, and regulatory complexity, and it is obvious that this must be determined or governed by an increasing number of genes. In attaining this complexity, accompanied by varying degrees of specialization, there has been a parallel specialization of chromatin. Some chromosomes have become specialized in being concerned exclusively, or in part, with sex determination, and there are many kinds of chromosomally based sex-determining mechanisms. Specialization is also seen in that chromatin can divide into two kinds: euchromatin containing the genes that govern readily recognizable traits, and heterochromatin, which may be inactive genetically, regulatory in some as yet undetermined way, or concerned with their formation of rRNA in the nuclear-organizing region, a subdivision of specialization within heterochromatin.

The amount of DNA in some cells cannot, however, be explained totally by an increase in the number of structural genes, that is, those genes determining proteins for enzymatic or structural purposes (Table 11.6). For example, the lungfish, *Amphiuma* and *Necturus* among the Amphibia, and *Vicia faba* among the legumes

Table 11.6
Haploid chromosome numbers and DNA values per haploid cell and per chromosome (given in terms of numbers of nucleotides) in a group of representative organisms, with a range of figures of mitochondria, chloroplasts, and viruses included for comparative purposes. (After Sparrow, Price, and Underbrink, 1972.)

Organism	*Chromosome number*	*DNA/cell*	*DNA/chrom.*
mitochondria	1	$5.2 \times 10^4 - 1.0 \times 10^6$	Same
chloroplasts	1	$1.8 \times 10^5 - 2.8 \times 10^7$	"
tobacco necrosis satellite virus (RNA)	1	1.3×10^3	"
DNA pox viruses	1	5.3×10^4	"
Bacillus subtilis (bacterium)	1	1.3×10^7	"
Escherichia coli (colon bacterium)	1	$9.0 \times 10^6 - 3.4 \times 10^7$	"
Hemophilus influenzae (influenza bacterium)	1	4.0×10^7	"
Anacystis nidulans (blue-green alga)	?	1.2×10^7	?
Oscillatoria linosa (blue-green alga)	?	1.6×10^{10}	?
Spirogyra setiformis (green alga)	4	7.0×10^9	1.8×10^9
Euglena gracilis (green flagellate)	45	5.8×10^9	1.3×10^8
Aspergillus nidulans (black mold)	8	8.8×10^7	1.1×10^7
Neurospora crassa (Pink bread mold)	7	8.6×10^7	$1.2 - 10^7$
Osmunda cinnamomea (cinnamon fern)	22	9.6×10^{10}	4.4×10^9
Pinus strobus (white pine)	12	8.4×10^{10}	7.0×10^9
Lilium longiflorum (Easter lily)	12	1.1×10^{11}	9.2×10^9
Chrysanthemum sp.	18	1.0×10^{10}	5.6×10^8
Drosophila melanogaster (fruit fly)	4	1.7×10^8	4.2×10^7
Gryllus domesticus (cricket)	11	1.1×10^{10}	1.0×10^9
Esox lucias (pike)	9	1.7×10^9	1.9×10^8
Protopterus sp. (lungfish)	17	1.0×10^{11}	1.4×10^{10}
Rana pipiens (leopard frog)	13	1.4×10^{10}	1.1×10^9
Necturus maculosus (mudpuppy)	12	1.7×10^{11}	1.4×10^{10}
Boa constrictor	18	3.5×10^9	1.9×10^8
Gallus domesticus (chicken)	39	2.3×10^9	5.7 c 10^7
Mus musculus (mouse)	20	6.5×10^9	3.2 c 10^8
Homo sapiens (humans)	23	6.0×10^9	2.6×10^8

have many times greater amounts of DNA per nucleus than their very close relatives. Part of this may be due to gene redundancy, that is, genes of particular kinds may be represented in the genome by more than one copy. Structural genes are generally represented by nonredundant genes, but those for the several kinds of rRNAs (particularly) and tRNAs have been shown to be highly redundant, sometimes with thousands of copies of the same gene present in a single chromosome. How this has come about is not known, but it is an obvious device for maintaining a rich supply of RNAs for biosynthetic purposes. In addition to these redundant genes, there is also in some species DNA whose function is not known but whose presence can account for substantial amounts of the DNA in each cell. In the newt, *Triturus cristata*, for example, a "spacer" DNA has been recognized; each block consists of seven nucleotides and is represented a million or more times in the same genome.

The eukaryotic genome is a complex one, and not all of the evolutionary innovations that have occurred are understood. But the evolutionary potential residing in the eukaryotic cell can be dramatized by pointing out that it took a far longer period of time for the eukaryotic cell to evolve from a prokaryotic ancestor(s) (about 2 billion years) than for the earliest eukaryotic cell to evolve into humans (about 900 million years). Prokaryotic species show very little differentiation of cells, the heterocysts of the blue-green algae being one of the few kinds of differentiated cells. Multicellularity is also absent. The blue-green algae exhibit a pseudomulticellularity in the form of filaments or flat sheets of cells, but the cells are held together by their gelatinous sheaths and possess no true intercellular connections through plasmadesmata or pits. Evolutionary change in the prokaryotic system took the form of varied metabolic pathways, whereas the eukaryotic system retained, for the most part, a common aerobic metabolism but showed great potential in exploiting aspects of form and size. This was made possible by the acquisition of multicellularity, intercellular communication, and the potential for cellular differentiation as a developmental phenomenon, features which in turn required the evolution of genomic control systems. We are just beginning to realize the intricacies and complexities of this kind of cellular behavior.

Bibliography

Bogorad, L. 1975. Evolution of organelles and eukaryotic genomes. *Science* **188**, 891–898.

Calvin, M. 1975. Chemical evolution. *American Scientist* **63**, 169–177.

Cohen, S. 1973. Mitochondria and chloroplasts revisited. *American Scientist* **61**, 437–443.

Ehrensvard, G. 1962. *Life: Origin and Development.* University of Chicago Press, Chicago.

Fox, S. W. 1975. Looking forward to the present. *BioSystems* **6**, 165–175.

Fox, S. W., and K. Dose. 1972. *Molecular Evolution and the Origin of Life.* W. H. Freeman & Co., San Francisco.

Gibbs, M., ed. 1971. *The Structure and Function of Chloroplasts.* Springer-Verlag, New York.

Kirk, J. T. O., and R. A. E. Tilney-Bassett. 1967. *The Plastids.* W. H. Freeman & Co., San Francisco.

Lima-de-Faria, A., ed. 1969. *Handbook of Molecular Cytology.* North-Holland Publishing Co., Amsterdam. An excellent reference volume dealing with all aspects of organellar structure, function, and origins.

Margulis, L. 1970. *Origin of Eukaryotic Cells.* Yale University Press, New Haven.

———. 1971. Symbiosis and Evolution. *Scientific American* **225**, 48–57.

——— 1974. On the evolutionary origin and possible mechanism of colchicine-sensitive mitotic movements. *BioSystems* **6**, 16–36.

Miller, S. L., and L. E. Orgel. 1974. *The Origins of Life on Earth.* Prentice-Hall, Inc., Englewood Cliffs, N. J.

Nass, S. 1970. The significance of the structural and functional similarities of bacteria and mitochondria. *International Review of Cytology* **28**, 55.

Pickett-Heaps, J. D. 1969. The evolution of the mitotic apparatus: an attempt at comparative ultrastructural cytology in dividing cells. *Cytobios* **1**, 257.

———. 1974*a*. The evolution of mitosis and the eukaryotic condition. *Biosystems* **6**, 37–48.

———. 1974*b*. *Structure, Reproduction and Evolution in Some Green Algae.* Sinauer Associates, Stamford, Conn.

Raff, R. A., and H. R. Mahler. 1972. The Non-symbiotic origin of mitochondria. *Science* **177**, 575–582.

Raven, P. H. 1970. A multiple origin for plastids and mitochondria. *Science* **169**, 641.

Sager, R. 1972. *Cytoplasmic Genes and Organelles.* Academic Press, New York.

Uzzell, T., and C. Spolsky. 1974. Mitochondria and plastids as endosymbionts: a revival of special creation? *American Scientist* **62**, 334–343.

Appendix A

Textbooks and monographs of general cytological interest

BITTAR, E. E., ed. 1973. *Cell Biology in Medicine.* John Wiley & Sons, New York. Despite the title, the papers cover a wide variety of basic cell areas from both a biochemical and a structural point of view, making it a useful compendium of recent findings.

BLOOM, W., and D. A. FAWCETT. 1968. *A Textbook of Histology* (9th ed.). W. B. Saunders Co., Philadelphia. A standard and widely used book on human histology, written primarily for medical students; contains a great deal of well-illustrated material on the entire spectrum of vertebrate cell types.

BRACHET, J., and A. E. MIRSKY, eds. 1959–1964. *The Cell.* Volumes 1–6. Academic Press, New York. Collections of articles by specialists on a broad variety of cell topics. Somewhat out of date, but still useful for review purposes.

DEROBERTIS, E. D. P., F. A. SAEZ, and E. M. F. DEROBERTIS, JR. 1975. *Cell Biology* (6th ed.). W. B. Saunders Co., Philadelphia. The most recently published of the general textbooks of cytology; much emphasis is given to techniques in addition to the usual structural and biochemical aspects.

DU PRAW, E. J. 1968. *Cell and Molecular Biology.* Academic Press, New York. A general textbook of cell biology, with much emphasis on the more recent biochemical findings and the relation of these to the structural aspects of the cell.

FAWCETT, D. W. 1966. *The Cell: An Atlas of Fine Structure.* W. B. Saunders Co., Philadelphia. A collection of superb electron micrographs, each described in detail. The great majority of illustrations are of animal cells.

HAMKALO, B. A., and J. PAPACONSTANTINOU, eds. 1973. *Molecular Cytogenetics.* Plenum Press, New York. A collection of research articles concerned with chromosomal organization, behavior, and regulation in both prokaryotes and eukaryotes.

KENNEDY, D., ed. 1958–1965. *The Living Cell.* W. H. Freeman & Co. San Francisco. A collection of articles from the *Scientific American,* grouped, with editorial commentary, into the following categories: Levels of Complexity, Organelles, Energetics, Synthesis, Division and Differentiation, and Special Activities. Somewhat outdated, but still useful.

LEDBETTER, M. C., and K. R. PORTER. 1970. *An Introduction to the Fine Structure of Plant Cells.* Springer-Verlag, New York. Comparable in excellence of electron microscopy with the Fawcett volume, but dealing exclusively with plant materials.

LEHNINGER, A. L. 1975. *Biochemistry* (2nd ed.). Worth, New York. An excellent and comprehensive textbook of biochemistry, widely used and recently updated.

LIMA-DE-FARIA, A., ed. 1969. *Handbook of Molecular Cytology.* North-Holland Publishing Co., Amsterdam. An excellent reference volume dealing with the biochemistry and ultrastructure of prokaryotic and eukaryotic cells.

STERN, H., and D. L. NANNEY. 1965. *The Biology of Cells.* John Wiley & Sons, New York. An introductory textbook dealing with the foundations of biological thought—the cell theory, the gene theory, etc.—and then the body of biology that takes off from these generalizations. The emphasis is largely on the physical-chemical aspects of cell behavior.

SWANSON, C. P. 1957. *Cytology and Cytogenetics.* Prentice-Hall, Inc., Englewood Cliffs, N. J. Parts of the volume are much out of date, but that dealing with the evolution of cellular systems is still valid.

WHITE, M. J. D. 1973. *Animal Cytology and Evolution* (3rd ed.). Cambridge University Press, London. The emphasis in this volume is on the evolution of chromosomal systems; technically detailed and difficult. A comprehensive treatment of one facet of cell biology.

WOLFE, S. L. 1972. *Biology of the Cell.* Wadsworth Publishing Co., Inc. Belmont, Cal. An introductory textbook that, in addition to information on the structure and function of cells, contains much detail on specific cellular techniques. The electron micrographs are excellent.

Appendix B

Reviews containing occasional articles of cytological interest

Advances in Cell and Molecular Biology. Academic Press, New York. Volume I appeared in 1971.

Advances in Genetics. Academic Press, New York. The first volume was published in 1947, with additional volumes appearing nearly every year since then.

Annual Review of Biochemistry. Annual Reviews, Inc., Palo Alto, Cal. Volume I appeared in 1932; has continued since then.

Annual Review of Plant Physiology. Annual Reviews, Inc., Palo Alto, Cal. Volume I appeared in 1950.

Cold Spring Harbor Symposia on Quantitative Biology. Cold Spring Harbor Laboratory, Cold Spring Harbor, N. Y. The following volumes are of cytological interest, containing articles by leading authorities:
Volume XXXI. 1966. *The Genetic Code.*
Volume XXXIII. 1968. *Replication of DNA in Microorganisms.*
Volume XXXIV. 1969. *The Mechanism of Protein Synthesis.*
Volume XXXV. 1970. *Transcription of Genetic Material.*
Volume XXXVIII. 1973. *Chromosome Structure and Function.*

Appendix C

Journals containing articles of cytological interest

American Journal of Human Genetics
American Scientist
Canadian Journal of Genetics and Cytology
Cell and Tissue Research
Chromosoma
Cytobios
Cytogenetics
Developmental Biology
Experimental Cell Research
Genetical Research
Genetics
Hereditas
Journal of Cell Biology
Journal of Cell Science
Journal of Cellular Physiology
Journal of Genetics
Journal of Heredity
Journal of Medical Genetics
Journal of Membrane Biology
Journal of Molecular Biology
Journal of Submicroscopic Cytology
Journal of Ultrastructure Research
Nature
Nucleus
Proceedings of the National Academy of Science
Protoplasma
Science
Scientific American

Index

D

E

F

G

H

I

K

L

M

N

O

P

R

S

T